Physical Science

A Mastery-Oriented Curriculum

CENTRIPETAL
PRESS

Austin, Texas
2017

Physical Science

A Mastery-Oriented Curriculum

John D. Mays

CENTRIPETAL
PRESS

Austin, Texas
2017

Published by Centripetal Press

centripetalpress.com

CENTRIPETAL
PRESS

Printed in the United States of America

Second printing, with new preface, 2019

ISBN: 978-0-9981699-4-1

Centripetal Press is an imprint of Novare Science & Math LLC.

Cover design by Nada Orlic, http://nadaorlic.info/

For a catalog of titles published by Centripetal Press, visit centripetalpress.com.

REVIEWER

This text was carefully reviewed for technical accuracy and clarity of expression by

Chris Mack   Adjunct Faculty, University of Texas at Austin
PhD, University of Texas at Austin, Chemical Engineering
MS, Electrical Engineering, University of Maryland
BS degrees in Physics, Electrical Engineering, Chemistry, Chemical Engineering, Rose-Hulman Institute of Technology

Any errors or ambiguities that remain are the responsibility of the author.

ACKNOWLEDGEMENTS

Special thanks to my brother, Jeffrey, for suggesting to me the fundamental structural idea behind this book. One small comment led to a revolution in my thinking about writing science textbooks. Thanks also to Gerald Tilma, who provided great skill in his work as copy editor. Thanks as always to Dr. Chris Mack, who loves to read and comment on what I write, and whose comments always lead to endless trains of thought. Any errors, omissions, or cloudy thinking that remain are my own fault.

Finally, thanks to my faithful wife Neda, who brings me a stunning work of art to eat for lunch every day. This book was built on salad.

Contents

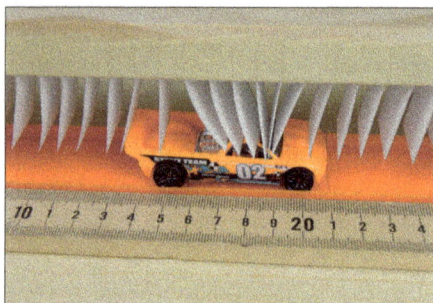

The book you have before you is clearly not a typical middle school physical science text. It is not two inches thick, it does not weigh five pounds, and it does not attempt to cover every topic in all of the physical sciences. These are not accidents.

I believe that science education, including textbook design, needs a major overhaul. The design philosophy behind this text represents my idea of where to start. My philosophy of pedagogy—and thus textbook design—is based on three core principles—*Mastery*, *Integration*, and *Wonder*. I summarize these principles this way:

MASTERY

The norm for classes in contemporary schools is what I call the *Cram–Pass–Forget cycle*. Students cram for tests, pass them, and then forget most of what they crammed in just a few weeks. Teachers across the nation know exactly what this looks like because they see it day after day. This cycle is a waste of time for teachers and students. Instead, students should *learn*, *master*, and *retain* what they have learned. Realizing this in the classroom requires both the teacher and the students to make significant changes in the ways they approach the tasks of teaching, testing, practicing, and studying. For textbooks, a mastery orientation requires that the curriculum cover fewer topics, and cover them deeply. When students engage with content, their efforts should be directed at comprehension, practice, and retention of important concepts, not mere exposure to masses of material they will not remember. Methods that promote the Cram–Pass–Forget cycle must be replaced with more effective methods.

In a mastery-oriented learning environment, students notice that the workload is demanding, and initially some of them resist. But all the students learn far more than they do otherwise. They typically remember the material for years, and most of them—even those who aren't much interested in school—eventually come to recognize the value of such a learning experience. Some recognize this immediately, some only after they go off to college and find themselves ahead of their peers because of things they remember from many years before.

An example of how a mastery approach is realized in this text is in the use of unit conversions in computational exercises. Unit conversions and the SI System of units are introduced in Chapter 8. After that, all computations in Chapters 8, 9, 10, and 13 require students to perform unit conversions and to maintain their familiarity with the SI System units and prefixes.

For more information about how mastery concepts should be implemented with this text, see the "Notes on Using this Text" section below, and my book *From Wonder to Mastery: A Transformative Model for Science Education*, available in early 2021 at centripetalpress.com.

INTEGRATION

Effective science instruction requires that a number of content areas not usually represented adequately in curriculum materials be deeply embedded in the curriculum. These content areas include scientific process, basic epistemology, mathematics, scientific history, and English language usage. Typically, science classes do not place the necessary emphasis on these areas, and as a result, students fail to have a coherent and effective learning experience. Science teachers must think deeply about how their courses address the need for these and other key areas of integration and make adjustments to curriculum, teaching methods, assessments, and coordination between science courses.

An example of how integration is realized in this text is that the questions and exercises in every chapter and in every experiment require students to write out answers in complete sentences. This helps develop their ability to write well and to articulate scientific principles accurately. Another example is the introduction of graphical (and thus quantitative) presentation of data in the reports students write for the experiments. A third example is the way the discussion in the text builds toward Chapter 7, where students engage with different categories of knowledge and with the distinction between scientific claims and truth claims.

WONDER

Science and mathematics provide us with unique ways of seeing the world. The world is a stunning place, full of surprises and jaw-dropping phenomena. But today there are many factors standing in between our students and the world they might otherwise become fascinated with. Safety concerns interfere with kids playing and exploring outdoors. Liability concerns make it hard to find a decent chemistry set. And unfortunately, it is common today for young people grow up spending most of their time indoors with digital media. The natural draw of nature for budding scientists is now commonly missed. Many people have never seen the night sky in an area that is completely dark, and have no idea of the stunning beauty of the heavens at night. Most kids have not hiked and camped in the forests, and have not learned to listen to the sounds made by animals, insect, and trees.

Additionally, the environmental challenges we face today from pollution, resource exploitation, and especially climate change require a new generation of people who care about the earth. But people usually do not care about what they do not love, and they do not love what they do not know about. Helping students to know the natural world has never been more important than it is today. Only if they know the world will students begin to love it, and only then will they be motivated to take care of it.

We cannot correct all these deficiencies in a textbook. But wherever possible, I try to point out amazing facts about the world we live in. My hope is that the more kids learn about how amazing the world is, the likelier it becomes that their natural passions and interests will kick in and lead them into the adventure of scientific study.

LAB JOURNALS AND LAB REPORTS

The overall goal of experiment documentation for middle school students is to continue laying the foundation for writing full lab reports—from scratch—when they get to high school. The target is for students to begin writing full lab reports in 9th grade, and the standard for such reports is presented in my book *The Student Lab Report Handbook* (also available on our website, centripetalpress.com). Toward that end, I believe lab reports at the middle school level should focus on students describing what they did, presenting their results, and engaging with the questions.

Part of integrating English language development into science instruction is accomplished through the use of lab journals. Each of the 12 experiments included in this text requires students to document their work in a lab journal. Details about using lab journals are in the section entitled "Getting Started with Experiments," as well as in the first chapter of *The Student Lab Report Handbook*.

In each of the Experimental Investigations, I pose questions for students to engage with as they consider their results and observations. I leave the specifics of the student reports up to the individual instructor; requirements should be based on the preparedness and background of the students in a given class.

Several of the Experimental Investigations require students to calculate the "percent difference." This is the same quantity most secondary science teachers refer to as the "experimental error." For my reasons for using different terminology, please refer to Appendix B.

CONDUCTING EXPERIMENTS

The instructions written in the Experimental Investigations are rather brief. To keep students from getting lost in the details, I did not include some of the information instructors need to have at hand. I do assume that students at this level are supervised by an adult for the experimental work.

Full details are available at no charge to instructors using this text. Simply contact us by email or phone and we will send it to you. Ask for the *Physical Science Experiment Resource Manual*. The manual is supplied electronically, and includes detailed materials lists, parts costs and sourcing information for all supplies, sample experimental results, and more detailed information about how to conduct each experiment effectively. The Experiment Resource Manual is also included on the Resource CD for this text (available at our website).

Five of the 12 experiments require construction of special wooden or metal parts. We recognize that this may pose a difficulty for some students, primarily those who do not have access to tools for cutting wood and metal. Accordingly, we offer an experiment "parts kit" containing the special parts for these five experiments. (Note: The kit includes only the unique wooden and metal parts that require cutting and assembly. Ordinary materials and supplies that can be sourced elsewhere are not included.)

NOTES ON USING THIS TEXT

One of the major motivations behind the design of this text is for students to be delivered from the Cram–Pass–Forget cycle alluded to earlier. Compared to other texts, this book is small. An appropriate amount of content has been assembled for students to master in a single year of study.

But for mastery to be realized, certain practices need to be a regular part of the classroom experience. Regular review, rehearsal, and practice of older material must occur alongside the study of new material. Otherwise, students simply forget things a few weeks after being tested on them.

Here are a few specific practices that should be regularly present in classes using this text. These practices simply represent good teaching, and should be part of any well-structured course.

1. *Class Discussion*   Discuss concepts, ask questions, and give students opportunities to express concepts in their own words. Use the "Learning Check" questions at the end of each section and the Chapter Exercises as questions to stimulate discussion. All these questions should be discussed in class. Take time with this type of discussion and don't hurry. This is why the book is mod-

estly sized—so there will be adequate time for extended discussion and deep engagement with the material.

2. *Group Study*   Divide students into groups for teamwork on answering some of the more difficult questions in the Chapter Exercises. Then have teams take turns reporting their answers to the class. Stimulate discussion between teams about the merits of different answers to a particular question, and seek a class consensus on a well-formed answer.

3. *Review Discussion*   Go through questions again from past chapters. It is also fun to create competitions among teams in which teams score points by giving good answers to review questions. Some kind of activity like this should occur at least once every other week throughout the duration of the course.

4. *Review Computations*   Once the computational exercises begin (Chapter 8), students should experience a steady stream of weekly review computations for the rest of the course. Once a new computation has been introduced, students should see new sets of problems containing similar computations to work as review every week. Correct answers should always be furnished with new review problems sets. Review problems should make frequent use of unit conversions using the metric prefixes from Table 8.4 and the conversion factors from Table 8.6. The *Resource CD* for this text contains Weekly Review Guides for this purpose.

5. *Enrichment Activities*   Although concepts are covered thoroughly in the text, understanding is enhanced and memory is strengthened when students engage with the content in activities outside the text. Classes should incorporate full-length instructional videos, classroom demonstrations, hands-on play with models (such as for modeling molecules), internet images, You Tube videos, games, student presentations, projects, and other activities. Teachers should emphasize activities designed to make students *active* participants in the learning process, rather than simply passive observers. To assist with this, we have compiled a list of relevant short videos, images, and other resources, organized chapter by chapter. Look for the Tips and Tools link under the Extras tab on our website.

In addition to the comments here, more extensive information about the organization of chapter content and recommendations for teaching with this text are available on the Resource CD. Teachers are strongly encouraged to study that information in order to make the most of the pedagogical philosophy behind the design of this text.

It seems to me that everyone should love the world of science and technology. After all, science is about studying this fascinating world that we live in. Technology is the process of applying scientific principles to nature to solve particular engineering problems. These are really fun things to do.

Long before I was a science teacher, I was an engineer. I suppose I was in training my whole life to become an engineer. When I was a kid, I was always exploring the natural world. With my friends, we explored the woods and creeks near our home. We caught crayfish with a piece of bacon on the end of a string. We made dams in the gutters during rainstorms. We searched for interesting rocks and insects. We learned how to take care of a tortoise and how to pick up a snapping turtle without getting our fingers snapped off!

I was also learning how to build things (still one of my favorite things to do!). I built tree houses and model cars (a great number of each). I went to construction sites and talked to the workers to learn about what they were doing. I took my bicycle completely apart and put it back together. (It was hard to pedal after that—I didn't know how to tighten the wheels and pedals properly!) My friends and I built wooden go-carts (from scratch) that we could race down the hilly streets where we lived. When I was 10, I even built an elaborate work bench in our garage especially for building model cars.

I was also learning how to solve particular technical problems. As a Boy Scout who went camping every month for five years, I learned how to start a fire when the wood is soaking wet, how to start a fire without matches or a lighter, how to prevent our food from being destroyed by bears, how to sharpen a knife, how to carve things, how to make a lanyard and tie good knots, how to purify water so we didn't get sick, and how to find my way through a forest with a map and compass.

I could go on and on, but the point is this. It is really fun to explore the way the world works and to apply scientific principles to solve engineering problems. These are pretty much what this book is about. We are going to explore some of the fundamental principles of physics and chemistry—that's what the phrase *physical science* means. And in the 12 experiments, we see how these principles can be applied to the real world. The experiments all involve building or setting up some kind of apparatus using common materials, and that is both fun and interesting. And most of the experiments require you take make careful measurements—something that is challenging because it is not so easy as it sounds!

I hope you have a lot of fun with this study. Once you get hooked on exploring nature, the world just gets more fascinating and wonderful every year—and you can almost never get bored! If you do get hooked and want to become a scientist or engineer, that would be super. The world will never run out of problems to solve, and helping solve them is work that is open to just about anyone who is willing to put their mind to it—maybe even you!

And now for a note on how this science course is organized. I assume if you are reading this that you are around 11–14 years old. You are probably in 6th, 7th, or 8th grade. You probably love science and love studying about the amazing world we live in. I love these things too, and I love teaching students about this wonderful world.

I am sure that you have already been studying topics in science for several years. But now that you are in middle school, it's time for your studies in science to become a bit more organized. Do you like knowing how things fit together in the "big picture"? When you study a complicated topic in history or math, do you like to start from the beginning and build up your knowledge in a careful, organized way? For myself, I know that if I understand how things fit together—starting with the basics—it helps me understand and remember things better.

In this book, this is what we are going to do. When we talk about the "laws of nature," the laws of physics and chemistry, we can talk about the world of very tiny things, like atoms and molecules. We can also talk about the world of larger things like balloons, rockets, and machines. We can even talk about the world of *very* large things like planets and galaxies. For you to have a solid understanding about how the laws of nature work, we need to study *both* the small and the large. So in this book, we begin with some things science has discovered about the microscopic world. This will help you understand the fundamental principles that govern how the everyday world works.

Now we are ready to begin. Enjoy your study of our fascinating world!

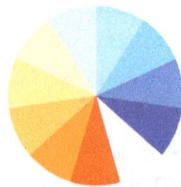

CENTRIPETAL
PRESS

Physical Science

A Mastery-Oriented Curriculum

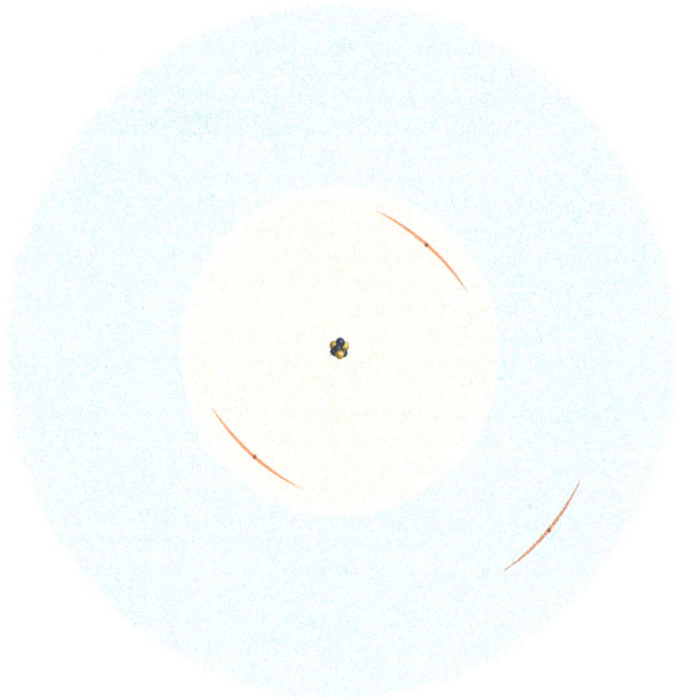

The drawing above is a depiction of a lithium atom. In the center is the atomic nucleus, containing four neutrons and three protons. In the much larger spherical regions surrounding the nucleus are the atom's three electrons, two in the inner region and one in the larger outer region. To make the nucleus visible at this scale, it is drawn about 2,500 times larger than it should be. If the nucleus were drawn to scale for a diagram of this size, it would be 1/300th the size of the period at the end of this sentence.

OBJECTIVES

After studying this chapter and completing the exercises, you should be able to do each of the following tasks, using supporting terms and principles as necessary.

1. Name and briefly explain the three basic things the universe is made of.
2. Describe how the particles in atoms are organized.
3. Describe each of the three basic subatomic particles.
4. Describe the atomic models developed by John Dalton, J.J. Thomson, Ernest Rutherford, and Niels Bohr.
5. Describe the key features that the quantum model of the atom added to correct and complete the Bohr model.
6. State the contributions of Democritus and James Chadwick to atomic theory.

VOCABULARY TERMS

You should be able to define or describe each of these terms in a complete sentence or paragraph.

1. atom	6. mass	11. order
2. charge	7. matter	12. proton
3. electron	8. neutron	13. shell
4. energy	9. nucleus	14. subatomic particle
5. ion	10. orbital	15. volume

1.1   The Three Most Basic Things

What are the pages of this book made of? Paper, of course, but what is paper made of? The answer is that paper is made of the fibers from various kinds of plants, including trees. But what are these fibers made of at the most basic level? You probably already know the answer—*atoms*. The material stuff in the every day world is made of atoms, parts of atoms and a few other strange particles we can't see.

Matter is just our word for substances made of particles that have *mass* and take up space (have *volume*). The matter we normally encounter is made of atoms. There are different ways the atoms can be arranged, such as crystals and molecules. There are also different forms matter can take, depending on how hot or cold it is. We will discuss these things in more detail later. Our point here is that matter is one of the three basic ingredients that form the universe we live in.

Matter is one of the basic things the physical universe is made of. But matter is not all there is

All matter is substance that has mass and volume.

Energy holds everything together and enables any process to happen.

in the physical world. Going back to the pages of this book—how did the pages get here? How were they fashioned, printed and bound? And thinking even more deeply, what holds the pages together? Why don't their atoms fly apart, like spray paint coming out of a can?

The answer to these questions relates to *energy*. The pages of this book were fashioned into their present form through the use of energy. The machines that cut the trees, the factory that made the paper, and all the people involved in making the paper and the book used energy to do their work. But thinking more deeply again, the atoms in the pages are sticking together because of the energy in their attractions for each other. The atoms *themselves* are held together by energy. Nothing anywhere can happen without energy being involved, and energy itself is what holds everything together.

Some scientists are content to say that matter and energy are the two basic ingredients of which the universe is made. However, there is one more basic ingredient that I think should not be left out. I am going to call this ingredient *order*.

For over 100 years, scientists have marveled at the orderliness present at every level in the structure of the universe. Stars aren't just distributed randomly in space; they are organized into spiral galaxies. At lower temperatures, protons, neutrons, and electrons are tightly organized into very predictable structures in atoms. Molecules in our bodies are part of a colossal protein factory that processes our food, carries energy to our muscles, flushes out waste, operates our many different organ systems, and builds new cells as we need them.

The orderliness of the universe allows us to model the behavior of nature using mathematics. We call these models the "laws of nature" or the "laws of physics." (As we will see in Chapter 7, a better term for these laws is theories.) But the fact that we can model nature with mathematics is simply amazing.

Some of the order around us comes from human intelligence. We build houses and machines, write books and music, craft paintings and sculptures. But as mentioned above, there is order at large in the universe, and there appears to be deep mathematical structure in the universe.

Order is present everywhere in nature.

Many believe—even some nonreligious scientists and philosophers—that the order observed throughout the universe indicates that there is a purpose for the universe and for creatures like us who live in it. Some believe that we create meaning and purpose for ourselves; but others believe that since we humans express desires, intentions, and purposes, these thoughts about purpose must come from a larger purpose for the universe that comes from outside the universe—a power that we speak of as transcendent.

There is much food for thought here! The universe is a stunning and wonderful place. So far as we know, human beings are unique creatures. We know of no other species that engages in language, reason, humor, art, love, and religious belief the way we do. There is so much in the world around us to be excited about and interested in. How can life ever be boring? There is so much to learn, so much to wonder about!

To summarize this first section, all material objects—all matter—is made of atoms. Energy is present in nature, holding everything together and enabling everything to happen. And order is present in the laws of physics and chemistry and in the use of matter and energy to make things. We will consider energy and order in more depth in later chapters. Our task for the rest of this chapter is to consider atoms in more detail.

Learning Check 1.1

1. Describe the three basic things the universe is composed of.
2. Give an example of how order is evident in nature.
3. In the list below, some things are clearly the result of the order resulting from human intelligence, and some are the result of the order embedded in the laws of nature. Explain which is the case for each item.

> › an artist's painting
> › the arrangement of pieces of confetti on the floor at a party
> › the shape of a wadded up piece of paper
> › the design of the pages in a book
> › the arrangement of all the leaves in a tree
> › the sound of a cello string when it is plucked
> › the arrangement of the keys on a computer keyboard

1.2   Atoms

Chances are that you have learned about atoms before, so you may already have an idea of how they are put together. In this section, I am going to describe some basic scientific facts about atoms. Later, we will take a brief look at how scientists figured these things out.

Atoms are much too small to see. In order to see an object with our eyes, the object has to be big enough to reflect light waves into our eyes. But atoms are much smaller than the waves of visible light, so they do not reflect the waves, and we cannot see the atoms. What we know about atoms we have *inferred* from thousands of experiments. To infer something is to figure it out from the evidence. If I go out to my car and find the

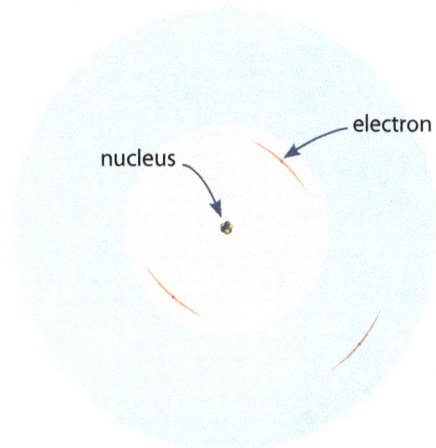

Figure 1.1. An atom with its nucleus of protons and neutrons, and electrons in a much larger region surrounding the nucleus. The nucleus is actually much smaller than shown in the picture.

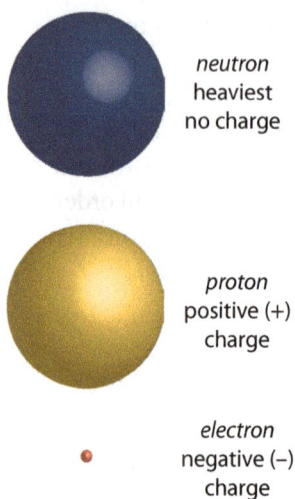

neutron
heaviest
no charge

proton
positive (+)
charge

electron
negative (–)
charge

Figure 1.2. The three subatomic particles.

back bumper smashed, I infer that someone hit my car, even though I was not there to see it.

Figure 1.1 is a diagram of a small atom, the way scientists currently understand them. In the center is the *nucleus* (which is actually much smaller than shown in the figure). Within an atom are three different types of *subatomic particles*, depicted in Figure 1.2. There are two kinds of subatomic particles in the nucleus: protons and neutrons. These particles have almost the same weight, although neutrons are a tiny bit heavier. Protons have a property called *charge*. This property is responsible for electricity and everything electrical in nature. There are two kinds of charge, which we call positive and negative. (Benjamin Franklin was the first to call charge by these terms, back in the 18th century.) The charge on protons is positive. Neutrons have no charge.

A third particle inside the atom is the electron. Electrons weigh about 2,000 times less than protons. This means that their weight almost does not matter. But what does matter is their charge, which is exactly the same strength as protons, but negative. The electrons buzz around in a sort of layered cloud around the nucleus. More on electrons in a moment.

Other than the nucleus and the electrons, the rest of the atom is *completely empty space*. This is actually a bit mind boggling, so here is an example to help you visualize this. Figure 1.3 shows an engraving of the ancient sports stadium in Rome called the Coliseum. The tiny figures on the ground near the center of the Coliseum are people. Imagine that one of those people has a flower pinned to his lapel. If the nucleus of an atom were the size of the head of that pin, the cloud where the electrons are would be the size of the entire Coliseum! Everything else in the atom is empty space—nothing in there, not even air. (Of course, air is also made of atoms.)

Finally, when we speak of an atom alone by itself, we typically assume the atom is *electrically neutral*. This means that there is no net charge on the atom. The only way this can be is if the atom has an equal number of protons and electrons so that their charges balance out.

Figure 1.3. Engraving of the Roman Coliseum by Giovanni Piranesi.

Scientists, Experiments, and Technology

Although atoms are too small to be seen, there are technologies that can make images of atoms we can study. The image to the right was made by a scanning tunneling microscope (STM) and the individual atoms are imaged as little circles. The STM uses beams of electrons reflecting off objects to construct an image of the object's structure at a very small scale. This image shows the surface of a sample of gold. The atoms inside a sample of gold are arranged in a regular, repeating pattern—rows of atoms without gaps. But at the surface, gold atoms can have a gap between the rows that occurs after every five rows of atoms.

In recent years scientists have been learning how to construct materials with very specific arrangements of atoms. A fascinating example is carbon nanotubes, represented in the computer image to the right. These are hollow tubes of carbon atoms with walls one atom thick. The diameter of a nanotube is about seven times the diameter of the carbon atom itself, and the tubes are extremely strong. An STM image of a nanotube is shown to the right.

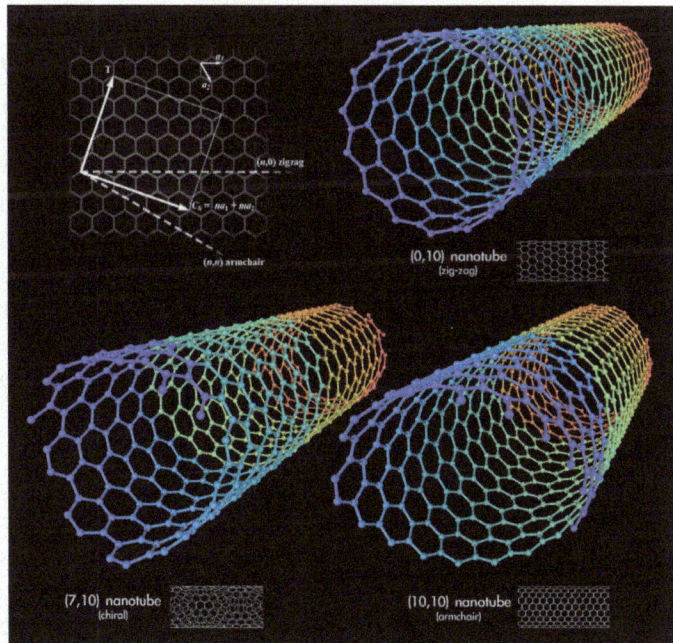

It's pretty easy for most atoms to gain or lose electrons. When they do, they are not electrically neutral any more. They have a net charge, either positive or negative. Atoms with a net charge like this are called *ions*. Ions with opposite charges attract each other. In later chapter, we will see that this is one of the main reasons atoms stick together to form chemical compounds.

Learning Check 1.2

1. Explain why scientists can claim that atoms exist, even though atoms cannot be seen.
2. Describe the locations of the three types of subatomic particles found in atoms.
3. Describe which subatomic particles have charge and which do not. For those that do, identify which kind.
4. Compare the weights of the three subatomic particles.
5. Explain what ions are and how they form.

1.3   Electrons

Let's talk just a bit more about the electrons and how they are arranged inside the atom. First, you need to know that electrons are *weird*, and it is hard to say just exactly what they *are*. Even though we refer to them as particles, they are certainly not hard little things like pellets or B-Bs. Electrons sometimes act like tiny particles, but they also sometimes act like *waves*. This is hard for everyone to understand, but that's just how it is. We all have to live with the strange properties of electrons and just try to understand them the best we can. Another thing about electrons is that there is no way to know precisely where they are and how fast they are going at the same time. This is why I drew the red streaks around the electrons in Figure 1.1. By showing them as sort of smeared, I am trying to show the uncertainty we have about where they are or how fast they are moving.

Figure 1.4. The first five orbitals in an atom. Each one can hold up to two electrons.

In an atom, every electron has a very specific amount of energy. The electrons are arranged in the atom according to how much energy they have. The clouds they buzz around in are called *orbitals* or *shells*. Electrons with the same amount of energy go in the same orbital, but only two electrons can go in each orbital. The large spheres in Figure 1.1 represent the first two orbitals that every atom has. Figure 1.4 shows the shapes of the first five orbitals every atom has, beginning with the two spherical ones. After the first two, the next three are shaped in a double arrangement that looks like a thick hamburger bun. The orbital shapes get even weirder after that, as you may learn later when you take chemistry in high school.

Learning Check 1.3

1. State at least five facts about electrons.
2. Describe the shapes of the first five electron orbitals in atoms.

1.4   The Development of Atomic Theory

Our theories about atoms have been under development for a long time. Over the centuries, there have been scores of important scientists who contributed key insights to our present theory—or model—of the atom. Here we will look at a few of the most important developments along the way.

The ancient Greek philosopher Democritus is usually given credit for first imagining that matter is made of atoms (Figure 1.5). Democritus lived in the 5th century BCE, and proposed that everything was made of tiny, indivisible particles. The word atom comes from the Greek word meaning *indivisible*. For over 2,000 years after Democritus, nothing much happened to further our understanding of atoms. But then the scientific revolution began to take off, and major developments began to occur regularly.

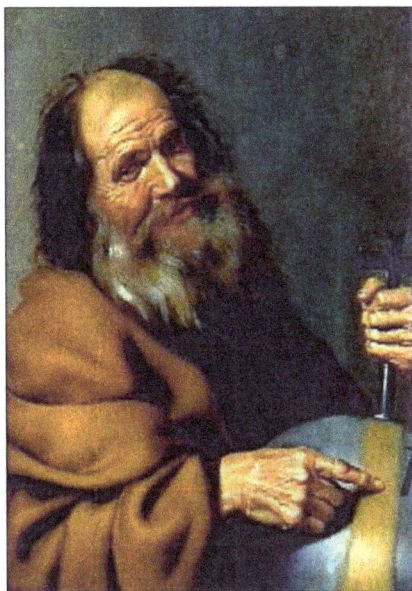

Figure 1.5. Greek philosopher Democritus.

In 1803, English scientist John Dalton (Figure 1.6) published the first fully scientific model of the atom. Dalton's theory included the idea that everything was made of indivisible atoms, as Democritus had said. Dalton went on to say that atoms combine together in whole-number ratios to form the compounds that different substances are made of. Dalton also proposed that atoms are not created or destroyed during chemical reactions, and that every atom of a given element is identical. All the points in Dalton's theory were either correct or partially correct, and Dalton's model was a major step forward.

Dalton's 1803 atomic model: indivisible particles.

Dalton's model was not correct about the notion that atoms are indivisible. The first news

Figure 1.6. English scientist John Dalton.

Figure 1.7. English scientist J.J. Thomson.

that atoms had smaller pieces inside them came from the work of another English scientist, J.J. Thomson (Figure 1.7). In 1897, Thomson performed a brilliant series of experiments that produced beams of electrons inside a glass tube. At the time, no one knew anything about electrons, but Thomson took the bold step of proposing that the beams he had produced were made of particles that came from within atoms.

Thomson's 1897 atomic model: the Plum pudding model—a cloud of positively charged material with thousands of negatively charged particles embedded in it.

As a result of his work, Thomson proposed a new atomic model, one that everyone now calls the *plum pudding model*. Now, most American students these days don't know much about plum pudding, so you can think of Thomson's model as the "watermelon model." As illustrated in Figure 1.8, Thomson modeled the atom as a cloud of positively charged material with thousands of negatively charged particles embedded in it, like a watermelon with its many seeds.

The next scientist to develop a new atomic model was New Zealander Ernest Rutherford, (Figure 1.9).

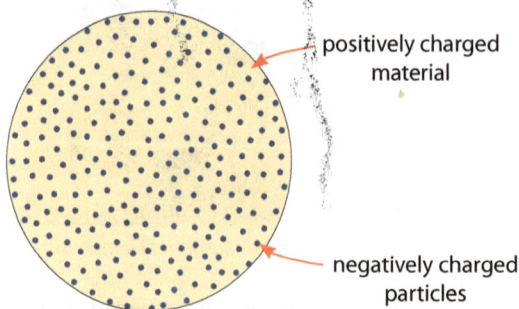

positively charged material

negatively charged particles

Figure 1.8. Thomson's "plum pudding" model of the atom.

In 1909, Rutherford was in England experimenting with firing small, positively charged particles at a thin foil made of pure gold. This work led Rutherford to conclude that all the positive charge in an atom is concentrated in the center, not spread out as Thomson had proposed. Rutherford called this central concentra-

Rutherford's 1909 atomic model: a tiny nucleus containing the positive charge and almost all the mass; negative electrons surrounding the nucleus; most of the atom is empty space.

tion of charge the *nucleus*. Rutherford also proposed that the electrons discovered by J.J. Thomson were outside the nucleus, surrounding it, and that most of the atom was empty space.

As you can see, with Rutherford's work we have come a long way towards understanding atoms, and we now have a general idea of how they are structured. The

The 1913 Bohr model: Building on Rutherford—electrons orbit the nucleus like planets. The energy of the electron determines its orbit, and only fixed energies are possible.

next development was put forward by Danish physicist Niels Bohr in 1913 (Figure 1.10). Bohr was the first to propose that the electrons were orbiting the nucleus like planets orbiting the sun. As depicted in Figure 1.11 on page 12, Bohr theorized that the orbits represented different "energy levels" for electrons. Lower energies were closer to the nucleus and higher energies were farther out. The lower-energy orbits would fill up first. The lowest orbit could hold two electrons. Orbits two and three could each hold eight electrons, and there were higher-energy orbits after that.

Figure 1.9. Physicist Ernest Rutherford, from New Zealand.

Bohr's model was very successful at explaining atomic behavior. However, it soon became clear that the electrons aren't exactly orbiting. Instead, an electron sort of zooms around—at extremely high speed—in a three-dimensional cloud defined by how much energy the electron has (as we saw back in Figure 1.4). And as we saw before, it is difficult even to think of electrons as particles at all, since they also have wave-like properties. One scientist said that since we don't really know what electrons are, we should just call them *slithy toves*. And when we talk about what they do, we can just say they *gyre and gimble in the wabe!*[1]

Our short history of the atomic model would not be complete without mentioning the discovery of one final important piece to the puzzle. In 1932, English scientist James Chadwick discovered the neutron. Scientists already knew that nearly all of the atom's mass was in the

The quantum model: Building on Bohr— electrons reside in orbitals of various shapes nested around the nucleus.

Figure 1.10. Danish physicist Niels Bohr.

nucleus, along with all the positive charge. But what they knew about mass and charge didn't match up until Chadwick demonstrated that there were electrically *neutral* particles, also in the nucleus, that had almost the same mass

---

[1]   In case you have forgotten, these terms are from the poem "Jabberwocky," in Lewis Carroll's *Through the Looking-Glass, and What Alice Found There.*

(slightly more) as the protons. With Chadwick's discovery our basic understanding of the atom was complete.

Scientists learned much more about atoms from experiments conducted throughout the 20th century. Our current model of the atom is called the *quantum model*. As I describe at the beginning of the chapter, the quantum model places the electrons in orbitals, rather than orbits. The quantum model describes the shapes of all the orbitals, and the rules governing which electrons go where among an atom's orbitals. These rules are at the heart of chemistry, which is all about how the electrons in atoms interact with each other. We will leave the rest of the details of the quantum model for another science course.

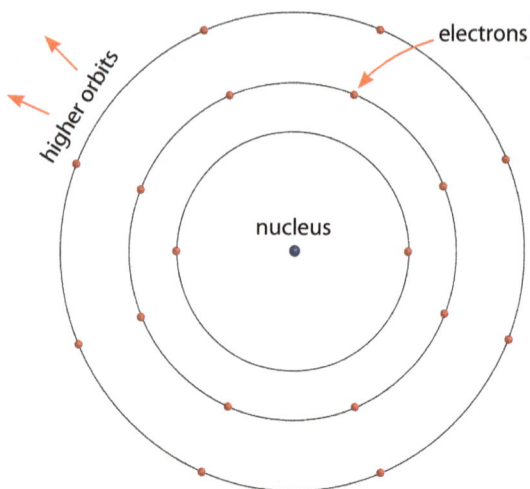

Figure 1.11. Bohr's planetary atomic model.

Learning Check 1.4

1. Describe the atomic models proposed by Dalton, Thomson, Rutherford, and Bohr.
2. Describe the additional features included in the quantum model.
3. Why did one scientist use the silly language of "Jabberwocky" to describe electrons?

Scientists, Experiments, and Technology

The particles used by Ernest Rutherford to explore the thin gold foil in his research are called *alpha particles*. Alpha particles (often written as α-particles) are a form of nuclear radiation. Each alpha particle contains two protons and two neutrons. Alpha particles are emitted naturally as radioactive substances go through the process called *nuclear decay*. The image on the opposite page depicts a large nucleus emitting an alpha particle during nuclear decay. When this decay happens, the alpha particles typically exit the atomic nucleus at a speed of 15,000,000 meters per second!

Alpha particles are sometimes used in common technologies such as smoke detectors. But I would rather tell you about a new technology being

Scientists, Experiments, and Technology (continued)

explored by researchers today. This technology is a cancer treatment called *unsealed source radiotherapy.*

The idea is to make use of the fact that though alpha particles will damage tissue, they do not penetrate the tissue very deeply. In unsealed source radiotherapy, small amounts of a radioactive substance are introduced into the body and directed near the site of a cancerous tumor. As the radioactive substance decays, the alpha particles it emits bombard the tumor and destroy it. Of course, the healthy tissue surrounding the tumor is also hit by the alpha particles. But because the alpha-particles do not penetrate the surrounding tissue very far, the healthy tissue is damaged only slightly. The damaged tissue will heal. The important thing is that the life-threatening cancer is destroyed.

Chapter 1 Exercises

Answer each of the questions below as completely as you can. Write your responses in complete sentences.

1. Describe J.J. Thomson's contributions to the development of the atomic model.

2. When scientists say that atoms are mostly empty space, what do they mean? (How empty are they, and what's in the empty part?)

3. Describe the particles found in the nucleus of atoms.

4. What determines where the electrons are in an atom?

5. Why was John Dalton's atomic model so important?

6. Describe the three basic ingredients the universe is made of.

7. What are some of the properties of electrons?

8. What are some examples from nature that demonstrate that order is an important aspect of the natural world?

9. Describe the atomic model proposed by Ernest Rutherford.

10. What are *orbitals*?

**Chapter 2
Sources of Energy**

At wind farms like this one near Fluvanna, Texas, wind energy is converted into electrical energy with enormous wind turbines. You can get an idea of how huge these wind turbines are by looking at the base of the third tower back from the front of the photo—the tiny black and white objects at the left are five grazing cows!

Wind farms are becoming hugely popular as a clean energy source. It is now common to see turbine blades being transported by big trucks out on the highway. Obviously, these machines need to be installed in locations where steady winds blow nearly all the time. One problem is that some of our best windy locations are out in the western deserts, far from population centers. Installing the electrical wiring necessary to connect turbines at these remote locations to the rest of the country has big practical limitations. Thus, wind turbines alone are not going to solve our energy problems.

OBJECTIVES

After studying this chapter and completing the exercises, you should be able to do each of the following tasks, using supporting terms and principles as necessary.

1. Describe four (possibly five) locations where the bulk of the energy in the universe is found.
2. Describe how the energy produced by the sun is generated and how it is transmitted to the earth.
3. Explain what is meant by the "dual nature of light."
4. Explain what a quantum of energy is.
5. List six regions in the electromagnetic spectrum, in order from longest wavelength to shortest.
6. List the six colors in the visible spectrum, in order from longest wavelength to shortest, and the wavelength range of visible light.
7. Describe six major forms of energy, and explain how each one is used to produce electrical power in the 21st century.
8. Describe some of the drawbacks of using fossil fuels for electrical power production.
9. Describe some of the drawbacks of using nuclear energy for electrical power production.

VOCABULARY TERMS

You should be able to define or describe each of these terms in a complete sentence or paragraph.

1. chemical potential energy
2. electromagnetic radiation
3. electromagnetic spectrum
4. energy
5. fission
6. fusion
7. gravitational potential energy
8. heat
9. hydroelectric power
10. kinetic energy
11. mass
12. nuclear energy
13. photon
14. photovoltaic cell
15. potential energy
16. quantum
17. quantized
18. solar cell
19. thermal energy
20. turbine
21. visible spectrum
22. wavelength
23. wave packet

2.1   What Is Energy?

As we saw in the last chapter, energy is one of the three main ingredients in the physical world around us. You have been talking about energy all your life. Certain foods give you energy. When you are tired you feel out of energy. Because of increasing pressure on our resources and on the environment, everyone these days is talking about saving energy and "green energy."

But even though talking about energy is common, defining energy is tricky. It is easy to talk about what energy *does*; it is difficult to say what it *is*. But this problem is not going to get in our way too much. We can still get a good handle on energy by studying the things we *do* know about it. Here are some of the things we know about energy:

- Energy is one of the three fundamental ingredients in the universe (matter, energy, and intelligence).
- All the energy there is in the universe was there in the universe at the time the universe began. This energy exists now, and much of it is available for our use.
- Energy is needed to accomplish any and every task.
- Energy exists in many different forms and can be transformed from one form to another.
- Energy obeys the law of conservation of energy.

We discussed the first of these points in Section 1.1. In the rest of this chapter and the next, we unpack the others.

Learning Check 2.1

1. Make a list of several tasks that require energy to perform. Include tasks that would be accomplished by machines, by animals, and by humans.
2. Identify three processes that occur in nature—but not involving animals or humans—in which energy is involved.

2.2   Where Is the Energy?

Approximately, 13.8 billion years ago, matter, energy, and the laws of nature came into existence. No matter how you think about it, this original event is so amazing it is beyond comprehension. Figure 2.1 is a diagram showing the present view among cosmologists (scientists who study the origin and development of the universe) about major events in cosmic history and how the universe has expanded since it began.

As for the matter in the universe, most of the visible matter consists of the atoms in the stars, and in gases found in and between the galaxies. Obviously, there is also matter in the planets, moons, and comets found in galaxies throughout the universe. I wrote "visible matter" because many scientists now think that there is also "dark matter" in the universe that we cannot see. (We don't know much about dark matter yet.)

Scientists, Experiments, and Technology

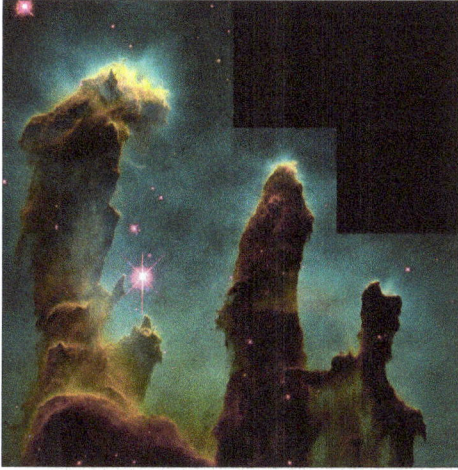

The amount of energy in the universe is beyond comprehension. Instruments now operating in space have provided some beautiful images for us to contemplate. The first two images here were produced by the Hubble Space Telescope. The first is the Eagle Nebula, a young formation of stars and star-forming gas and dust in the constellation Serpens. The

second is the beautiful Cat's Eye Nebula in the constellation Draco. The Cat's Eye has a very complex structure, and there is much about it that scientists do not presently understand.

The lower image is the Andromeda Galaxy. The Andromeda Galaxy is the nearest galaxy to our own Milky Way, even though it is 2.5 million light-years away. The image was produced by the European Herschel Space Observatory.

Astronomy is an ancient (the oldest) natural science, and was part of all the earliest known civilizations.

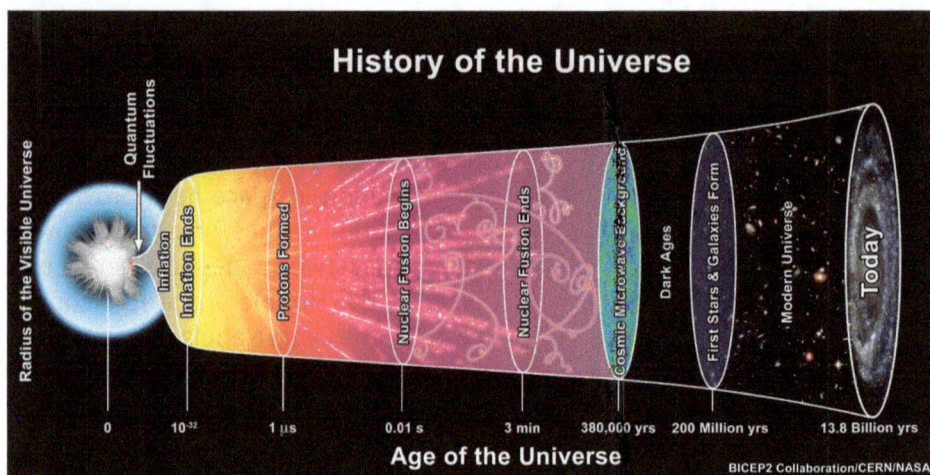

Figure 2.1. An illustration of the major events in cosmic history.

It is a bit tougher to identify where all the energy in the universe is. It is also kind of mind-blowing to think about all this energy—there is a lot of it. The energy in the universe can be thought of as residing in four main places, possibly five.

First, a lot of the energy in the universe is in the light that all stars are producing. The more general term for this energy is *electromagnetic radiation*, waves of pure, massless energy that penetrate every part of outer space. We will look more closely at electromagnetic radiation in the next section.

Figure 2.2. Using this large antenna scientists discovered the Cosmic Microwave Background in 1964.

The Cosmic Microwave Background is believed to be energy left over from the universe's sudden origin, also known as the Big Bang.

A second place where a lot of energy is found is the electromagnetic radiation in the so-called *Cosmic Microwave Background* (CMB). Using the huge antenna shown in Figure 2.2, in 1964 scientists discovered that the universe is full of microwaves, electromagnetic radiation that permeates space in every direction. The prevailing view among scientists about the CMB is that it is energy left over from the original creation event, also known as the Big Bang. In fact, the existence of the CMB and its particular structure is regarded as very strong evidence for the Big Bang. It is the radiation in the CMB that is responsible for the fact that the temperature throughout outer space is about three degrees above absolute zero.

A third location for the universe's energy is in the motion of all the galaxies as they rotate and rush away from each other in our expanding universe. The energy associated with moving objects is called *kinetic energy*. The galaxies are big and they are moving very fast, so there is a lot of kinetic energy in their motion.

Fourth, there is a great deal of energy in the gravity present in the universe. This may sound strange to say, so here is an analogy using a bow and arrow. It takes energy to pull back the string of a bow and bend the bow. If you pull back a bowstring, the energy it took do the stretching is stored in the bent bow. If the bowstring is released, the energy in the bow is released as well (and is used to shoot the arrow). Now, you know that gravity tries to pull things together, like a bent bow trying to straighten itself. Energy is stored when objects are held in a gravitational field, just like the energy stored in the bent bow that is ready to shoot the arrow. Like the energy stored in the bent bow, there is energy in all the gravitational attractions between the objects in the universe.

The energy in a gravitational attraction between stars and planets is like the energy stored in a bent bow.

Finally, there may be a fifth place where the universe's energy is located. Just as scientists suspect the existence of dark matter, they also suspect that the universe may be full of "dark energy." Dark energy is a fairly new theory. There are some very strong arguments for suspecting that dark energy exists. But we are still in the early days of this idea, and it will take some years before there is enough experimental evidence to back up this theory. Thus, whether dark energy actually exists or not remains to be seen.

Interestingly, we can even say that gravity is the cause of the light the stars produce. This is so interesting that I will take the time here to explain the process.

The energy produced by the sun—and that we depend on to live—is due to a continuous nuclear reaction going on inside it. This nuclear process, illustrated in Figure 2.3, is called *fusion*. It is the same process as the nuclear reaction that occurs when a nuclear hydrogen bomb is detonated. The sun is full of hydrogen atoms, and the nucleus of a hydrogen atom is a single, lone proton. Fusion reactions in the sun occur because the tremendous gravitational attraction in the sun pulls the hydrogen atoms in the sun together so forcefully that they stick together, or *fuse*. The sun's internal gravitational attraction causes the continuous fusion of hydrogen atoms, which in turn leads to a continuous release of energy in the form of *electromagnetic radiation* or *waves*. Some of these waves are visible to us as light. Other waves are invisible but can be felt, such as the heat you feel when you stand in direct sunlight.

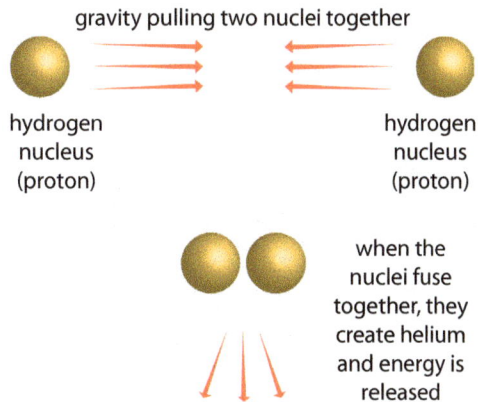

gravity pulling two nuclei together

hydrogen nucleus (proton)

hydrogen nucleus (proton)

when the nuclei fuse together, they create helium and energy is released

Figure 2.3. Fusion between two hydrogen nuclei releases energy.

At present, the only energy we can use from outer space is the electromagnetic waves coming to us, primarily from the sun. Our other energy sources are in the earth. We look at a number of these different energy sources next.

Learning Check 2.2

1. List the four (or five) major locations where we find the universe's energy.
2. How does the energy from the sun get to us here on earth?
3. What process in the sun is producing the sun's energy? Describe this process.

2.3   Sources and Forms of Energy

Studying the forms of matter and energy, and the ways energy and matter relate to one another, is a major part of what the physical sciences are all about. We can learn a lot about the different forms of energy by looking at the sources of energy we depend on. In this section, we look at six major sources of energy.

Electromagnetic Radiation By far the most important source of energy for life on earth is the energy from the sun. Light and heat from the sun make all our crops grow, which feeds everyone on the planet. It also feeds every animal on the planet, and many of these animals feed us too through their meat and their milk. Heat from the sun also warms the oceans, the land, and the atmosphere, making the temperature on earth just right for life to flourish.

In addition to the energy from the sun that grows our food and warms our planet, we can also capture this energy and turn it directly into electricity though the use of *photovoltaic cells*, or *solar cells*. Figure 2.4 shows a solar panel capturing the sun's light. This energy is converted to electricity by the photovoltaic cells in the panel, and stored in a battery during the day. Energy from the battery then powers the street light at night.

As solar technology continues to improve, solar cells are being used more widely to supply the energy we need. Figure 2.5 shows a power generating station in Fukuyama, Japan that uses a huge field of solar panels to capture enough energy to help power the city. Solar panels are still pretty expensive to install, and they obvi-

Figure 2.4. The solar panel at the top of the pole at this bus stop captures energy from the sun and uses it to power the light at night.

ously don't produce any electricity at night. But during the day, the electrical power they produce is free and inexhaustible.

Figure 2.5. Fukuyama solar power generating station.

As we have seen, the energy traveling to earth from the sun is in the form of electromagnetic radiation. It is usually difficult for ordinary people (like most of us) to get a grip on what electromagnetic radiation is. So if thinking about electromagnetic radiation makes your head spin, join the club. It makes *my* head spin, too. My advice is that you do your best to understand as much as you can about what it *is*, but that you focus more on how it *works*.

The weirdest thing I know about this form of energy is this: electromagnetic radiation behaves like *waves*, but it also behaves like *particles*. Because of this "dual nature," scientists have adopted the terms *wave packet* and *photon* to refer to a single particle of electromagnetic energy. Historically, the wave theory of light was fully developed before the modern-day idea of photons ever arose. Light does all the things other waves do, such as reflection, refraction and diffraction, which we will study later. Because of light's behavior as waves, which scientists already knew a lot about in the 19th century, the wave theory of light was worked out in full detail.

Light exhibits the properties of both waves and particles. For this reason, we say that light has a "dual nature."

But at the end of the 19th century, there were several important phenomena involving light energy that scientists did not understand. The solution to these problems emerged when the great German physicist Albert Einstein (Figure 2.6) came along in 1905 and announced that light came in tiny lumps, which we now call photons, but which the physicists also call *quanta*. (*Quanta* is the plural of *quantum*, which comes from Latin.)

This was a *major* discovery in the history of science. But Einstein's discovery was about more than just light. His discovery was that energy itself—in any

Energy is quantized. It comes in little lumps called quanta.

Figure 2.6. German physicist Albert Einstein.

form—is *quantized*. When something is quantized, that means it comes in quanta, or lumps. A good analogy to this is water, which seems smooth and continuous but which is actually divided up into separate molecules. The smallest piece of water is one water molecule. Analogously, energy seems like it is available in any imaginable quantity, but it is not. It is quantized—it comes in quanta, and the smallest piece of energy is one quantum of energy.

Even though we now know that electromagnetic radiation is quantized, we still use wave terminology a lot when discussing light. When we distinguish electromagnetic waves from one another, the most common way to do it is by referring to the wavelength of the waves. As illustrated graphically in Figure 2.7, the wavelength is the distance between the peaks in the wave. Electromagnetic radiation exists in a huge spectrum of wavelengths. Appropriately, we call this the *electromagnetic spectrum*. In Figure 2.8, I have shown the approximate locations of several of the more well-known regions in the electromagnetic spectrum, in order from longer wavelengths on the left to shorter wavelengths on the right.

Each of the names in the diagram—such as FM Radio—designates a region of wavelengths

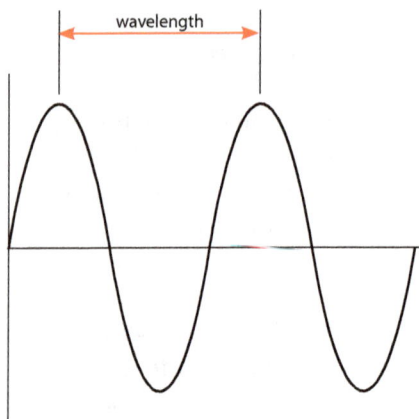

Figure 2.7. The wavelength of a wave.

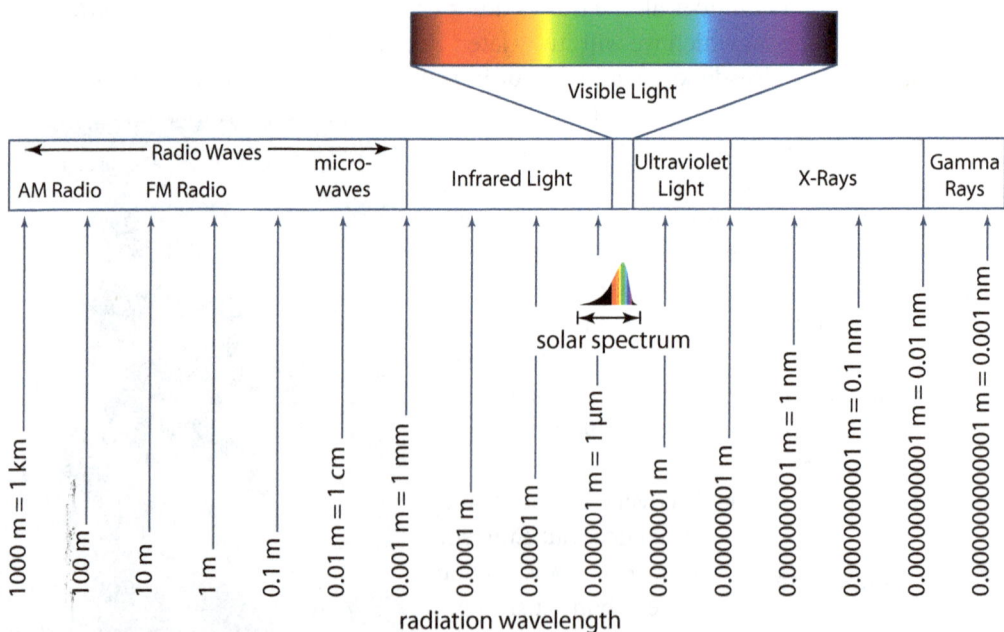

Figure 2.8. The electromagnetic spectrum.

in the electromagnetic spectrum. As you can see, the only difference between FM radio waves and X-Rays is that the wavelengths of X-Rays are about 40 billion times shorter than the roughly three-meter length of FM radio waves. X-Rays, as you may know, can be used to produce images on film of the bones inside our bodies, such as the elbow shown in Figure 2.9.

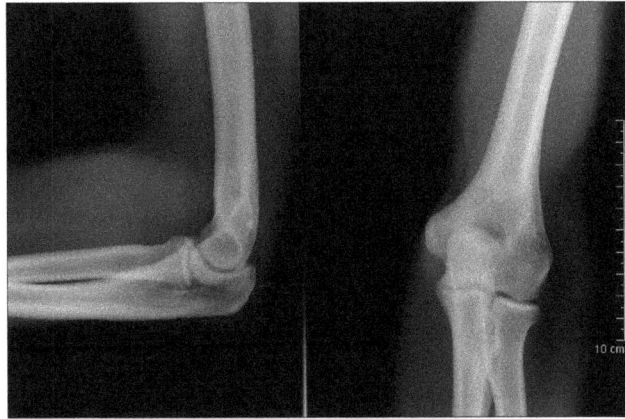

Figure 2.9. X-Rays are used to produce images like these two images of an elbow. The images produced by X-Rays are called radiographs.

In Figure 2.8, I have also shown the solar spectrum, the range of wavelengths produced by the sun. The sun's radiation is strongest in the middle part of the spectrum and tapers off toward the ends. The black region on the left end of the solar spectrum is infrared light—heat we can feel but cannot see. The most intense energy is in the *visible spectrum*, indicated by the colors from red to violet. On the right end is a short span of ultraviolet radiation.

From the diagram, you can see that the wavelength range of the electromagnetic spectrum is so wide that we really need to use the metric prefixes to talk conveniently about the different wavelengths. Table 2.1 lists the lengths and metric prefixes I used in making the spectrum diagram.

Length in Meters	Equivalent Length in Other Units	Length Written with Unit Symbol
1000 m	1 kilometer	1 km
0.01 m	1 centimeter	1 cm
0.001 m	1 millimeter	1 mm
0.000001 m	1 micrometer	1 µm
0.000000001 m	1 nanometer	1 nm

Table 2.1. Common metric length units and their symbols.

For talking about light, the most important unit in this table is the nanometer, one billionth of a meter. This is because the wavelengths of visible light are most conveniently expressed in nanometers. In Figure 2.10, the colors in the visible spectrum are shown. These are the wavelengths of light that our eyes respond to. Each different wavelength corresponds to a different color in the spectrum. The numerical values in the figure are the approximate wavelengths marking the boundaries

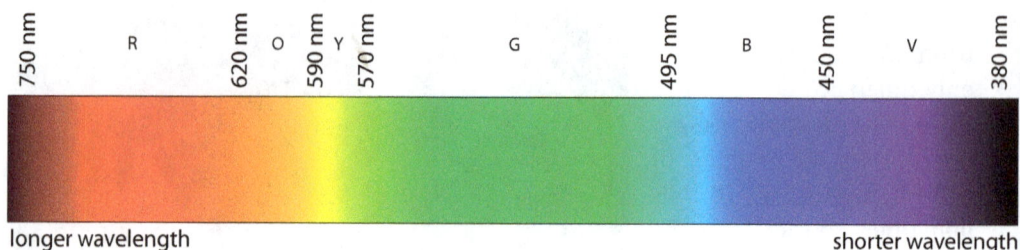

longer wavelength   shorter wavelength

Figure 2.10. The visible light portion of the electromagnetic spectrum showing approximate reference wavelengths.

between the six major colors. For example, orange light lies approximately in the range of 620 nanometers to 590 nanometers.

The colors in the visible spectrum are the same as the colors in a rainbow, and you should know the color sequence by heart: red—orange—yellow—green—blue—violet. The entire visible spectrum runs from wavelengths of about 700 nanometers (red) to 400 nanometers (violet).

There are two final points to make about electromagnetic radiation. First, the energy in electromagnetic waves gets higher as the wavelength gets shorter. This is why X-Rays and other short-wavelength radiation can be harmful to us. Longer wavelengths have lower energy, so we can have AM and FM radio waves all around us (and we do) without harm.

Finally, all electromagnetic waves travel at the same speed, the speed of light, which is 300,000,000 meters per second in air or in the vacuum of space. Electromagnetic waves travel at this speed regardless of their wavelengths, so this is the speed not only for visible light, but for radio waves and X-Rays as well. In water, light only travels at about 75% of its speed in air. In glass, light slows to about 66% of its speed in air.

Learning Check 2.3a

1. Explain the purpose of a photovoltaic cell.
2. Why is light said to have a dual nature?
3. What does it mean to say that energy and light are quantized?
4. Explain what the electromagnetic spectrum is.
5. Identify, in order by wavelength, eight regions in the electromagnetic spectrum, and six colors in the visible spectrum.
6. Why is shorter-wavelength electromagnetic radiation more harmful to humans than longer-wavelength radiation?

Kinetic Energy

I mentioned previously that the energy associated with moving objects is called *kinetic energy*. If someone lightly tosses a baseball at you and it hits you on the shoulder, it does not really hurt at all. But a baseball flying off the end of a bat is traveling very fast and has a lot more kinetic energy—that is why it hurts a lot if it hits you. So kinetic energy depends on *speed*.

Kinetic energy also depends on *mass*. I have already mentioned mass once or twice in this book, but so far we have not stopped to discuss it. So we need to pause here and talk about mass. Please now read the box below that explains the term mass. After you do so, return here to continue reading.

Kinetic energy is the energy associated with moving objects. Kinetic energy depends on both the object's speed and its mass.

As I said, kinetic energy also depends on mass. What do you think does more damage to a car in a parking lot: a runaway shopping cart moving at 10 miles per hour, or another car moving at 10 miles per hour? The shopping cart might put a little ding in a car door when it hits, but another car moving at that speed can crush

---

*We Pause Here to Talk About **Mass***

The term *mass* is a measure of how much matter there is in a given substance. Everything made of atoms has mass. Mass is different from *weight*. Weight is caused by gravity, and the weight of an object depends on where it is. Objects on the moon only weigh about 1/6 of their weight on earth, and in outer space, where there is no gravity, objects have no weight at all. But the mass of an object does not depend on where it is. The mass of a given object is the same, whether the object is on earth, on the moon, or in deep outer space. This is because an object's mass is based on the matter the object is made of.

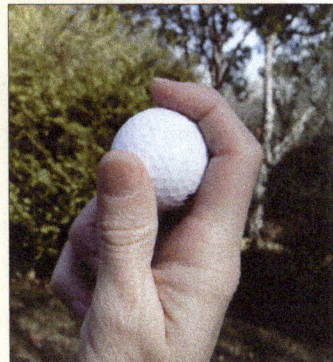

So consider this thought experiment involving a golf ball and a table tennis ball. Throwing a golf ball takes more force than throwing a table tennis ball. This is because the golf ball has more mass. Now, the golf ball also weighs more than the table tennis ball does. But imagine taking them both out into space. Out in space, both balls are weightless, because there is no gravity to give them weight. But in outer space, it still takes more force to throw the golf ball than it does to throw the table tennis ball, because the golf ball has more mass.

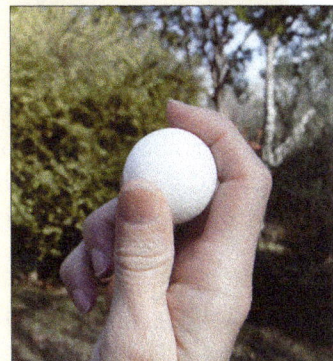

The more atoms there are packed into an object, the more mass it has. And since the different types of atoms themselves have different masses, an object made of more massive atoms has more mass than an object made of an equal number of less massive atoms.

Figure 2.11. The new wind turbines provide us with a lot of electrical energy.

the door in. The car has a much greater mass than the shopping cart, so even when moving at the same speed it has a lot more kinetic energy and can thus do a lot more damage.

In summary, every moving object possesses kinetic energy, and the amount of kinetic energy depends on both its speed and its mass.

Wind energy is an important source of energy. The energy is in the kinetic energy of the moving air molecules, and even though air molecules don't have much mass, when huge numbers of them move together we get a lot of energy. You have probably seen wind turbines like those in the photo of Figure 2.11. This photo of a wind turbine at a beach in the United Kingdom gives an idea of how huge they are. These turbines are driven by the kinetic energy in the wind—the moving air molecules—and they convert the wind's kinetic energy into electrical energy. This electrical energy is then transported in electrical wiring to different communities to help supply their electrical power needs.

Learning Check 2.3b

1. What is mass?
2. Consider throwing a golf ball and a table tennis ball. If you throw them at the same speed, which one has more kinetic energy, and why?
3. Describe one way we are able to use the kinetic energy found in nature to generate electricity.

Potential Energy

Several forms of energy go by the term *potential energy*. Potential energy is energy that is stored somehow, but which can be released from storage and put to use. A simple example of potential energy is the energy in the compressed spring of a dart gun. When the trigger is pulled, the potential energy in the compressed spring is released and converted into kinetic energy in the moving dart that flies out of the gun.

Two important types of potential energy that relate to our sources of energy are *gravitational potential energy* and *chemical potential energy*. We look at both of these next.

Gravitational Potential Energy   Recall from Section 2.2 that there is energy in the gravitational attraction between two objects. This energy is the source of *gravitational potential energy*. This form of energy has been put to great use around the world in the construction of large dams that use the gravitational potential energy in the water behind the dam to produce electricity. The water in a lake is being attracted toward the center of the earth, and if the water is allowed to fall, its gravitational potential energy is converted into kinetic energy as it picks up speed.

Figure 2.12 shows the water held by Hoover Dam, and Figure 2.13 shows the upstream side of the dam just after its completion in 1935, before the lake had filled.

Shown in both photos are the cylindrical housings that contain the turbines that drive the electric generators. Water from the lake enters the turbine housings and falls, converting enormous amounts of gravitational potential energy into kinetic energy in the spinning turbines, and then into electrical energy by the electric generators. After powering the turbines, the water is released into the bottom of the canyon on the other side of the dam, where it continues its journey down the Colorado River.

The term *hydroelectric power* is used to describe electric power generated by the water retained by a dam. Construction of a dam is usually a controversial undertaking because of the land that goes un-

Figure 2.12. Hoover Dam, completed in 1935 on the border between Arizona and Nevada. The turbines in the cylindrical housings are driven by the gravitational potential energy of the water. The turbines drive electric generators to produce electrical power.

Figure 2.13. The upstream side of Hoover Dam in 1935, just as the lake was filling. The four turbine housings can be clearly seen.

Figure 2.14. This unusual water tower in Brazil was named Varginha's Spacecraft, after an alleged UFO siting that occurred in Varginha. The energy required to pump the water up into the tower is stored as gravitational potential energy in the elevated water.

Gravitational potential energy is the energy in an object that is held in a gravitational field. If the object is released, the gravitational potential energy is converted into kinetic energy as the object falls.

der the lake and the interruption of river flow to those downstream who depend on it. Nevertheless, once finished, the electric power is produced without fuel and without pollution. Ultimately, the energy comes from the sun because it is the sun's heat that evaporates the water into the atmosphere, beginning the water cycle that continues in the rain and the water flow in the rivers.

Another common use of gravitational potential energy is in the storage of water in water towers, like the handsome Nave Espacial de Varginha, shown in Figure 2.14. Electric pumps are used to pump the water up into the tower. With the water up in the tower, its weight produces a steady pressure in the water lines that take the water to the homes and offices where it is used. By allowing the weight of the water to create the water pressure, the pressure is there all the time without any electric pumps running. The pumps can switch on periodically, as needed, to refill the tank when it gets low.

Chemical Potential Energy

Chemical potential energy is the term for energy stored in the chemical bonds holding together the molecules in various fuels or foods. Chemical potential energy is released during a chemical reaction that breaks down the compounds in the fuel or food, releasing energy in the process. An example of this occurring is when fuels are burned to release heat. Another example of chemical potential energy being released is when your body digests the food you eat. The digestive process breaks down the molecules in the food, releasing the energy the body needs to function.

Chemical potential energy is the energy locked into the bonds of molecules. This energy is released by chemical reactions such as the burning of fuel.

By far, the most predominant sources of energy since the beginning of the industrial era are fossil fuels—coal, petroleum, and natural gas. These three fuels are extracted from underground mines and reservoirs. As Figure 2.15 illustrates, the fuels are transported to every city in the world where they are used to generate electricity, fuel vehicles, and heat homes. One of the major uses of petroleum is in the production of gasoline, diesel fuel, and jet fuel. The majority of our transportation depends on these fuels. Petroleum is also the source of propane and heating oil, which are used in many areas for heating homes.

Fossil fuels, abundant and inexpensive, powered the Industrial Revolution of the 19th century, which forever changed how people live in modern cities. Fossil fuels continue to be the major energy source in the world today for producing electricity. In one process, coal is burned to boil water in a large boiler. The high-pressure steam from the boiler is then used to power a turbine, and the turbine powers a generator that produces electricity. (See the separate box on the next page describing how the steam turbine-generator system works.) In another process, natural gas is burned directly in a gas turbine to power the generator. We now have clean water, advanced medical technologies, easy transportation, and comfortable homes because of fossil fuels.

However, as helpful as these fuels have been in improving our standard of living, use of fossil fuels has come at a price. Burning fossil fuels pro-

Figure 2.15. Coal (top), petroleum (center), and natural gas (bottom) being transported and stored.

duces a lot of pollutants, including carbon dioxide (CO_2). The amount of CO_2 in the earth's atmosphere has been growing steadily for many decades, rising from a historical level of 280 parts per million (ppm) to over 400 ppm today. There is now strong scientific evidence that the rising level of CO_2 is the result of burning fossil fuels and is leading to changes in the global climate that could have very undesirable consequences for life on earth.

*We Pause Here to Talk About **Turbines***

We have talked about wind turbines and about steam turbines. Here we will look at what turbines are and how they are used to produce electricity.

At the left is a photo of a common pinwheel. You have seen these spinning as the wind passes over their blades. In the same way that wind makes a pinwheel spin, the wind makes the large wind turbines spin. The same principle is also at work in turbines driven by water at a hydroelectric station, or by steam, such as the one shown in the next photo. A steam turbine is encased in a steel housing, and when the high-pressure steam is injected into the housing, the steam acting on the turbine blades forces the turbine to spin.

The main turbine shaft is connected to an electric generator, which uses

magnetic principles to generate the electricity. The third photo shows a steam turbine (far right) connected to an electric generator (center) in a power station. On the left a motor is connected to the generator to get the generator spinning during start-up. After that, the turbine takes over.

The topic of "climate change" is very politically controversial. Some argue either that climate change is not happening, or that it's no big deal, or that it is happening, but that it is due to natural processes on earth and not due to human industrial activity. However, the huge majority of climate scientists agree that the known rise in atmospheric CO_2 is due to the burning of fossil fuels, and that unless we find ways of significantly reducing or even eliminating CO_2 emissions soon, disastrous changes in the weather patterns and in the levels of the oceans are coming over the next several decades.

Everyone agrees that development of new energy sources that are clean and renewable is very desirable.

Even though some people do not accept the scientific evidence for climate change, the fact remains that burning coal, oil, and gas at high rates all over the earth does pollute the environment. It is also a fact that the reserves of fossil fuels in the earth are limited and that sooner or later supplies will run low. For these reasons, everyone agrees that development of new energy sources that are clean and renewable is very desirable. Accordingly, development of alternative energy sources is a major part of scientific research today. The wind turbines and solar cells we have discussed are a big part of the drive for clean, renewable energy sources.

Besides the CO_2 emissions issue, there are several other drawbacks to the use of fossil fuels for electrical power. One is the risk of an oil spill or gas release during transportation in pipelines or ships, or at the well where the oil or gas is pumped out of the ground. A large oil spill, of which there have been many, can cause serious damage to the environment. There have been two major oil spills in the United States in recent years. In 1989, the Exxon oil tanker *Valdez* ran aground in Alaska's Prince William Sound and leaked over 10 million gallons of crude oil into the ocean. Exxon's clean-up crews worked heroically for months to clean up the mess, but it took years for the environment to recover. More recently, in April 2010 an explosion on the British Petroleum oil rig *Deepwater Horizon* in the Gulf of Mexico caused oil to flow continuously into the gulf for three months from a broken wellhead on the sea floor. Before the well was finally sealed, some 210 million gallons of oil had leaked into the Gulf, causing extensive environmental damage and economic loss. In the satellite photo of Figure 2.16, the oil from the BP spill appears white on the water's surface because of the way it reflects the sunlight.

Figure 2.16. Oil on the water in the Gulf during the 2010 BP Deepwater Horizon oil spill.

Releases of natural gas into the atmosphere during drilling operations are also harmful. Natural gas consists primarily of methane, and methane is far worse in its effect on the climate change problem than CO_2 is.

Another drawback to the use of fossil fuels, especially of crude oil (petroleum), is that in America, as in many other countries, we depend heavily on oil supplies from oil-producing foreign countries. Thus, there is always the risk that our oil supplies could be cut off due to an event such as a war or a natural disaster.

Learning Check 2.3c

1. When talking about different forms of energy, what does the term "potential" mean?
2. What is gravitational potential energy?
3. How can the water in a lake behind a dam be used to generate electricity?
4. Why is water stored in water towers?
5. What is chemical potential energy?
6. Describe three drawbacks associated with using fossil fuels for electrical energy production.
7. Explain how fossil fuels are used to produce electricity.

Thermal Energy

Thermal energy is a term we use to describe the energy in a substance that has been heated. For example, if you put a pan of water on a gas stove and light the burner, the chemical potential energy in the gas is released by the burning (a chemical reaction). This chemical potential energy from the gas is released in the form of heat, and the heat flows into the water in the pan, heating up the water. Now that the water is hot, it contains more thermal energy than it did when it was cool.

Thermal energy is a natural energy source. Far below the surface of the earth, the earth's interior is extremely hot—about 6700°C, scientists estimate. This is hotter than the surface of the sun! Most of the energy heating the interior of the earth is coming from nuclear decay of radioactive elements, as I described in the box at the end of Chapter 1. In mountainous areas, there are often channels underground

Figure 2.17. This geothermal power plant in Steamboat Springs, Nevada uses the heat from inside the earth to generate electricity.

that allow the molten rock deep in the earth to come close to the surface. This heat near the surface is the energy source driving volcanoes and geysers. In these mountainous areas, the hot material underground is close enough to the surface that it is possible to drill down and tap into it. The thermal energy in the ground can be then used to generate electricity. The energy obtained from these kinds of plants, such as the one shown in Figure 2.17, is called *geothermal energy*.

Geothermal energy production has been around since the early 1900s, but geothermal projects became more popular in the late twentieth century as energy costs began to rise and environmental concerns began to become more prominent.

Learning Check 2.3d

1. What is the source of earth's thermal energy?
2. Why might the use of geothermal energy not be practical in many parts of the world?

Nuclear Energy We have already seen (Section 2.1) that the nuclear process known as fusion releases a lot of energy. This is the process going on inside the sun and the other stars. Fusion occurs when protons are forced to join together. Fusion occurs in the sun and in nuclear weapons, but so far scientists have not succeeded in using fusion in a controlled way to generate power we can use. (A nuclear explosion is an *uncontrolled* nuclear reaction. Not very useful for generating electricity!)

However, there is another nuclear process, *fission*, that we do know how to use. Fission, illustrated in Figure 2.18, is the splitting apart of the nucleus in a large atom, such as uranium. Just as with fusion, energy is released when fission occurs.

Inside a lump of uranium, the fission, or splitting, of one atom can result in a nuclear chain reaction. A neutron is fired into the nucleus of a uranium atom, shattering the nucleus. Energy is released, and large chunks of the nucleus

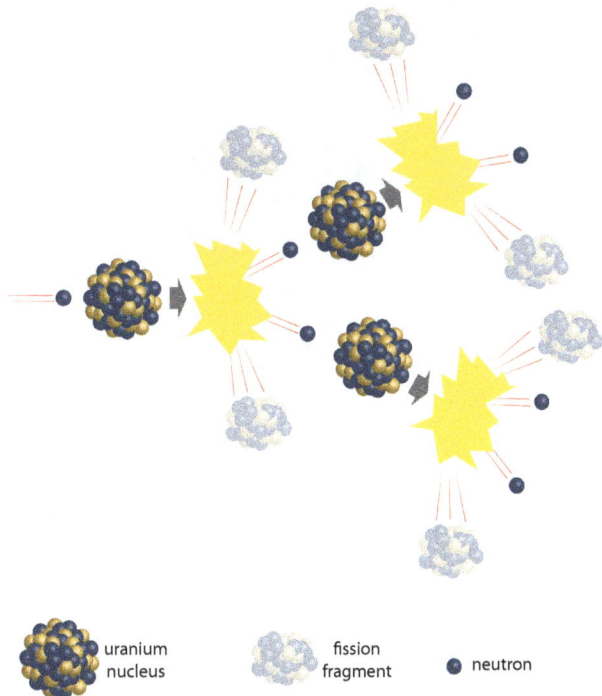

uranium nucleus   fission fragment   neutron

Figure 2.18. A fission reaction is a chain reaction that begins when a neutron shatters the nucleus of a uranium atom. Neutrons coming out of the shattered nucleus hit other nuclei, shattering them, and continuing the chain reaction.

called "fission fragments" fly apart, along with a few lone neutrons. These neutrons collide with other nearby nuclei, shattering them. Each time a nucleus splits, energy is released and more neutrons come flying out to continue the chain reaction. This process can be controlled so that the chain reaction continues at just the right rate, without dying out and without getting out of hand.

Just as with the burning of fossil fuels, nuclear power stations use the heat created by the nuclear fission reactions to produce steam, and the steam is used to drive a turbine connected to an electric generator. There are 390 nuclear power plants in the world, like the one shown in Figure 2.19, with another 65 plants under construction. Nuclear power stations currently supply 13% of the world's electricity, and about 19% of the electricity in the U.S.

Nuclear power has always been very controversial for several reasons. First, the waste produced is deadly radioactive because the nuclear decay process in the fuel continues after the fuel is no longer useful in a reactor. Not only is the waste deadly, it remains so for tens of thousands—or even millions—of years. Much debate and research is devoted to finding technologies to reduce or reprocess waste, but right now the main waste disposal strategy in the U.S. is to store the waste permanently in controlled storage locations.

Figure 2.19. Sequoyah Nuclear Generating Station, in southeast Tennessee.

The second point of controversy is the threat of terrorists or rogue nations obtaining nuclear material and using it to make weapons. The more nuclear power stations there are, the more fuel production and waste disposal there is, and the higher the risk is that nuclear materials will fall into the wrong hands.

The third point is the possibility of an accident or failure at a nuclear power station that could release radioactive material into the atmosphere. This has happened several times, although the worst accidents have all happened outside the United States. A disastrous accident, the worst in history, occurred in Chernobyl, Ukraine, in 1986, and another bad accident occurred in Fukushima, Japan, in 2011. To date, the only serious incident in the United States was the Three Mile Island incident in Pennsylvania in 1979. Since then, safety standards and regulations for nuclear power facilities have been very stringent. Authorities believe these strict regulations will prevent accidents in the future, but many activists do not trust the authorities and are completely opposed to nuclear power generation.

These issues are at the center of political debate about energy in the U.S. today. Growth in population and industry means that more power generation capability is needed—more than can be met by solar and wind technologies at the present time. But building more fossil fuel generating plants means adding to the global CO_2 emissions, whereas building more nuclear facilities entails all the risks I have just described. The same controversies and political debates surround the construction of new power generating facilities in other countries.

It seems clear that conservation—using less energy and producing less waste—must be part of any successful energy strategy. Through conservation and new technologies, we may be able to rein in our energy consumption and reduce pollution. Our goal must be to live within the long-term resource capabilities of planet earth. At present, we are not doing so.

Learning Check 2.3e

1. Explain the difference between the nuclear processes of fusion and fission.
2. Explain how nuclear reactions can be used to produce electricity.
3. Describe three drawbacks associated with nuclear power.

Chapter 2 Exercises

Answer each of the questions below as completely as you can. Write your responses in complete sentences.

1. Describe the four (or five) locations in the universe where most of the energy in the universe is found.
2. Explain how gravity is the hidden cause behind the heat and light produced by the sun.
3. Explain what electromagnetic radiation is, and describe the electromagnetic spectrum.
4. In Section 2.3, we discussed six major forms of energy. Make a chart of these six forms of energy. In your chart, include the following for each one: (a) a definition, (b) a description of the source of the energy, and (c) a description of the process by which the energy is made available to supply the energy needs of homes, buildings, vehicles, or machines.
5. Identify the three common forms of fossil fuel, and describe three drawbacks associated with the use of fossil fuels.
6. Describe three drawbacks associated with the use of nuclear power.
7. Consider this. Reducing waste not only reduces pollution, it reduces energy consumption as well. See if you can identify several reasons for this, and put them together in a paragraph.
8. Describe some of the main political issues involved in building new electrical power generating stations.

Getting Started with Experiments

The Art of Experimental Science

Experimental research is one of the things that makes science so interesting and so much fun. Science is a lot more than just learning things in books. Throughout the history of science, new discoveries have been made and new theories have been tested in the laboratory. If you are the type of person who loves fooling around with parts, wires, wood, and chemicals, then the experiments in this book will be right up your alley. If you are the type of person who would prefer to stay inside where it is air conditioned and drink tea, then the experiments will give you an opportunity to get your hands dirty. Who knows, you may find you love doing experiments! History is full of people—both men and women—who helped in a lab when they were 13, and ended up becoming experimental chemists or physicists. It could happen to you.

In the next few pages we will look at some of the important things to keep in mind while doing experimental work. I will conclude this introduction with a tutorial on preparing scientific graphs.

Safety

Safety is a major concern in any science laboratory, whether that lab is in a classroom, a research facility, your kitchen, or your garage. Here are some standard safety rules you should know and always follow:

1. Always wear safety goggles or safety glasses when heating substances on a hot plate or over a flame.

2. When handling hot substances or apparatus, use tongs or wear thermally protective gloves.

3. Use great care when handling glassware. As I always say, there are three ways to break something—improper procedures, silliness, or carelessness—and all are bad in a lab!

4. Wear protective eyewear and gloves when handling hazardous materials.

5. Always work under the supervision of a responsible and knowledgeable adult when using sharp tools, hot plates, flames, or chemicals.

6. Make sure you have a phone in your work area in case you ever need to call for help.

7. Always follow written procedures, and don't take short cuts. Do not revise procedures to suit yourself without consulting with a responsible and knowledgeable person who knows about the kind of work you are attempting to perform.

8. Never taste things in a science lab unless your instructor directs you to.

9. Always keep long hair tied back out of the way, and don't wear loose, blowsy, or baggy clothing while working in a lab.

10. Make sure you have adequate ventilation.

11. Make sure you have a fire extinguisher in your work area.

12. Exercise care in everything you do, pay attention, and avoid horseplay.

Care and Accuracy

It is common for those new to experimental work to underestimate just how careful one needs to be in order to get accurate results from a science experiment. Students' results are sometimes so inaccurate that they are useless, often requiring the students to do the work over again.

I have already mentioned that carelessness can be a safety hazard. But carelessness can also result in equipment that doesn't work properly, results that don't turn out correctly, or data that are useless. If your experimental data aren't any good, then you have wasted your time, and possibly your teammates' time as well.

So from the very beginning, make it your goal to follow directions carefully, to assemble apparatus with care and patience, and to record measurements with as much accuracy as possible. Resist the temptation to consider hastily or carelessly performed work as "good enough." Developing a passion for care and accuracy makes a big difference in the quality of your results. And when your results are superior, you learn more and you end up finding scientific experiments much more satisfying.

Doing It Over Is Okay

Sometimes, even when you are being as careful as you know how to be, experiments don't turn out the way they are supposed to. Welcome to the world of the scientist! How many thousands of light bulbs was it that Thomas Edison made before he found one that worked? Life in the science lab means sometimes things don't work. So let me offer you this advice and encouragement. When things don't work out right, just try to figure out how to improve your method and do the experiment, or part of it, over again. This may not be convenient; everyone is busy. But it is the right thing to do. If you have no idea what went wrong, then get some advice from someone who can help you figure it out.

Keeping a Lab Journal

Every practicing scientist maintains a *lab journal*, a written record of everything he or she does in the lab. As a science student, it is very important that you learn how to maintain your own lab journal, and that you faithfully document your work in it. Here's why lab journals are important in the real world:

- When work is being passed from one researcher to another, the journal is a record of what has been done in the past and how it was accomplished.
- When particular methods, equipment, or procedures used in the past need to be used again, the details are all in the lab journal.
- When people become famous, apply for patents, win awards, and so on, all the background information folks need in order to verify the work or write the scientist's biography is all there in the lab journal!

You may not invent something that needs to be patented, but you will need to write reports on your experiments, and to do that you need a record of what you did, who helped, when it happened, and what equipment you used. You also need a place to record your data. So when you do experimental work, always document everything in your lab journal. (Your teacher may even grade you on how well your journal is kept.)

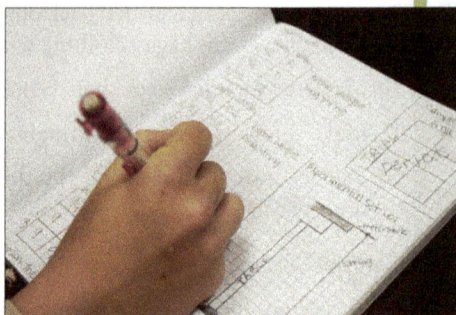

Here is some advice about lab journals:

1. Keep your lab journal very neat and well organized. Don't doodle in it, draw in it, or mess it up.
2. Don't use a spiral notebook. Use a bound composition book with quadrille (graph) paper. (Quadrille paper makes it easy to set up tables and graphs.) Acceptable journals are the National 53-108, Mead 09100, and others available online and at office supply stores.
3. Put your name on it in case you misplace it.
4. For every experiment you work on, enter the following information:

   - the date (always enter the date again every day you work)
   - the names of team members working with you (enter these also every day you work, so you have a record of who is there and who is not each time you meet)
   - a *complete* list of all equipment, apparatus, materials and supplies you use in conducting the experiment
   - the manufacturer and model number for any electronic equipment you use
   - tables with *all* your data, with the original units of measure
   - calculations or unit conversions you perform as part of the experiment
   - observations or notes about anything that happens that you may need to write about in your report or remember later, including records of work that has to be repeated and why
   - methods or procedures you use, and the reasons for using them

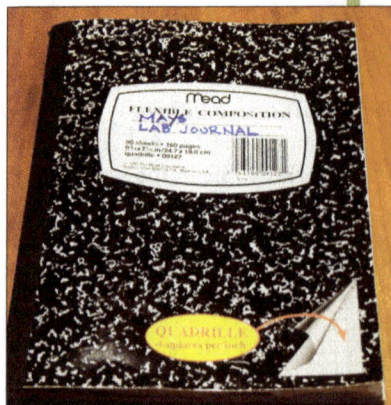

There are other items you can enter in your journal that become more important as you get into high school and college (such as sources, contacts, and prices for special chemicals or parts you have to order), but the list above should cover the things that you need to worry about for now. Take pride in maintaining a thorough lab journal. Make it a habit always to have it with you when you work in the lab, and always to document your work in your journal.

Scientific Graphs

Graphs are extremely important in reporting scientific information. Often, a graph is the best way to present scientific information so that the reader can examine it and understand it most easily. Because scientific graphs are so important, many of the experiments in this text require you to prepare a graph for displaying your experimental results.

Depending on what grade you are in, you may have already learned about using Cartesian coordinates in graphs in your math class. If so, then learning how to prepare a proper scientific graph will be easy. But here I assume that some of the students using this book haven't studied graphing yet and I describe how to prepare scientific graphs in some detail. Even if you already learned how to make graphs in your math class, there are still many specific details about scientific graphs that you need to know. So read this section carefully.

We begin by considering an example experiment and some data from that experiment. We will use these data to illustrate how to set up and format a scientific graph.

You may know that automobiles use water to keep the engine cool. You may also know that to keep the water from boiling in the summer or freezing in the winter, a product called *antifreeze* is mixed in with the water in the car's engine. I tested mixtures of water and antifreeze to see what the boiling point would be with different amounts of antifreeze mixed in. What I found is shown in the table to the right. (You probably know that plain water boils at 100°C. But you may not know that common thermometers are only accurate to +/– 1°C. This means that the reading might be off by one degree too high, or one degree too low. That is probably the explanation for why my boiling point was recorded as 101.0°C with plain water.)

Amount of antifreeze, by volume (%)	Boiling point (°C)
0	101.0
10	102.8
20	104.3
30	105.1
40	106.6
50	110.0

In this experiment there are two quantities we call *variables*. These variables are the amount of antifreeze in the mixture, and the boiling point of the liquid. Notice that it makes sense to think of one of these variables as depending on the other. I select the amount of antifreeze I put in the mixture and the boiling point that results depends on my selection. In the context of graphing scientific information, the variable the scientist selects values for is called the *independent variable*. In my example experiment, the amount of antifreeze in the mixture is the independent variable. The variable that depends on the scientist's selection is called the *dependent variable*, and in the example this is the boiling point.

A basic graph is a grid with a horizontal line across the bottom and a vertical line down the left side, as shown at the top of the next page. The two lines are called *axes*, and they are used as scales for locating the values of the variables for each data point (each row in the data table).

On a graph, we associate the independent variable with a scale marked out on the horizontal axis of the graph. We associate the dependent variable with the vertical axis of the graph. The first thing we have to do to set up the graph is decide what scales to use.

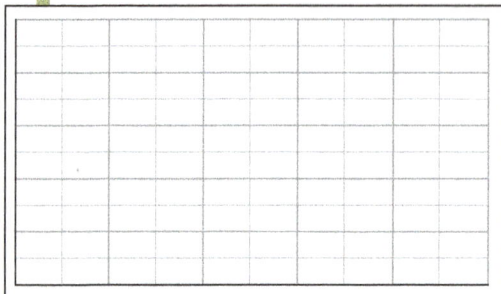

Look again at the data values in the data table. As you see, the values for the independent variable, the antifreeze concentration, range from 0 to 50. We want to pick a scale that comfortably covers this range of values without having a great deal of excess space on either end. I am going to choose a scale that starts at 0 and goes up to 60, just a bit higher than the highest value.

From the table, the values for the dependent variable, the boiling point, range from 101.0 to 110.0. So I will scale my vertical axis from 100 to 112. When I label the axes with these two scales, the graph looks like the illustration below. (Most of the time in math your axis scales probably start at zero. Well here's a little secret: they don't have to!)

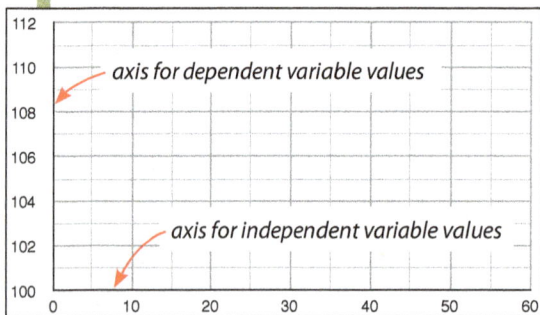

The next thing we need to do is label each of the axes. The label must contain the variable associated with that axis, and the units of measure that go with it. In the third illustration I have added these labels. Notice that I placed the units of measure in parentheses. This is one of the standard methods of formatting the units and is the method you should use for now. Notice also that there are no capital letters in my labels. This is traditional for scientific graphs, and is the formatting I prefer for my students (although some scientific publications are now capitalizing the variable names).

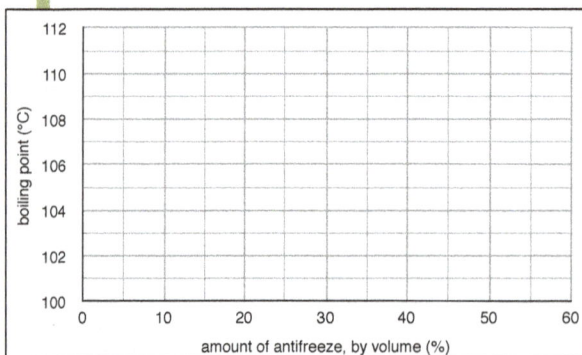

Now we are ready to locate each of the data points from the table on the graph. To do so, do the following for each of the data points (rows) in the data table. First, find the value of the independent variable on the horizontal axis, and using a ruler and pencil, draw a light vertical line from there up into the graph. Next, locate the value of the dependent variable on the vertical axis, and, using a ruler, draw a light horizontal line from there into the graph. Where these two lines meet is where you place some kind of symbol to represent that data point. In the next illustration, I depict this process for the third data point using dashed red lines. The first two data points are already shown on the graph.

Repeat the data point location process until all the data points from the data table are accurately located on the graph. Then connect each of the data points with a straight line, drawn with a ruler. The completed graph looks like the final illustration below.

There are a few final points to make. First, the type of graph I have shown how to make in this tutorial is called an *x-y scatter plot*, or simply *scatter plot*. There are many other types of graphs, but this one is the most important to know about. Scatter plots are used all the time in science to present data and other types of scientific information.

Second, if you are going to place your graph into a report, then your graph needs a title. Standard formatting for titles is to capitalize only the first letter of the first word, and to place a period at the end.

Third, until you get to some very fancy math in college, chances are that the scales on your graphs need to be *linear*. This means that the scales need to be marked in round numbers and regularly spaced. On my graph, the horizontal axis scale is marked in tens, spaced two lines apart. The vertical graph is marked in twos, two lines apart. The regular spacing is crucial in order for the graph to display the correct relationships between the data points.

Fourth, you don't always have to connect the dots in a graph, but it is common to do so.

Fifth, if you are displaying more than one data set on the same graph, you need to use different symbols or colors for the dots and lines of the different data sets. We run into this in Experimental Investigation 2.

Finally, in high school you should begin learning how to prepare graphs on a computer. For now, your instructor may prefer that you concentrate on learning how to draw nice looking graphs by hand. (I think that is appropriate for middle school students.) But when drawing graphs by hand, make sure you a) use a ruler for the axes and for connecting data points, b) make the graph as neat as you are capable of making it, and c) locate your data points very accurately.

Now it's your turn!

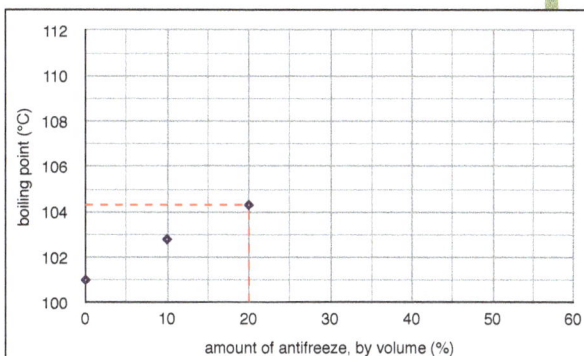

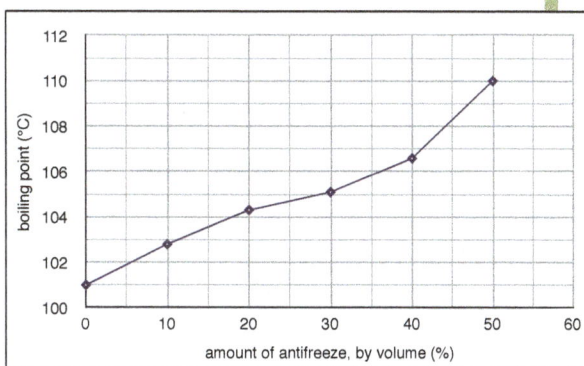

Experimental Investigation 1: Kinetic Energy

Overview

- Using a Hot Wheels car and a ramp of Hot Wheels track, release the car from a standard height repeatedly, adding weights to the car each time to increase its mass. Measure the car's mass before each run.
- *The goal of this experiment is to determine how the kinetic energy of the car varies as its mass changes, assuming the car's speed at the bottom of the ramp stays the same.*
- Use an assembly of friction flaps and a measuring rule to slow the car and measure how far it travels while stopping. Use the stopping distance as a measure of the kinetic energy the car has at the bottom of the ramp.
- Verify that the speed of the car is not affected by mass by releasing two cars with different masses together.
- Collect mass and distance data, and prepare a graph of stopping distance versus mass.

Basic Materials List

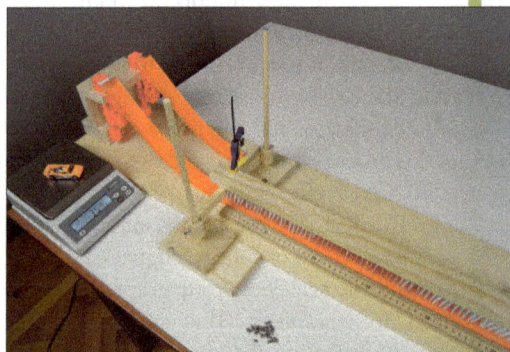

- Hot Wheels cars and track
- lead weights (split-shot fishing sinkers)
- mass balance and measuring rule
- friction flap assembly of 3 × 5 index cards, and apparatus for adjusting its height
- apparatus for mounting the track

In the discussion of kinetic energy, you learned that the kinetic energy of a moving object depends on both the object's mass and its speed. In this investigation, we examine how increasing an object's mass increases the energy the object is carrying in its motion. This experiment will be fun because we get to use Hot Wheels cars! Your setup for this investigation is similar to the one shown above.

We start the car rolling by allowing it to roll freely down a track formed into a ramp. This is how the car gets its kinetic energy—the gravitational potential energy at the top of the ramp is converted into kinetic energy at the bottom of the ramp. Our setup allows us to increase the mass—and the kinetic energy—of the car without increasing its speed. We do this by releasing the car over and over from the same height, so that its speed at the bottom of the ramp is always the same (more on this in a moment). But each time we release the car, we increase its mass just a bit by adding small lead weights to the car. To enable this, we use a car with a cargo bed to hold the weights.

Note to Instructors

Full details on materials, preparations, and procedures for all the *Experimental Investigations* are available from the publisher (see Teacher Preface).

To get an idea of how much kinetic energy the car has at the bottom of the ramp, we make a device that uses friction to slow the car in a stopping zone. Our friction stopper uses flaps of index card to slow the car, and we gauge the amount of kinetic energy the car has by measuring how far into the stopping zone the car goes before the friction flaps bring it to a stop. The photo to the left shows a car being stopped by the friction flaps. The measuring rule next to the track allows the experimenter to measure how far the car travels into the stopping zone.

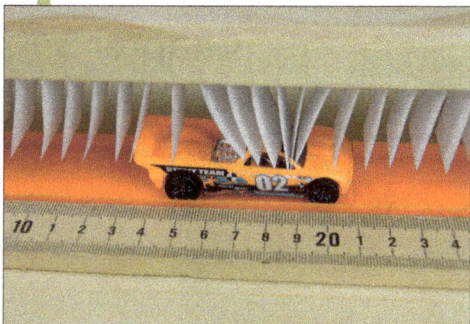

Let's go back for a minute to my comment that every time the car is released from the same height it will be going the same speed at the bottom of the ramp. Back in the 17th century, Galileo demonstrated that falling objects all accelerate at the same rate. This means that *any* object released from a given height above a table will be going the same speed when it reaches the table, regardless of its mass. Accordingly, our car will always be going the same speed at the bottom of the ramp, even as we add weight to it.

But rather than take Galileo's word for it, as part of this experiment you need to verify this for yourself. In the photo, you can see two ramps side by side. Before placing the friction flaps over the track, make a few trials with two cars released from the same height. Use two identical cars, and place weights in one of them to make it about 10–15% heavier than the other. As you do this, see if you can verify Galileo's discovery. In order to make this verification, you need to work out a way to release the two cars at exactly the same time.

For collecting experimental data, you use a single car. Adjust the height of the friction flap assembly so it is level, and so it stops the car before the car reaches the end of the track, even when the car is full of lead weights. Release it 3–5 times from the same height, adding another bit of weight to it each time. Vary the weight from a minimum (car empty), up to at least 130% of the car's empty weight. For each run, measure the car's mass and how far it travels into the friction/stopping zone, and record these data in a table in your lab journal.

Analysis

An important part of the analysis is to prepare a graph of stopping distance (vertical axis) versus mass (horizontal axis). Your instructor will help you with setting up your graph and labeling it properly. In your report for this experiment, address the following questions.

1. Were you able to verify that the car's speed at the bottom of the ramp didn't change, even as mass was added to the car? Describe how you did this.
2. How does the stopping distance relate to the car's kinetic energy? Use your graph to help address this question. Explain what caused the stopping distance to increase.
3. Where does the kinetic energy of the car go as the car stops?
4. What does the shape of your graph tell you about the relationship between kinetic energy and mass?

Sir Benjamin Thompson, also known as Count Rumford, was an American-born British physicist and inventor. In the late 18th century, Thompson studied the heat produced during the process of boring cannon (drilling the hole down the center of a cannon after it is cast), and concluded that the heat was produced by the motion and friction of the drilling machine. His theory about heat was completely opposed to the prevailing view at the time, which held that heat was a substance inside objects. His ideas were foundational for the formulation of the law of conservation of energy a century later.

OBJECTIVES

After studying this chapter and completing the exercises, you should be able to do each of the following tasks, using supporting terms and principles as necessary.

1. State the law of conservation of energy and give examples of its application.
2. Explain how friction on a moving object affects the forms of energy present in a given situation.
3. Explain Einstein's principle of mass-energy equivalence, and use the principle to explain the source of the energy in nuclear reactions.
4. Describe the three ways heat can transfer energy from one substance to another.
5. Explain how the heat radiating from a warm object can be used to determine the object's surface temperature.
6. Define *work*, and give examples of situations when work is or is not being performed.
7. Use the concept of internal energy to explain how a substance retains energy within its atoms or molecules.

VOCABULARY TERMS

You should be able to define or describe each of these terms in a complete sentence or paragraph.

1. conduction
2. conservation of energy
3. convection
4. fluid
5. heat
6. heat transfer
7. internal energy
8. mass-energy equivalence
9. mechanical equivalent of heat
10. radiation
11. thermal equilibrium
12. work

3.1 The Law of Conservation of Energy

In our energy study last chapter, we saw many examples of energy in one form being converted into another form. Nuclear energy produced by fusion reactions in the sun produces electromagnetic radiation. The gravitational potential energy in dammed-up water turns into kinetic energy as the water falls, and then electrical energy as the water spins a turbine-generator. Chemical potential energy in the molecules of fuel is converted into thermal energy in steam, and then kinetic energy in a spinning steam turbine, and finally into electrical energy by a generator.

There are dozens of examples like this of energy in one form being transformed into another. It was during the 19th century that scientists began to solidify the theory known as the *mechanical equivalent of heat*. According to this theory, mechanical forms of energy (such as kinetic energy) and heat are the same thing—energy. One can be converted into the other, but they are simply different forms of the same thing. At the time, this was a big discovery because for centuries scientists thought that heat was some kind of weightless gas—called *caloric*—that passed from warmer substances to cooler ones.

As usual in scientific research, many experiments exploring this issue were conducted by many scientists, but over time the caloric theory just didn't hold up. Heat was not a substance or gas flowing out of hot objects. Later in this chapter, we will explore the nature of heat in more detail—how energy is stored inside substances, and how it is transferred from one substance to another. For now all we need to note is that heat is just a form of energy, like kinetic energy and the other forms of energy we have studied.

Once the experimentation of the 18th and 19th centuries had made evident the principle of the mechanical equivalent of heat, scientists soon discovered the *law of conservation of energy*, one of the most fundamental laws in physics. This important law is stated in the following box.

> *The law of conservation of energy:*
> **Energy can be neither created nor destroyed, only changed in form.**

To illustrate this principle, let's look at two examples. The first example is a kid sliding down a metal slide, as depicted in Figure 3.1. At the top of the slide, the kid is at rest. His body has gravitational potential energy because of his elevated position in the gravitational field of the earth. As he slides down, the gravitational potential energy converts into kinetic energy as the kid picks up speed. Also, the friction between the kid's pants and the slide causes heating, and heat is released into the atmosphere. This is because friction on moving objects always causes heating, which always releases energy into the environment. When he reaches the bottom, he has no gravitational potential energy left. The energy he started with has all been converted into heat and kinetic energy.

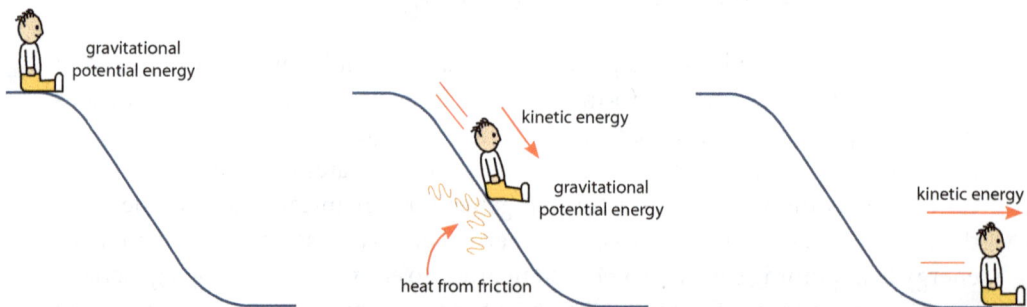

gravitational potential energy

kinetic energy

gravitational potential energy

heat from friction

kinetic energy

Figure 3.1. Energy transformations as a kid slides down a slide.

According to the law of conservation of energy, none of the original gravitational potential energy is lost or destroyed, and no new energy is created. The only thing that happens is that the gravitational potential energy the kid starts with converts into different forms of energy. We can even express this in the form of an equation, like this:

$$\boxed{\text{original gravitational potential energy}} = \boxed{\text{final kinetic energy}} + \boxed{\text{heat produced by friction}}$$

This is a sort of accounting equation. It states the kind of energy we start with and then accounts for where it all goes. The equation says that all the gravitational potential energy the kid has at the beginning is equal to the sum of the kinetic energy he has at the end and the energy released as heat on the way down.

Our second example is to track the energy involved in an exploding firecracker, as illustrated in Figure 3.2. Before the explosion, there is chemical potential energy in the chemicals in the firecracker. With the explosion, this chemical potential energy is converted into several different forms of energy, including energy in the light, heat, and sound from the explosion, as well as the kinetic energy in the flying debris as the firecracker blows apart.

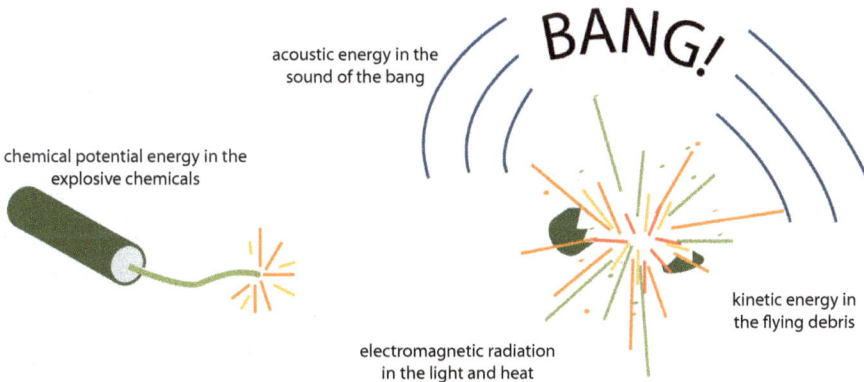

acoustic energy in the sound of the bang

BANG!

chemical potential energy in the explosive chemicals

kinetic energy in the flying debris

electromagnetic radiation in the light and heat

Figure 3.2. Energy transformations in an exploding firecracker.

As in the previous example, when the firecracker explodes no energy is created or destroyed. The original energy is all there, just in different forms. Writing this in an equation,

$$\boxed{\text{original chemical potential energy}} = \boxed{\text{kinetic energy in flying debris}} + \boxed{\text{heat and light}} + \boxed{\text{acoustic energy}}$$

With one exception, which we examine in the next section, current scientific theory now holds that the law of conservation of energy applies to every process and in every part of the universe. It is one of the fundamental laws of physics. In the next section, we take the law of conservation of energy one step further.

3.2   Mass-Energy Equivalence

From the previous chapter, you know that it was Albert Einstein that first theorized that light and energy are quantized. In 1905, when he published his famous paper on the subject, Einstein also published a paper demonstrating that mass and energy are related—equivalent, in fact—and that one can be transformed into the other! This theory is now called *mass-energy equivalence.*

The equation Einstein discovered is one of the most famous equations in the history of science:

$$E = mc^2$$

In this equation, E stands for energy, m stands for mass, and c stands for the speed of light, which is 300,000,000 meters per second. The equation basically says that there is an equivalent amount of energy associated with any given amount of mass. If you take a certain mass, in kilograms, and multiply it by the square of the speed of light (a very large number), the result is the equivalent amount of energy associated with that amount of mass.

Mass and energy are related, and one can be converted into the other.

Remember our discussions about fission and fusion in Chapters 1 and 2? In each case, energy is released. Einstein's equation enables us to understand why. When two protons fuse together during fusion, their mass after they stick together is a tiny bit less than the sum of the masses of the two protons before they fuse together. What happens to the rest of the mass? It is transformed into energy, just as Einstein's equation predicts. The same thing happens in fission. After a large atomic nucleus has been split apart, the sum of the masses of all the pieces is less than the original nuclear mass. The difference is converted to energy. This is truly amazing. It is also amazing that Einstein formulated this theory long before any experiments could be performed to confirm it.

So far as we know, the only exception to the law of conservation of energy presented in the previous section is when mass is converted to energy during a nuclear reaction. But if we take the mass-energy equivalence into account using Einstein's equation, then the law of conservation of energy applies across the board, without exception. Scientists refer to this as the *conservation of mass-energy.*

Learning Check 3.2

1.   Where does the energy come from in the fission reactors we use to produce electricity?
2.   What does *mass-energy equivalence* mean?

3.3   Heat and Heat Transfer

In the context of science, *heat* is a technical term with a very specific meaning. The term heat applies specifically to energy that is in the process of being transferred from a warm object to a cooler one. In fact, it is incorrect to refer to an object or substance as containing or possessing heat. You can say that an object has kinetic energy, and you can refer to the internal energy of a substance (which I explain later in the chapter), but heat is a term that refers to energy that is in transit—on its way from one place to someplace else. In this section, we look at the three ways this happens.

No doubt you already know that heat always flows from hotter objects to cooler objects. (We are not going to talk about thermodynamics much in this course, but in case you are wondering, it is the second law of thermodynamics that says heat always flows from the warmer object to the cooler one and never the other way around.) Heat continues to flow from the warmer object to the cooler one until they reach the same temperature. When they are the same temperature and heat flow between them ceases, the objects are in a state called *thermal equilibrium.*

Objects are in thermal equilibrium when they are at the same temperature, and no heat is flowing between them.

There are three ways heat can transfer from one substance to another. First, we have already seen (Section 2.3) how heat can travel as electromagnetic waves. Both the light and heat from the sun travel to earth as electromagnetic radiation. When we refer to heat transfer by *radiation*, this is what we mean.

We can see light, of course, but we cannot see the heat from a hot object like the sun because heat is in the infrared region of the electromagnetic spectrum. In Figure 3.3 is a sketch I drew to suggest how heat can radiate from a hot object, such as a horseshoe being heated in a blacksmith's forge. The red wavy lines are supposed to suggest the heat radiating from that hot horseshoe, but as I said, we cannot see the heat. We sure can feel it on our skin, though!

Heat transfer by radiation occurs when energy moves by infrared electromagnetic waves.

The second way heat transfer happens is by a process called *conduction*. Conduction occurs in solids, and of all solids metals conduct heat the most readily.

As with radiation, conduction is a very familiar process to all of us. You know that if you put a metal object, such as an iron frying pan, on a fire the handle of the object soon gets hot.

Scientists, Experiments, and Technology

All warm objects emit infrared radiation into the environment. In fact, by detecting the wavelength where the strongest radiation is, we can calculate the temperature of the object. This is because at any temperature, a warm object radiates a range of wavelengths, and the radiation is strongest at a particular wavelength for a given temperature. The hotter an object is, the shorter the wavelength of the peak in the radiation the object emits.

In the graph shown to the left, the black curves show the radiation emitted by a hot object at three different temperatures. The 5000 K curve is at about the surface temperature of the sun. The 3000 K curve is a bit below the temperature of the glowing metal filament in a traditional light bulb. Note that the peak in the curves moves to longer—more infrared—wavelengths as the temperature gets lower.

In the images below, the upper photo was made with visible light. The lower photo was made with infrared light, and then the different infrared wavelengths were translated to visible wavelengths so we can see the radiation patterns. The temperature scale in the lower photo associates the colors in the infrared photo to the temperature.

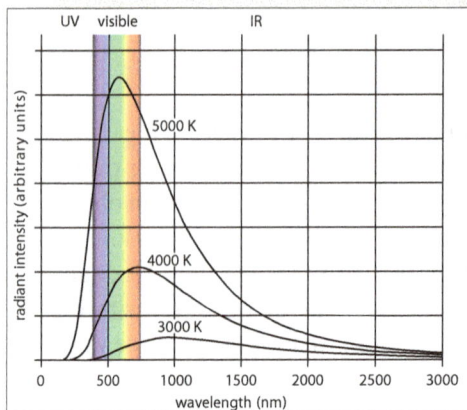

Notice that the black bag is opaque to visible light, but transparent to infrared. The reverse is true of the man's glasses. The color of the bag indicates that it is nice and cool—about 75°F. Notice also that the man's facial hair is the same temperature as his face, so it appears as the same color in the infrared photo.

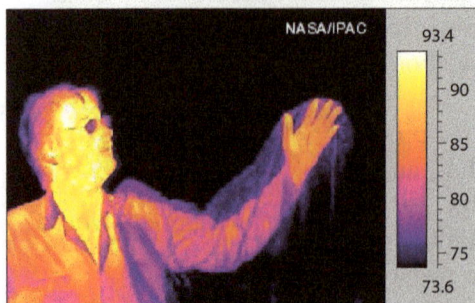

The relationship between the surface temperature of an object and the radiation it emits gave birth to a great new temperature measurement technology a few years ago. The hand-held device uses a lens to capture the infrared radiation from an object and display the object's surface temperature.

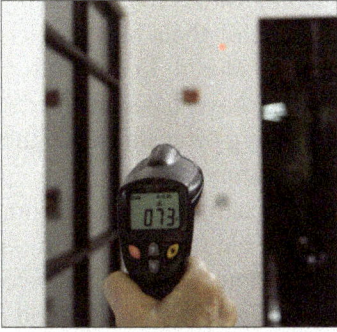

These devices have built-in laser pointers to make it easy to point the device at a distant object, but the laser is not part of the temperature sensing at all. In the photo to the left, a maintenance employee is measuring the temperature of the walls in a building to help assess how the building's air conditioners are performing. You can see the laser spot on the wall in the background.

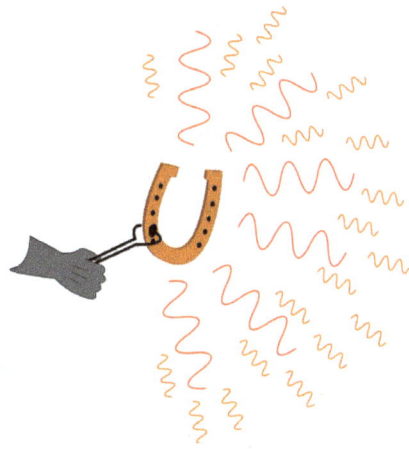

Figure 3.3. Heat transfer by radiation.

This happens even if the handle isn't over the fire. The reason is that when an object gets hot, the atoms inside the object vibrate very vigorously. In a metal, these atomic vibrations in the hot part of the object are transferred to other nearby atoms so that they begin vibrating vigorously, too, which means they are hot, too. The sketch in Figure 3.4 illustrates this process.

In the sketch, the blue spheres represent the atoms in the metal. I drew them in a regular, geometric pattern because in metals that's how the atoms are arranged. Underneath the metal, there is a heat source, such as a gas flame on a stove. As the heat from the flame warms the atoms in the metal, they begin vibrating much faster. These vibrations are transferred to neighboring atoms, which warms them so they begin vibrating faster, too. By this process of transferring vibrations, the heat spreads throughout the solid metal, although the atoms near the flame are always vibrating the most vigorously, and so are the hottest.

The third way heat transfer can occur is by *convection*. Convection occurs in *fluids* (liquids and gases). In fluids, atoms are free to move around, mix and mingle, and bump into each other. In hot fluids, the atoms are moving faster and in cool fluids the atoms are moving slower. As illustrated in Figure 3.5, if a hot fluid mingles with a cool fluid, the hot, fast atoms begin colliding with the cooler, slower atoms.

Heat transfer by conduction occurs when vibrating atoms in solids cause nearby atoms to vibrate, too.

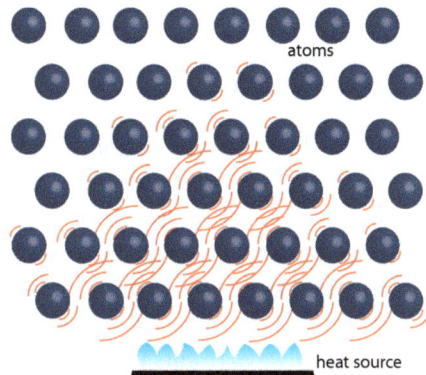

Figure 3.4. Heat transfer by conduction.

Figure 3.5. Heat transfer by convection. Red arrows represent fast, hot particles. Green arrows represent slower, cooler particles.

This mingling causes the hot atoms to transfer some of their kinetic energy to the cooler atoms. In the end, the cooler atoms have been warmed up by having energy transferred to them from the hot atoms.

A good example of convection at work is in the old radiators commonly used to heat homes in the days before central heating systems (and still in use in many older homes, particularly in the northern U.S.). Hot water is pumped through the inside of the radiator, making the iron radiator warm to the touch. When air molecules collide with the hot surface, they pick up kinetic energy from the rapidly vibrating atoms in the metal. These hot, fast moving molecules then gradually work their way through the room, colliding with the slower, cooler molecules and exchanging energy so the cooler molecules begin moving faster, which means they get hotter.

Heat transfer by convection occurs when hot and cold particles in a fluid mix and collide, causing slower, cooler molecules to gain kinetic energy and speed up.

Learning Check 3.3

1. Why is it incorrect to say that an object has heat in it?
2. What is thermal equilibrium?
3. How does conduction work?
4. How does convection work?
5. Write a paragraph explaining how we can determine the surface temperature of an object by sensing the infrared radiation the object is emitting.

3.4   Work

In the previous section, you learned that the term heat describes a process in which energy is transferred from one object to another. The term *work* is used to denote another such process. Work, like heat, is a technical term with a very specific meaning. Work is a *mechanical* process of transferring energy from one machine or person to another. When work is performed, a force is applied to an object, and the object is moved a certain distance in the direction of the force. This process always results in energy transfer.

Figure 3.6 shows four examples of work being performed. First, when an archer stretches back a bowstring, he does work on the bow. This results in potential energy stored in the bent bow. When the archer releases the bow, the bow does work on the arrow by pushing the arrow very hard for a few inches. The result of this work is that the arrow gains kinetic energy. In the second example, in the upper right, some boys are pushing a car up a hill. We say that the boys are *doing work* on the car. The result is that the car now has gravitational potential energy, since it is at the top of the hill. In the third photo, a white car is giving a push start to a red race car. The race car was at rest, and now is moving, so the white car did work on the red car, and the red car now has kinetic energy. Finally, a construction crane hoists

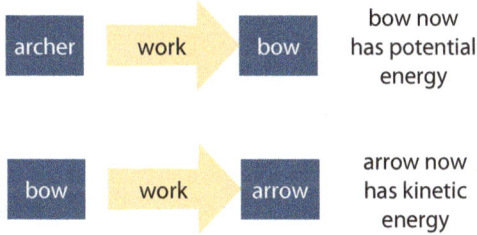

archer → work → bow   bow now has potential energy

boys → work → car   car now has gravitational potential energy

bow → work → arrow   arrow now has kinetic energy

white car → work → red car   red car now has kinetic energy

crane → work → bucket   bucket now has gravitational potential energy

Figure 3.6. Four examples of work, in which energy is transferred from one machine or person to another by applying a force and moving a distance.

a bucket of concrete up in the air. The crane does work on the bucket, and now the bucket has gravitational potential energy.

There is one more thing to notice about work. When work is done, energy transfer occurs. For this to happen, the force doing the work and the distance involved must point *in the same direction* or along the same line. If they are at right angles to one another, no work is done.

When work is performed, an applied force causes an object to move in the same direction.

This is illustrated in Figure 3.7. When lifting a bucket, a person does work on the bucket. Energy is transferred in this process because some of the energy in the person's body (chemical potential energy) is transferred to the bucket, which now has gravitational potential energy. But during the process of carrying the bucket from one place to another, *no work is done on the bucket*. The bucket is not being raised, so it is not gaining gravitational potential energy. And it is not being accelerated, so it is not gaining kinetic energy (except a little nudge at first to get it going, but that is not what I am talking about here). The bucket is not gaining any energy because the upward force applied to hold up the bucket is at right angles to the distance the bucket is being moved.

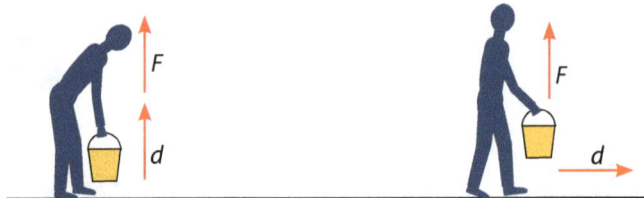

Figure 3.7. In raising the bucket (left), work is done on the bucket. In moving the bucket horizontally, no work is done on the bucket.

In this second case, does the person do any work at all? Yes, *but not on the bucket*. To perform the task of carrying the bucket, the energy transfer is heat from the person's body into the air. (This is a very complex process, and we cannot say what the work is done *on* without getting into the biology of the human body, which we are not going to do.) In short, since the bucket gains no energy, no work is done on it.

Learning Check 3.4

1. Make up four examples of your own that illustrate one machine or person doing work on another.
2. If you press as hard as you can against a brick wall until you begin breathing hard and sweating, did you do any work on the wall? Explain your answer.

3. In the photo to the left, food items are slowly moving along on a conveyor belt. Is the conveyor belt doing work on the food? Explain your answer.

4. At the right, baggage is moving up a conveyor belt into an airplane. The conveyer loads the baggage on the ground, and transports it up into the airplane. Is the conveyor belt doing work on the baggage? Explain your answer.

3.5   Internal Energy

When describing convection in Section 3.3, I noted that the particles in a hot fluid are moving faster than the particles in a cooler fluid. The fact that hotter particles move faster than cooler particles is very significant, and relates fundamentally to the energy in a substance.

Every substance is made of atoms, and the atoms of every substance are always in motion. In a solid, the atoms vibrate. In a fluid, the atoms are moving around at high speed. This means the atoms in any substance always have kinetic energy. This kinetic energy in the atoms in a substance enables a substance to absorb energy by being heated. The hotter the substance is, the faster its atoms move and the more kinetic energy the atoms have.

The *internal energy* of a substance is the sum of all of the kinetic energies of the individual particles (atoms or molecules) in the substance. Now, even though the number of particles in any ordinary quantity of matter is colossal, scientists have mathematical techniques that allow them to compute this sum, and thus to determine the internal energy in a substance. The hotter a substance is, the faster its particles are moving and the higher its internal energy is.

Internal energy is the sum of all the kinetic energies of the particles in a substance.

Now that you know about internal energy, let's consider the entire picture about the energy in an object as it warms up, how the energy is stored in the object, how the energy is transferred into and out of the object, and so on. Figure 3.8 is a photograph of one side of a hollow brick wall, with bricks on one side of the wall warming in the sun. The other side of these bricks is the inside of the wall, where it stays cool and dark. Let's use these warming bricks to consider how heat flows through the wall. Refer to the diagram in Figure 3.9 as we go.

First, the energy warming these bricks is in the infrared electromagnetic radiation arriving at the bricks from the sun. You might think that some of the energy warming the bricks would be from convection in the heated air, as the hot air molecules hit the bricks and transfer kinetic energy to the atoms in the bricks. But bricks in the direct sun become much hotter than the surrounding air, don't they? This means that the bricks are actually warming the air, rather than the other way around. On the left of the diagram, the electromagnetic radiation from the sun is shown arriving at the outer surface of the bricks.

Okay, so the energy in the infrared electromagnetic waves is absorbed by the atoms on the bricks' surface, increasing their thermal energy. What does this mean the atoms do? They vibrate faster—because the radiant heat has been transferred to the atoms in the bricks. Since they are very hot now, they are vibrating vigorously. Then, since the bricks are solid, the atoms in the next layer of brick begin vibrating faster, too, and so on, one layer after another, as the energy works its way by conduction to the inside of the brick. The heat conducts its way to the cool side of the bricks, on the inside of the wall. Now, since the bricks are warmer than the cool air inside the wall, electromagnetic energy radiates from the bricks into the inside of the wall, warming it up a bit. I drew the waves inside the wall shorter (vertically) than the waves hitting the bricks from the sun to show that the heat ra-

Figure 3.8. Bricks in the sun on a warm day.

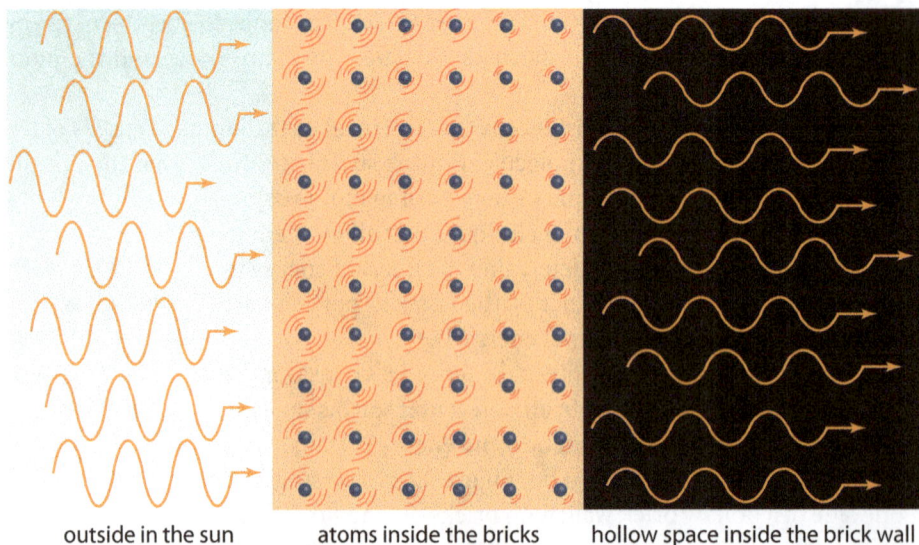

outside in the sun   atoms inside the bricks   hollow space inside the brick wall

Figure 3.9. Energy transfer through a hollow brick wall. Energy arrives at the bricks, passes through the bricks by conduction, and radiates out the cooler side. (Note: Since the bricks are hotter than the surrounding air, some energy also radiates from the bricks to the left, back out into the air.)

diating inside the wall is less intense. I also made the waves radiating inside the wall to have longer wavelengths than those arriving outside. The reason for this should be clear from our study of electromagnetic radiation earlier in this chapter (see the box on page 50). Lower temperature means longer wavelength radiation.

As we end this section, there is one final point I wish to make about internal energy. If you are thinking that internal energy sounds a lot like the thermal energy we studied in Chapter 2, you are right. The two are very similar and some books treat them as synonyms. The energy in a substance from being heated, which is our definition for thermal energy, is simply the increase in the substance's internal energy that occurs from the heating.

But there is more than one way to increase the internal energy of a substance, so some authorities make this distinction between thermal energy and internal energy: the term internal energy refers specifically to the sum of all the kinetic energies in the particles of a substance. The internal energy can be increased two ways: by heating or by compressing the substance into a smaller volume. By contrast, the term thermal energy refers specifically to the portion of the internal energy that is due to heating.

Learning Check 3.5

1. Imagine an unopened can of your favorite soft drink sitting in the sun. Explain what is meant when referring to the internal energy of the liquid inside that can.
2. Think about your can of soft drink again as it sits in the sun. Explain how the liquid in the can gets hot. Don't be vague; use the terms and concepts we have encountered in the last two sections.
3. Think again about the brick wall we discussed in this section. Consider what changes in Figure 3.9 if the sun is suddenly covered up by a thick black storm cloud. Redraw the diagram, and explain what is going on in this new situation.
4. Building on the previous question and your new diagram, assume the sun never comes back out before nightfall. How long does the situation in your new diagram continue? What if it stays cloudy the next day and the overnight temperature holds steady with no sun for many days. What happens?

3.6   Summary: Where Is the Energy?

In this chapter and the last, we have covered the basics about energy. Now, based on what I have presented, if you think about it, there are really only a few basic ways that energy can be present in substances or in space. Just by way of reviewing and summarizing our study of energy, let's review these possibilities.

The first is in the random motion of the atoms in a substance. If it is a solid, they are vibrating. If it is a liquid or a gas, they are flying around (kinetic energy). The

Figure 3.10. Internal energy: The sum of the kinetic energies of the moving particles.

Figure 3.11. Kinetic energy (large scale): Energy in the overall motion of an object.

sum of all these kinetic energies is the substance's internal energy. The internal energy of a substance depends on its temperature. The hotter it is, the faster its particles move, and the higher its internal energy is. The internal energy of a gas is suggested by Figure 3.10.

The second way a substance can possess energy is if the substance as a whole is moving, so that as a whole it has kinetic energy. The substance could be moving in a line, rotating, orbiting or some combination of these. If so, the substance as a whole has kinetic energy. This is depicted in Figure 3.11.

The third way a substance can have energy is if the substance is in a field, so that it has potential energy. As suggested by Figure 3.12, we studied the gravitational potential energy present with objects in a gravitational field. The object could also be in some other kind of field, such as a magnetic field or an electric field. If so, then the object may have some kind of potential energy associated with that kind of field. Chemical potential energy is actually due to the energy in electrical fields around protons and electrons in substances. (We will look more closely at electric and magnetic fields later.)

Fourth, as Einstein discovered, there is energy associated with the mass in a substance, and during a nuclear reaction some of the mass in an atom is

Figure 3.12. Gravitational potential energy: The energy that will be released by objects in a gravitational field if they are no longer held apart.

Note: This is the famous and beautiful "Earthrise" photo taken by Apollo 8 astronaut Bill Anders. The moon is in the foreground. On Christmas Eve, during the mission, the Apollo 8 crew read the first 10 verses from the book of Genesis on a national television broadcast. At the time, that broadcast was the most-watched TV program ever.

converted directly into energy, as depicted in Figure 3.13. If we were able to tear apart the nuclei in the atoms of a substance, massive quantities of energy would be released. But the only way to get at *all* this energy would be literally to destroy the substance by tearing apart its very atoms and breaking them down into individual protons and neutrons.

Finally, energy is present in the electromagnetic radiation found throughout outer space. This includes the infrared, visible, and ultraviolet radiation coming from the sun and stars, as well as the Cosmic Microwave Background radiation.

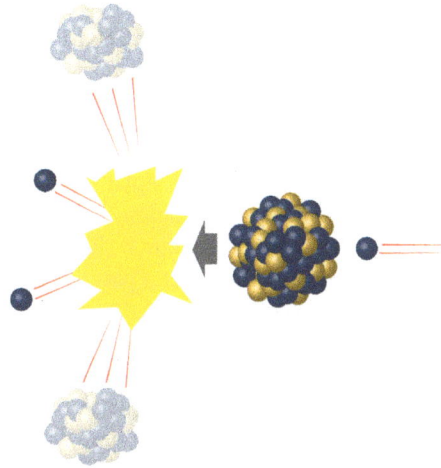

Figure 3.13. Nuclear energy: Some of an atom's mass is converted to energy during nuclear reactions.

Chapter 3 Exercises

Answer each of the questions below as completely as you can. Write your responses in complete sentences.

1. Make up a situation in which you start with one form of energy, and some event occurs, causing the energy to be transformed into at least three other forms of energy. Use the law of conservation of energy to account for all the energy in your example.
2. Imagine you are camping, and you are sitting inside your tent in the sun on a hot day. Your tent is dark blue, so not much light comes in, and it is dark in there. Why does it get hot inside your tent? Explain how this happens.
3. Suppose there is a small gasoline engine in the top of a big warehouse. The engine is attached to some pulleys and is used to hoist cartons up to the second and third floors for storage. Explain how work is involved in getting a carton up to the third floor.
4. Consider again the previous question, and suppose you carry a gasoline can full of fuel up to the third floor and fill up the engine's gas tank. To do this, you must do work on the can of fuel. Does this work energy get converted into the gravitational potential energy of the cartons hoisted to the upper floors? Why or why not?
5. Consider a glass of ice water sitting in a windowless room in a house. The temperature in the room is 72°F. Describe how energy from the room finds its way into the ice cubes in the ice water to melt them.
6. Consider again the glass of ice water in the previous question. Does the idea of thermal equilibrium apply at all here? Explain your answer.

Experimental Investigation 2: Heat Transfer by Conduction

Overview

- Assemble four containers of water. In a beaker, maintain boiling water at 100°C. In three Styrofoam cups place small amounts of room-temperature water. Place thermometers in each container.
- Connect the heat source (hot water) to the room-temperature water with pieces of copper, aluminum, and plastic wire or cord all the same diameter and length. *The goal of this experiment is to compare the ability of these three materials to conduct heat.* The amount of heat a material can transfer by conduction is called its *thermal conductivity*.
- Record the temperatures of the water in all four containers every 20 minutes until *thermal equilibrium* is reached.
- Collect temperature and time data, and prepare graphs of temperature versus time for the water in the four different containers. Plot the three curves for the cups on the same set of axes, and the temperature for the hot water beaker on a separate graph.

Basic Materials List

- hot plate
- digital thermometers (4)
- Pyrex beaker, 600 mL
- graduated cylinder, 100 mL
- Styrofoam cups and lids
- duct tape
- Styrofoam ice chest (2) and small cardboard box
- samples of copper and aluminum wire, and plastic cord

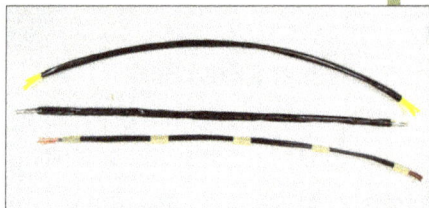

We expect metals to conduct heat better than plastic. This is why metal tools like tongs and frying pans sometimes have plastic handles. But how much better do metals conduct heat? And is there a difference in the thermal conductivities of different metals? In this experiment, we examine these questions scientifically.

To do this, we need a heat source, which will be a large beaker of water on a hot plate, maintained at a constant, hot temperature. Then we need a place for the heat to go. For this, we use three small Styrofoam cups of water. By monitoring the water temperature in the three cups, we have a way of *quantifying* how much heat is conducted into the water in each of the cups. Third, we need materials to conduct heat from the heat source to the water in the cups. For two of these materials, we use pieces of aluminum and copper wire. This way we can compare the thermal conductivities of these two common metals. For the third cup, we need a plastic material shaped like wire. The nylon monofilament line used in grass trimmers ("string trimmers") works well.

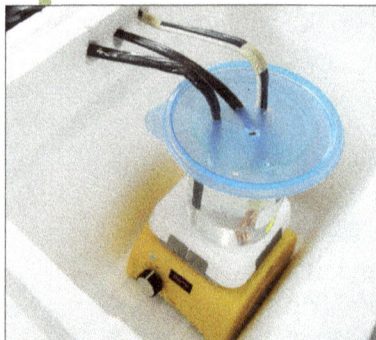

To prepare the heat conductors, cut equal lengths of copper wire, aluminum wire, and nylon line. You need four pieces of each material, and they need to be long enough to reach comfortably from the hot water beaker into the Styrofoam cups. Wrap the four pieces of each material together in duct tape to serve as a thermal insulator, as shown in the first photo. On each end of each pair of conductors, leave one inch of exposed material. Make sure the exposed part is the same length on each material so we have a valid comparison of thermal conductivity.

The setup is illustrated in the next two photos. A beaker containing about 500 mL of water is on a hot plate in a Styrofoam ice chest. This keeps the water in the beaker at a steady temperature of 100°C without warming up the room. Tape the insulating cups to a cardboard box inside a second ice chest. Use the graduated cylinder to measure out 100 mL of tap water into each of the cups, making sure that each cup has the same amount of water in it. Punch two holes in each lid, one for a thermometer and one to allow one of the conducting materials to enter the water. Label the cups so you don't forget which one is which. (The cardboard box is there to elevate the cups to the same height as the hot water beaker, so the thermal conductors can be as short as possible.)

To run the experiment, place the tap water in the cups 2–3 hours in advance so the water is at room temperature when the experiment begins. Insert the three conducting materials into the cups and the beaker through holes in the sides of the ice chests. Record the four starting temperatures in your lab journal. Switch on the hot plate and continue to record all four temperatures every 20 minutes until thermal equilibrium is reached (about four hours). You will know the system is at thermal equilibrium when all temperatures are more or less holding steady.

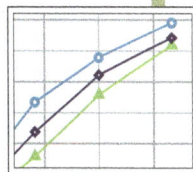

Analysis

An important part of the analysis is to prepare graphs of the four data sets, using the horizontal axis for the time and the vertical axis for the temperature. Plot the data for the three small cups on the same axes (same graph). The water in the cups begins at around 22°C, so you probably want to scale the vertical axis from 20°C up to a bit above the highest cup-temperature reading you have. Since you are plotting three sets of data on the same graph, you need to use a different symbol or color for each data set, as illustrated in the sample graph to the right.

The hot water temperature is supposed to be steady at 100°C. This is a lot higher than the cup temperatures, so it needs to be on a graph by itself. Your instructor will help you with setting up your graphs and labeling them properly.

In your report for this experiment, address the following questions:

1. From studying the graphs, how do the thermal conductivities of the three materials compare?
2. Why do you think the temperatures from the hot water beaker are recorded and plotted on a graph? What purpose does that serve?
3. Why are the three conductor materials wrapped in duct tape?
4. Why is it important that the three different materials are cut to the same length?
5. Imagine you are designing a part to go inside a machine, which we will call part U. Part U connects to three other parts, called a, b, and c. If you do not want heat to transfer between part U and parts a, b, and c, what kind of material should you consider using for making part U?

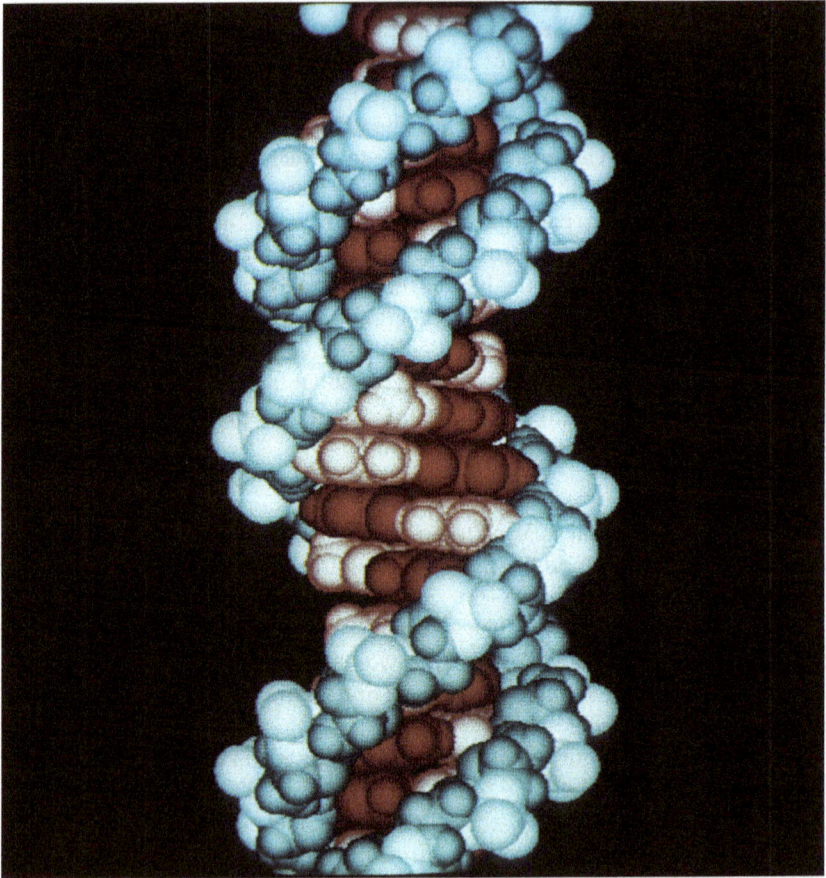

DNA is an enormous molecule found in the cells of all living organisms. DNA is referred to as a macromolecule, *because it is assembled from groups of smaller molecules called sugars, phosphates, and nucleotides, all woven together in a spiraling shape known as a* double helix. *The chains of paired nucleotides in human DNA (shown in red in the image above) can be up to approximately 220 million pairs long. The sequencing of the molecules in DNA contains a sophisticated multi-layered code. Embedded in this code is the genetic information that specifies and governs the biological functioning and physical traits of the organism.*

4.1 Why Are There Laws of Nature?

It is so common to hear people refer to the "laws of nature" that this phrase usually goes by without a second thought. But if you stop and think about it, the fact that laws of nature exist is actually quite remarkable. According to the standard scientific description of the origin of the universe, the entire universe began with an inconceivable explosion of energy, an event known as the *Big Bang*. From the moment the Big Bang happened, the laws of physics were present to shape the emerging matter and energy over 13.8 billion years into the stars, galaxies, and solar systems we see today. How could it be that from such an explosion a set of unwritten, mathematical principles would spontaneously emerge, and that these principles would consistently govern the behavior of matter and energy forever after?

Let's think about this for a moment. You know what explosions are like; you've probably seen thousands of them in movies. They are massively chaotic, with material flying everywhere and light and heat radiating out from the center of the explosion. But even in the midst of the chaos of an explosion, like the one shown in Figure 4.1, the motion of all the fragments, the patterns of light and heat, and the new compounds formed by the chemical (or nuclear) reactions involved are all governed by the laws of physics and chemistry. And apparently these laws were there from the beginning of the universe, governing the expansion of electromagnetic radiation from the moment of the Big Bang.

What do we mean by the "laws of nature"? Well, in the most general sense this phrase refers to everything in the entire domain of science, from the laws of electricity and magnetism, to the chemistry of exploding gunpowder, to the chemical system of detection and communication ants use to swarm a lump of sugar, to the patterns of behavior exhibited by a hunting tiger. The plain fact is, there *are* laws that govern how nature works. If there weren't, the matter and energy spewing out from the Big Bang would have had no plan or program for what to do next, and there would have been no regulating principles to bring order to the expanding mess. It would be just a mess of stuff, nothing else.

Figure 4.1. Even the chaos of explosions is governed by the laws of physics and chemistry.

Consider also that in order for us to do science at all, there must be predictable laws governing the way nature works. Without predictable laws, nature would not submit itself to scientific analysis. But if that were the case, we wouldn't be here to analyze it because the existence of life itself depends on predictable, orderly laws governing the way matter and energy interact.

The title of this section is, "Why Are There Laws of Nature?" Philosophers and theologians have expressed three different views in answer to this question. All agree that the laws of nature are fascinating, beautiful, and wonderful. Those who love science also agree that the laws of nature are interesting to study. But people differ as to why the laws of nature exist. Let's explore this question here briefly.

Science depends on the regular behavior of the world around us. If there weren't laws of nature, there would be no science. There would be no people, either.

Some people say there is no particular reason for the laws of nature. According to this first view, the laws of nature came into existence by chance. If this is true, then we need not seek a reason for the existence of the laws; we just accept them and go on to study them. A second view holds that there is some unknown natural principle that explains why the laws of nature exist. According to this view, there is a natural explanation for the existence of the laws of nature, but we have not discovered this principle yet. At this time, no one knows what such an unknown natural principle might be. The third view is held by those who believe in a divine being (God) who created the universe. According to this view, the laws of nature exist because they were created as part of nature by the creator, according to the purposes the creator had in mind at the creation.

Our purpose in this chapter is to explore some of the many fascinating consequences of the fact that the laws of nature do exist. As I wrote in the first chapter, the order in nature is one of the three basic things that exists (along with matter and energy). We will not further explore the reason why the laws of nature exist. We must leave that question to the philosophers and theologians. I encourage you to ask your parents or teachers about this if you are interested in thinking more about it. For the rest of the chapter, we will focus on what we can say about the orderliness we find in nature.

Learning Check 4.1

1. What is meant by the phrase "laws of nature"?
2. What are some examples of evidence for the existence of laws of nature?
3. Describe three possible explanations for why laws of nature exist.

4.2   Order and Structure in Nature

It has been said that all the basic laws of physics that govern the physical behavior of the known universe—which are expressible in mathematical equations—can be written on a single sheet of paper. Isn't that astonishing? Here are a few examples of these laws. In Figure 4.2 are the famous "Maxwell's Equations." These four equations were published by Scottish physicist James Clerk Maxwell in 1864, and they describe all of classical electricity and magnetism, including radio waves and light.

Everything we know about gravity is summarized in the general theory of relativity, shown in Figure 4.3. German physicist Albert Einstein published this theory in 1916. Maxwell's equations are difficult to learn and understand, but they are nothing compared to the difficulty of the mathematics behind this equation of Einstein's. And yet the main operating principle of all the gravity in the universe can be expressed in a single equation!

The behavior of particles such as protons and electrons at the atomic or

$$\nabla \cdot D = \rho_f$$

$$\nabla \cdot B = 0$$

$$\nabla \times E = -\frac{\partial B}{\partial t}$$

$$\nabla \times H = J_f + \frac{\partial D}{\partial t}$$

Figure 4.2. Maxwell's equations describe all electricity and magnetism.

$$R_{ab} - \frac{1}{2}Rg_{ab} = \frac{8\pi G}{c^4}T_{ab}$$

Figure 4.3. The primary equation of Einstein's general theory of relativity, a geometrical description of space and time that accounts for gravity.

$$i\hbar\frac{\partial}{\partial t}\Psi = \hat{H}\Psi$$

Figure 4.4. The general Schrödinger equation, describing the behavior of systems at the quantum (atomic) level.

65

$$\sigma_x \sigma_p \geq \frac{\hbar}{2}$$

Figure 4.5. An expression of the Heisenberg uncertainty principle, describing limits on what observers like us can know about objects at the quantum level.

quantum level is summarized by the "Schrödinger equation," shown in Figure 4.4. This equation, published in 1926 by Austrian physicist Erwin Schrödinger, has dominated physics in the 20th and 21st centuries (along with Einstein's equation).

And the limits on what it is possible to know about any quantum system are summarized by the Heisenberg uncertainty principle, shown in Figure 4.5.

Of course, there are a few thousand other equations that come up in the study of physics, but it is amazing that they really all boil down to those shown here and a few others. The breathtaking complexity of the matter and energy we see in nature can be summarized so simply and elegantly! Not only that, but we can discover these laws and understand them! It is wonderful that the universe—and the powers of the human mind—are this way.

Now, just as amazing as the simplicity and comprehensibility of nature's mathematical structure is the fact that nature has any mathematical structure at all. Many scientists have wondered about this. Several have written comments such as this statement by a Nobel Prize winning physicist: "the miracle of appropriateness of the language of mathematics for the formulation of the laws of physics" is "something bordering on the mysterious."

Many scientists have agreed that the mathematical structure of nature is a mystery that seems miraculous.

I have shown you some of the equations in physics that illustrate the order and structure embedded in nature. Now I want to show you some beautiful examples how this mathematical structure is evident just by *looking* at nature.

Just check out the stunning, three-dimensional arrangement of cubes in the pyrite crystal of Figure 4.6. This is not just mathematical; it looks like someone has been *playing* with mathematics!

Figure 4.6. Pyrite crystals exhibiting an effect known as "twinning."

Now, we all know about the planets in our solar system. The very creative image in Figure 4.7 shows them to scale, next to one another for size comparison (and for just gazing at their beauty.) Thousands of astronomers have labored at working out the mathematics of the planetary movements. All the planets travel in elegant, predictable orbits around the sun, and their orbits are all *elliptical*, which means they are shaped like this ⬭, and can be characterized with equations that look

Figure 4.7. The beautiful planets (left to right, back to front): Jupiter, Saturn, Uranus, Neptune, Earth, Venus, Mars, and Mercury.

like $ax^2 + by^2 = 1$. All the planetary moons travel in ellipses around their companion planets as well. (Some cool asides: the famous Great Red Spot on Jupiter is about three times the size of the earth. It is a giant storm, and we have known about it here on earth for 350 years. Saturn's rings are made almost entirely of water ice, and average about 20 meters thick. The outer diameter of the rings is about the same as the distance from earth to earth's moon.)

Now let's talk about the fruit fly, commonly used in biology classes all over the world to study the mathematical laws of genetic inheritance (which we won't go into). It is such a simple little bug, common and easily squished. But check out the magnificent geometry built into the eye of a fruit fly shown in Figure 4.8, an image captured by a scanning electron microscope (SEM). Figure 4.9 is another image of an eye captured by an SEM, the eye of an antarctic krill. (Krill are small ocean crustaceans related to shrimp, with bodies about 3/4 inch long.) Perfect hexagons— imagine that!

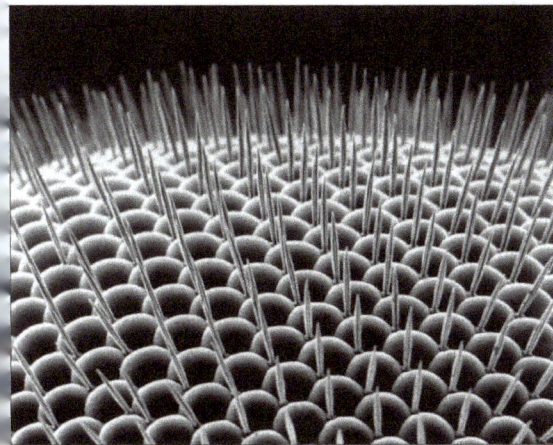

Figure 4.8. A scanning electron microscope image of the eye of a fruit fly.

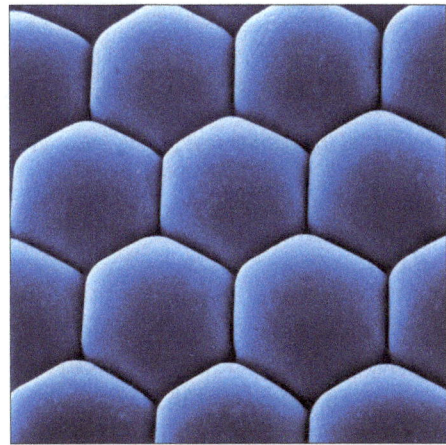

Figure 4.9. A scanning electron microscope image of the eye of an antarctic krill.

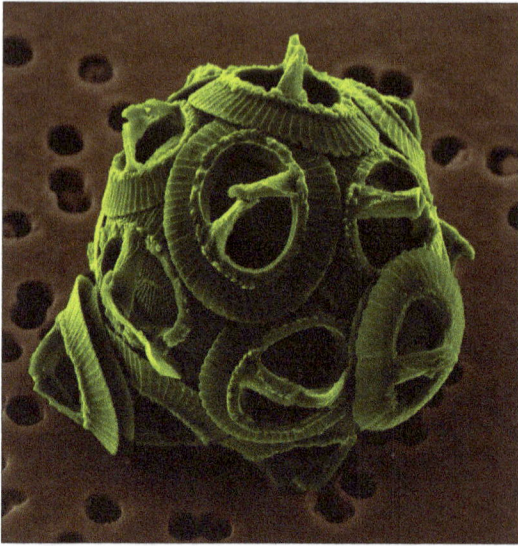

Figure 4.10. A scanning electron microscope image of a coccolithophore, a single-celled algae that is abundant in the oceans.

Figure 4.11. This bobtail squid is only a few centimeters long, but its light emitting capabilities allow it to hide its own silhouette.

We could go on endlessly with these examples, but let's look at just two more beauties. Figure 4.10 shows another SEM image, this time of a single-celled member of the algae family. If you can believe it, this thing is only about 10 micrometers in diameter, which is only about 20 times the wavelength of visible light. I just can't get over the rich, velvety beauty of this image. And mathematical structure? This thing looks like it was built in a watch factory! (By the way, scientists don't yet know what those round things are for, but I am confident they serve some purpose.)

Finally, in Figure 4.11 we have one of the world's cutest marine animals, the bobtail squid! (I'm sure you will love this: they are also known as dumpling squid, or stubby squid.) This little guy is only a couple of inches long. Don't you love the spiraling pattern of colors? (Why do the colors *spiral*?) But listen to what this critter has going on: it can create its own "invisibility cloak." No, really. This squid has a mantle called the "light organ," which is home to a glow-in-the-dark species of bacteria that live underneath it. By means of a fantastic optical system composed of a color filter, lens, and reflector built-in to the squid's mantle, the bacteria emit light beneath the animal. From below, this light mimics the appearance of the light hitting the top of the squid. The net effect of this superb optical imaging system is to mask the squid's own silhouette and protect the squid from predators! Imagine a squid and some bacteria having such elegant control over electromagnetic radiation. (You might also be interested to know that in exchange for the lighting service, the squid feeds the bacteria on sugar and amino acids.)

Oh wait—I almost forgot about the Fibonacci sequence, found all over the place in pine cones, sunflowers, flower stems, honeybees, and nautilus shells. On second thought, I will leave this one for you to look up yourself.

Learning Check 4.2

1. Describe some equations scientists use to model the laws of nature.
2. Describe some examples in which the mathematical structure in nature is evident from its appearance.

4.3   The Remarkable Universe

For several decades now, scientists have been aware of quite a number of coincidences associated with the existence of complex life on planet earth. Let's take a look at a few of these remarkable features.

Conditions Necessary for Life There are many conditions that must be satisfied for complex life as we know it to exist. Here are some of them.

1. We need a lot of water to regulate temperature and transport nutrients.

   The earth's surface is over 70% water (Figure 4.12). The water cycle involves a complex system of oceans, evaporation, cloud transport, rainfall, lakes, and rivers. This large amount of water is necessary to regulate the temperature of the planet. Also, because minerals can dissolve in water, the water cycle works as a global transport system to distribute the nutrients needed by plant and animal life. Current scientific thinking is that without the oceans, the temperature swings from day to night would be so drastic that complex life could not exist.

   Figure 4.12. 71% of the earth's surface is covered in water, one of the main necessities for complex life to exist.

2. We have to be just the right distance from the sun.

   Within our *solar system*, earth is positioned in what some scientists have called the "Goldilocks zone": not too hot, not too cold, but just right. If earth's orbit were a bit smaller or a bit larger, the oceans would be unstable and life probably could not exist.

3. We need an atmosphere to trap heat.

    Our atmosphere (Figure 4.13) is important because we breathe it. But it serves another purpose as well: to trap heat from the sun and help regulate the surface temperature of the earth. Without it, daily temperature fluctuations would be so high that complex life could not exist.

Figure 4.13. The earth's atmosphere, as seen during sunset from the International Space Station.

4. We need a large moon.

    The tilt of the earth's rotational axis is the cause of our seasonal climate variation, which helps keep the equator from getting too hot and the poles from getting too cold. Moderate temperatures are not only necessary for complex life to exist, but moderate temperatures also make sure earth's water does not get too hot and boil away, creating the kind of harsh environment found on Venus. Our large moon stabilizes the tilt of earth's axis, preserving the balance necessary for life.

5. Numerous physical constants must have just the right values.

    There are numerous constants found in the equations of the laws of physics. These include the gravitational constant, the masses of the fundamental particles, the electrical charges on the fundamental particles, and others. Scientists have noted that for many of these constants, if their values were only slightly different the physics and chemistry of the universe would not support complex life such as ourselves. Many scientists refer to this as the *fine-tuning* of these physical constants.

6. We need a magnetic field around the planet to shield us from the harmful solar wind.

    The liquid iron core of the earth creates a magnetic shield around the earth called the *magnetosphere*. The magnetosphere (Figure 4.14) provides protection for our planet from the "solar wind," a spray of

Figure 4.14. This artist's rendering of Earth's magnetosphere illustrates the way it protects us from the harmful effects of the solar wind.

high-energy protons and electrons coming from the sun. Without the iron core and the resulting magnetic shield, it is likely that the solar wind would rob earth of its atmosphere, its water, and its life.

Other Remarkable Observations

As we have seen, the conditions on earth are wonderfully appropriate for supporting life. But there are other remarkable observations that should be included on any list of things that make our world amazing. Here are a few more notable features of our world.

1. The universe had a beginning.

    Current scientific theory holds that the universe had a beginning. But science has not always held this. Big Bang theory was not formulated until after 1929, when astronomer Erwin Hubble discovered that the galaxies are all accelerating away from one another. Prior to that, many scientists held a "steady-state" view of the universe, in which the universe has no beginning and has always existed. Today, scientists recognize that the universe began 13.8 billion years ago, it has been expanding ever since, and its rate of expansion is increasing. In other words, the galaxies are accelerating away from each other.

2. All nature exhibits exquisite balance.

    It is unfortunate that many of us these days spend most of our time indoors. But those who have spent a great deal of time outdoors studying the forests, mountains, rivers, and oceans are often struck by the delicate balance maintained by the ecosystems of the planet and the life that lives in them. The creatures are all exquisitely adapted to their environments, displaying peculiar physical features that enable them to flourish. Further, the creatures all seem to depend on one another and on the proper functioning of earth's water cycle, day-night cycle, tides, and seasons. The balance in nature is one of its most remarkable features.

3. Our bodies exhibit incredible functionality.

    I hope the human body amazes you, because it is indeed amazing. Have you seen images showing the astonishing way babies develop in the womb, with a beating heart after 21 days? Have you thought about how wonderful it is that when we receive a minor wound, our blood coagulates (a highly complex process) to prevent us from losing too much blood? Aren't you amazed at the stereoscopic vision provided by our eyes that enables us to perceive in three dimensions, in color, and over a range of light intensity that no camera can match? Aren't you dazzled by the ability of the human immune system to fight off disease and infection? Consider that the human brain is understood to be the most complex object in the entire universe, enabling our amazing capabilities in mathematics, music, language, sculpture, athletics, engineering, medicine, and on and on. Sometimes when viewing beautiful works of art, such as Michelangelo's sculpture called the *Pietà*, or Hans Holbein's extraordinary sketches of the hands of Erasmus (Figure 4.15), one can find oneself actually weeping from sheer amazement. Thinking about the incredible ability people have to create such things is mind-blowing.

4. DNA is an extremely sophisticated, multi-level coding system.

We are still in the early stages of learning about human DNA, but already we know that the genetic coding in DNA conveys the entire set of instructions that specifies and regulates how the human body—the most sophisticated system in the universe—develops and functions. Amazingly, the coding in DNA operates at more than one level. There are codes embedded in the matrix of nucleotides in DNA, but there are also codes on top of those codes! Unraveling the coding embedded in human DNA is one of the hottest fields of scientific research today.

Figure 4.15. Our brains and hands are wonderful, and enable us to create beautiful works of art, such as this sketch by Hans Holbein the Younger (c. 1523).

5. We perceive that the earth and the heavens are beautiful.

Why do we perceive mountains, sunsets, forests, flowers, animals, and human beings as beautiful? Why do we perceive works of art as beautiful? In a universe that seems mostly hostile to life, it is remarkable that life exists in such abundance on earth, and that we humans with all our mental and emotional faculties are here in the midst of it. In fact, our perception of beauty is just one of the many emotions humans feel. The human range of emotion is unique. While scientists debate the degree to which animals experience emotion, there is no doubt about the amazing range of emotions humans experience. The fact that we perceive things as beautiful again points to the wonderful order amidst the complexity of the world around us. If our surroundings were not ordered in a harmonious way, we might still feel emotions such as fear, anger, and frustration, but we would hardly be expected to perceive beauty, love, peace, and joy. We are well adapted to the world we live in, and the orderliness of the laws of nature allows us to perceive many things around us as beautiful and good.

Conditions Necessary for Exploration We have looked at order and structure in nature. We have noted many other remarkable features of nature, including our wonderful human characteristics. One notable human characteristic is our thirst for knowledge and our desire to understand the world—that is, our yearning to explore. Well, as it turns out, many of the features of earth that enable complex life to exist also happen to make exploration of the universe possible. We conclude this section with a few examples of this amazing circumstance.

1. The moon is just the right size and distance from earth to produce solar eclipses.

    I wrote above about how important the size of the moon is for enabling complex life to exist on the earth. The moon's size is also exactly right for producing perfect solar eclipses, like one shown in Figure 4.16. Because of the moon's size and distance from earth, during an eclipse the main disk of the sun is obscured, shutting off its blinding light but still allowing us to see the sun's corona, a sort of atmosphere around the sun. Studies of the corona using images from eclipses were the key to discovering the elements of which the sun is composed, thus unlocking our understanding of the solar fusion process, solar history, the history of the universe, and in fact the entire field of astrophysics.

Figure 4.16. Our moon is just the right size to produce total solar eclipses, like this one observed from the Apollo 12 spacecraft in 1969.

2. Our galactic location provides a great view of the heavens.

    Our solar system is in a *galaxy* called the *Milky Way*. As shown in the artwork of Figure 4.17, the Milky Way Galaxy is shaped like a bulging disk with four main spiral arms extending out. I did not

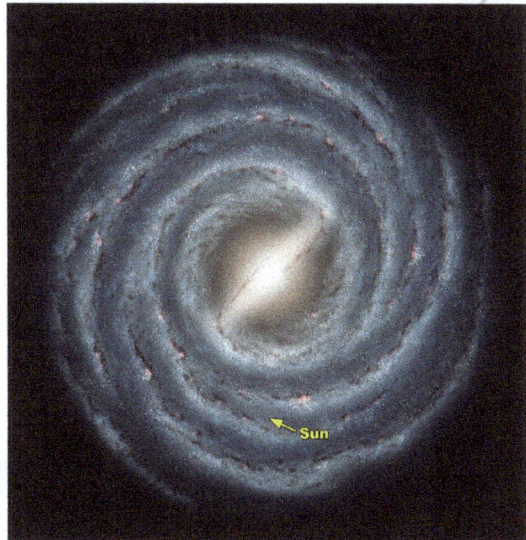

Figure 4.17. The position of earth in the galaxy is not only safe, but it is perfect for astronomical exploration.

mention previously that our solar system's location in the galaxy happens to be one of the safest places in the galaxy, but so it is. The main disk area is a very hazardous region because lots of star formation is going on, generating a lot of radiation hazardous to life. We are outside the hostile environment of the main disk, in a relatively safer place. But we also happen to be located in between two of the galaxy's main spiral arms. Inside the arms there is a lot of dust. If our solar system were located there, our view of the heavens outside our solar system and outside our galaxy would be obscured. But since we are located in between two arms, we have a great view outside the galaxy. As a result, we are able to see distant galaxies and study their features.

3. Our atmosphere is transparent.

If our atmosphere were like the atmosphere of Venus, we wouldn't be doing much astronomical research, at least not from the planet's surface. Venus is all clouds, and if you happened to be on the ground on Venus you would see nothing but thick cloud cover, 24/7. Fortunately for us, our atmosphere not only helps regulate the temperature by trapping the sun's heat, but it is transparent as well, enabling us to gaze out into the heavens. The layers in our lovely and very cooperative atmosphere are on fine display in Figure 4.18, an image captured by Expedition 22, the 22nd long-term endurance crew of the International Space Station. The image elegantly captures the silhouette of Space Shuttle Endeavor, just before docking.

Figure 4.18. This beautiful image of Space Shuttle Endeavor was captured by the crew of the International Space Station. Behind the shuttle the layers of earth's atmosphere are visible.

So it just so happens that natural features essential for the existence of complex life also just happen to make our world suitable for scientific exploration. What an amazing coincidence! We are ready now to dig more deeply into the physics and chemistry of the world around us. But studying science is a lot more fun when we are aware of just how wonderful our world is. Studying and learning is always hard work, but studying science is a lot more rewarding when we pause to take note of the wonders that surround us, wonders we often overlook. This world is truly an amazing place and the best part of studying science is that we get to study these wonderful things every day!

Learning Check 4.3

1. Make a list of six necessary conditions for complex life, and briefly describe each of them in your own words.
2. Make a list of at least five aspects of life on earth that are amazing or remarkable. On your list you may include some of the features discussed in this section, or some other features that you have thought of yourself. In your own words, explain why each of the features you list makes earth a remarkable place.
3. List three features of earth that facilitate our efforts to explore the universe around us.

Chapter 4 Exercises

Answer each of the questions below as completely as you can. Write your responses in complete sentences.

1. Write a paragraph explaining the different views on why the laws of nature exist.
2. Write a paragraph explaining what we mean when we refer to the "order and structure" found in nature.
3. Imagine that one of your friends says to you that she doesn't find the world to be a very remarkable place at all, and that, in fact, she finds the world to be quite ordinary and even boring. Write a paragraph or two in response to this friend. In your paragraph, make use of some of the material from this chapter.
4. Back at the beginning of Chapter 1, I write that there are three basic things in nature—matter, energy, and order. Why is it appropriate to say that order is one of the most basic things in nature? Write a paragraph explaining how you would respond to someone who claims that only matter and energy are basic, and that order is not part of nature.

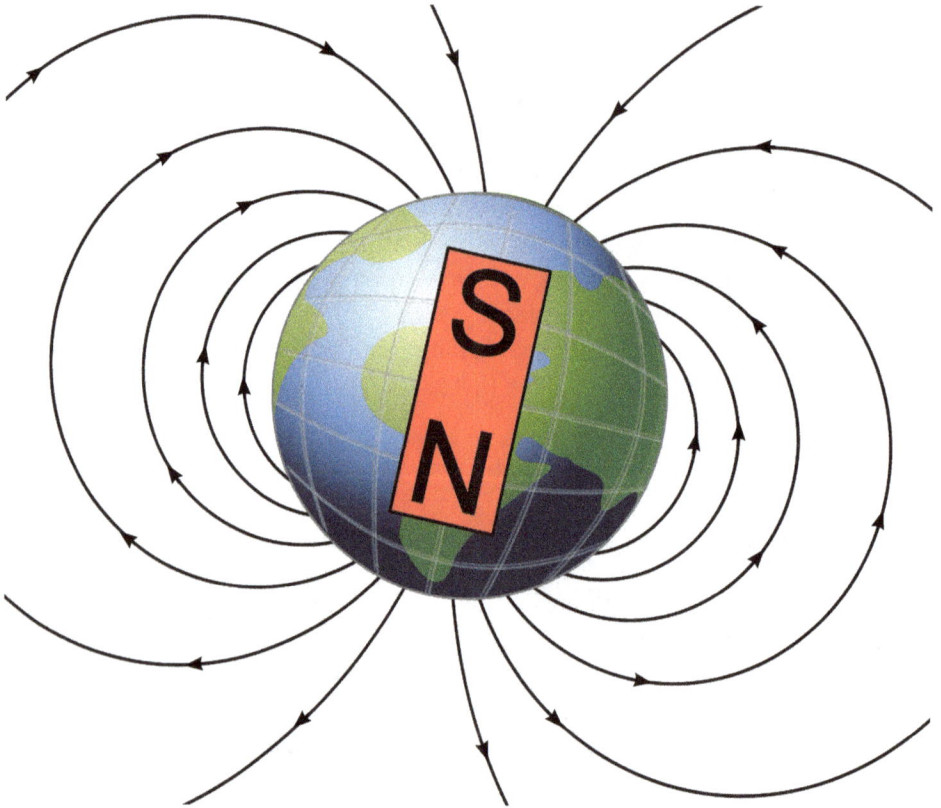

At the center of the earth is a massive volume of molten iron. Because of the rotation of the earth, this liquid iron produces a magnetic field that enshrouds the earth. We represent magnetic fields by curved lines pointing from the north pole of a magnet to the south pole. Contrary to what most people probably think, the earth's magnetic north pole is in the southern hemisphere, as shown in the sketch above. You can understand this easily by considering that the north pole of any magnet is repelled by other magnets' north poles, and attracted to other magnets' south poles. Since the north poles of all magnetic compasses point in a direction that is geographically north, we know that is the direction of the earth's magnetic south pole.

OBJECTIVES

After studying this chapter and completing the exercises, you should be able to do each of the following tasks, using supporting terms and principles as necessary.

1. Define the terms *force* and *field*.
2. List and describe the four fundamental forces found in nature.
3. Compare the strengths of the four fundamental forces.
4. State the key features of gravitational, electric, and magnetic fields.
5. Make a chart that shows the kinds of things that can cause and be affected by each of the three fields.
6. State when the gravitational theories of Isaac Newton and Albert Einstein were published, and briefly describe these two theories.
7. Explain how an electromagnet works.
8. Write a sentence stating Ampère's Law.

VOCABULARY TERMS

You should be able to define or describe each of these terms in a complete sentence or paragraph.

1. Ampère's Law
2. beta decay
3. electric field
4. electromagnet
5. electromagnetic force
6. electrostatic force
7. ferrous metal
8. field
9. field lines
10. force
11. general theory of relativity
12. gravitational field
13. gravitational force
14. interaction
15. inverse-square law
16. law of universal gravitation
17. magnetic field
18. monopole
19. nuclear decay
20. pole
21. radioactive
22. strong nuclear force
23. weak nuclear force

5.1   The Four Forces

When one object affects another object in any way, it is because forces are involved. A *force* is simply any push or pull. Magnets push against each other because of a magnetic force. The force of the earth pulling you toward it is a gravitational force. And the force that makes synthetic clothes cling together when they come out of the dryer is what we call an *electrostatic force*.

Although it may seem like there are a lot of different kinds of forces in nature, they actually all boil down to just four, and only two of them are noticeable to us in ordinary daily life. For reasons I explain a bit later, the four *fundamental forces* are also called the *fundamental interactions*.

The Gravitational Force

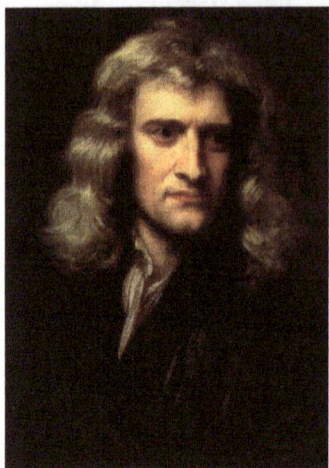

The force most familiar to all of us is the *gravitational force*. In 1687, English mathematician and scientist Isaac Newton (Figure 5.1) published his *law of universal gravitation*. In this new theory, Newton proposed that everything in the universe that had mass (that is, all matter) exerted a gravitational force on everything else in the universe that had mass. This theory represented a great amount of physical insight.

Newton's proposal was correct as far as it went, and was a major step forward in our understanding of planetary motion. For over 200 years, Newton's law stood as our best understanding of gravity, but actually it didn't come close to telling the whole story. That honor belongs to Albert Einstein's 1915 *general theory of relativity*, which I mentioned in the previous chapter. The general theory, which represents another profound physical insight, is our current understanding of how gravity works. Einstein envisioned gravity as *geometrical* in nature, and proposed that the presence of any object with mass warps the dimensions of space and time, which Einstein called *spacetime*,

Figure 5.1. English mathematician and scientist Sir Isaac Newton.

around itself. Figure 5.2 shows a way to visualize this spacetime curvature. With Einstein's theory, we now have a reason for why bodies exert gravitational attraction on one another. The force of gravity is not just some mysterious way things have of pulling on each other from a distance. The cause of gravitational forces is the curvature of spacetime.

Since, according to Einstein's theory, the spacetime around any object with mass is warped, this means that even light, though it is massless, is affected by it. If light is traveling through space and comes close to a massive object such as a star, its pathway in space changes direction as it passes the star. Einstein actually predicted that this happens,

Figure 5.2. A visual representation of the curvature of spacetime around the earth.

and he became an instant celebrity in 1919 when his prediction was confirmed by the analysis of the apparent positions of stars during a solar eclipse.

Einstein's theory of gravity is based on the curvature of spacetime around a massive object.

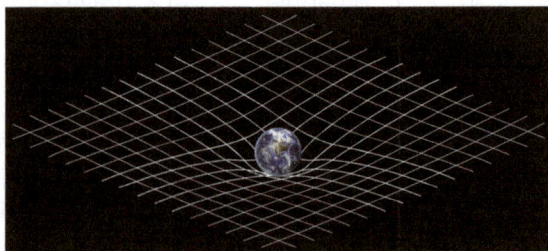

I would like to make one more point about gravity. This is about the mathematical relationship between the strength of the gravitational force and the distance between the objects involved. The math here may be a bit more advanced than what you have seen before. But hopefully, with your in-

*We Pause Here to Talk About **Inverse-Square Laws***

The force of gravitational attraction is what we call an *inverse-square law*. There are many inverse-square laws in nature (another example of the regular, mathematical structure in nature we considered in the previous chapter), so inverse-square relationships are very important in physics. This is why I want to introduce the subject to you, even though you may not have studied this kind of mathematics yet.

When a quantity like a force follows an inverse-square law, this means that the strength of the force varies inversely (decreases) in proportion to the square of the distance from the object causing the force. As an equation, this means the force of gravity looks like $F = k/d^2$. In this equation, F is the force of gravity and d is the distance between the two objects involved (that is, between their centers). The k represents a number that is not important for our discussion, so to simplify things let's just say $k = 1$. So now the equation is $F = 1/d^2$. An example will help to illustrate what this means.

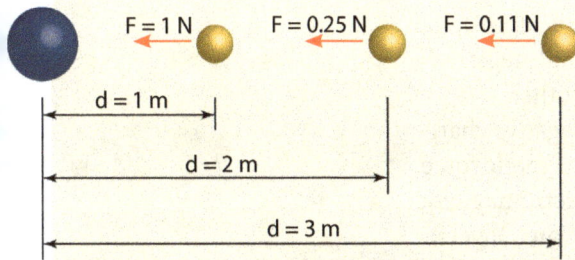

The figure shows a yellow object and a blue object. There is a gravitational attraction between them, as there is between any two objects. At a distance of 1 meter (1 m), the gravitational force on the yellow object is 1 newton (1 N). (In the metric system, forces are measured in units called newtons, named after Isaac Newton.) If the yellow object is moved to a distance of 2 m, which is twice as far away, the gravitational force on it drops, not by a factor of two, but by a factor of two squared, or four. So the force drops from 1 N to 1/4 N, or 0.25 N. At a distance of 3 m, the force is 1/9 N, or 0.11 N.

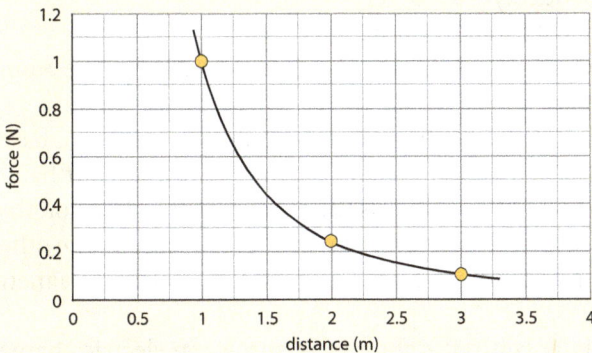

I have shown these force and distance values in the graph to the left. I have also drawn in a curve to show what the force is at other distances. The shape of that curve is the classic shape we associate with any inverse-square law.

structor's help, you will be able to follow it. I have placed it in the accompanying box about inverse-square laws.

The Electromagnetic Force

At the beginning of this chapter, I mentioned static electricity and repelling magnets as examples of forces. Both of these are actually examples of a single type of force called the *electromagnetic force*. Electricity and magnetism are very closely related. Electric current produces magnetic fields, and magnetic fields can produce electric current. This is why we consider electric and magnetic forces as simply different manifestations of a single fundamental force.

Here is something that may surprise you: except in the nucleus of an atom, nothing ever really "touches" anything else! No, I'm not crazy; allow me to explain. You know that all matter is made of atoms, and that atoms all have a tiny nucleus surrounded by a cloud of electrons. As illustrated in Figure 5.3, when one object, represented by the green-shaded atoms, gets near to another object, represented by the blue-shaded atoms, the electrons in their atoms, which are all negatively charged, begin pushing against each other.

The force of repulsion between negative charges is another example of the electromagnetic force. Like the gravitational force, the electromagnetic force follows an inverse-square law. The closer two objects get, the higher the repulsion between the electrons gets. The atoms never actually come in contact (if it were even possible for two electron clouds to come in contact). If two objects are suddenly forced very close together, the electromagnetic repulsion between them becomes extremely high—at the speed of light—and keeps the atoms in the objects apart from each other.

Nothing ever really touches anything else. The force of electric repulsion keeps them apart.

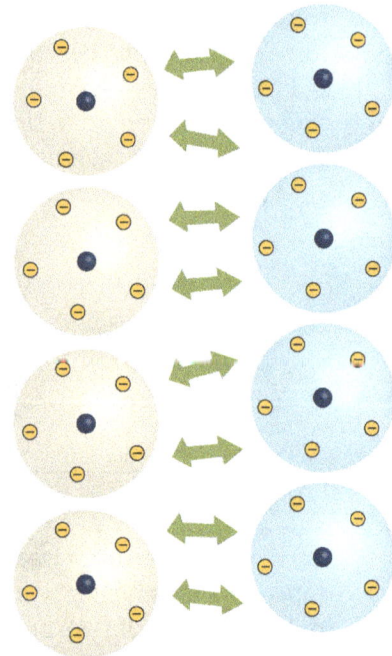

Figure 5.3. When two objects are near to each other, the electrons in their atoms repel one another.

So think about examples of things touching or hitting each other, like a baseball bat hitting a ball, an ice skater's skates pressing against the ice, or your hands holding a book. What is happening in each case is the same electron repulsion that Figure 5.3 illustrates. You are held to the earth by gravity. But basically *all* the other forces you experience on a daily basis are electromagnetic. The electromagnetic force dominates your life!

A good question to be asking yourself right now is this: when electric charges are close together, if the electromagnetic repulsion between them just gets higher and higher, then how do all the positively-charged protons stick together in the nucleus of atoms? The answer to that question is next.

The Strong Nuclear Force

The third fundamental force is the *strong nuclear force*, the strongest force there is. This force is 100 times as strong as the electromagnetic force. And get this: the strong nuclear force is 100,000,000,000,000,000,000,000,000,000,000,000,000 (that's 10^{38}) times as strong as gravity! *That's* what holds the protons together in the nuclei of atoms.

The strong nuclear force only operates inside the nucleus of atoms. Its reach is very short—only 0.0000000000000001 m at most. If two protons get this close together, the strong nuclear force grabs them and binds them together, releasing energy in the process.

The strong nuclear force is the strongest of the four fundamental forces.

This is exactly what happens in nuclear fusion, which we discussed in Chapter 2. In fact, the fusion process going on inside the sun involves all three of the forces we have discussed so far. First, the sun's immense gravity pulls hydrogen atoms toward each other at tremendous speed. The electromagnetic force of repulsion between the hydrogen atoms tries to keep them apart. But the huge amount of energy the atoms have acquired by being pulled toward the center of the sun causes them to blow through the massive electromagnetic force barrier and get close enough together for the strong nuclear force to kick in. When that happens, the protons in the two hydrogen atoms are fused together by the strong nuclear force to make the nucleus of a single helium atom, and energy is released. This is what is constantly going on in all the stars, and it is the source of the energy they produce.

The Weak Nuclear Force

As it turns out, the first three forces we have studied are not the only ones involved in the nuclear fusion in the sun. There is a fourth force in nature, the *weak nuclear force*, and this force participates in the process as well.

Back in Chapter 1, I mentioned a process called *nuclear decay*. The ongoing process of nuclear decay is what makes some substances *radioactive*. When nuclear decay occurs in an atom, some kind of transformation takes place in the nucleus. In *alpha decay*, which I mentioned in Chapter 1, an alpha particle (two protons and two neutrons) is ejected at high speed from the nucleus.

Beta decay, illustrated in Figure 5.4, is another type of nuclear decay. Beta decay is the process of a neutron spontaneously converting into three different new particles, a

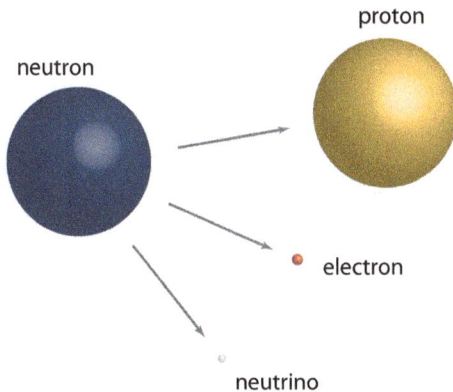

Figure 5.4. In beta decay, a neutron changes into a proton, an electron, and a neutrino.

Scientists, Experiments, and Technology

Marie Curie was one of the pioneers in the study of radioactivity, and is one of the most famous women in the history of science. Born in Warsaw, Poland, in 1891 she moved to Paris to work on her PhD at the age of 24. She remained there for the rest of her life.

The phenomenon of *radioactivity* was accidentally discovered by Henri Becquerel in Paris in 1896. Soon afterward, Marie and her new husband Pierre, also a scientist, were hard at work on trying to figure out the nature and cause of radioactivity, a term they coined. Radioactivity is nuclear radiation, which is due to nuclear decay, and can involve atomic particles as well as photons. (This is different from electromagnetic radiation, which is purely photons traveling as waves.)

In 1903, the Curies, along with Becquerel, shared the Nobel Prize in Physics for their work. Marie became the first woman to win a Nobel Prize.

In her further work, Marie discovered the elements radium and polonium, and she was the first to isolate pure radium. Pure radium glows constantly from the radiation it emits. Marie and her husband are pictured working with radium in the drawing at the left. Always proud of her origins, Marie named polonium after her home country. In 1911, Marie was awarded the Nobel Prize in Chemistry for her discoveries. She became the first person to win or share two Nobel Prizes, and she remains one of only two people who have won Nobel Prizes in two different fields.

In the early years of research into nuclear radiation, no one had any idea that radiation poses a health hazard, and Marie spent her life working with radioactive substances. This finally brought on the illness from which she died, and in 1934 Marie departed this world at the age of 66.

proton, an electron, and a strange, massless particle called a neutrino. This same process also happens in reverse, resulting in a proton converting into three different new particles, one of which is a neutron. (The other two are a neutrino and a positron. A positron is an electron with positive charge. These weird particles are not part of our everyday world, so it's okay if you've never heard of them.)

The weak nuclear force is the agent behind beta decay. That's its job, and that's really all you need to know about it. Its name comes from the fact that the weak nuclear force is 10 trillion times weaker than the strong nuclear force.

The weak nuclear force is the agent behind beta decay.

So how is the weak nuclear force involved in the sun's nuclear fusion? Here's how. The nucleus of a hydrogen atom is a single proton. The nucleus of a helium atom is two protons and two neutrons. Fusion is the process of fusing two hydrogen nuclei together to produce helium. See the problem yet? *Where do those two neutrons come from?* Answer: The weak nuclear force makes them from some other protons! Amazing.

There is one more interesting detail about the four forces that I want to mention. Forces are actually quite strange if you think about it. Take the electromagnetic force, for example. How do charges exert forces on one another without touching? What is actually *going on*? Einstein worried about this too, and called forces "spooky actions at a distance." As incredible as it may sound, according to our present understanding, what is going on is that at the subatomic level, protons, electrons and neutrons are actually *exchanging* zillions of other weird particles with each other. If you think this sounds weird, well, that's okay. Many of us do.

This is virtually impossible to imagine, but that's what's happening. This is why the four fundamental forces are now often called the four *fundamental interactions*. What's really happening when you kick a soccer ball is that the atoms in your shoe are interacting with the atoms in the ball. They exchange unimaginable numbers of particles like mad, and then they go their separate ways, each proud of the gifts they have received from the other atoms. (The whole thing is so bizarre that I can't help making wisecracks about it.)

Learning Check 5.1

1. Explain why nothing ever actually touches anything else.
2. Explain how an inverse-square law works.
3. Briefly explain Albert Einstein's idea of the curvature of spacetime.
4. Make a chart of the four fundamental forces. For each force, include a description of what the force does, and one or two special characteristics about the force.
5. Explain what happens in the process known as beta decay.
6. Briefly explain why the fundamental forces are also called interactions.

5.2   Three Types of Fields

When we talk about forces, it is natural for us to think of fields at the same time. After all, you've probably talked about gravitational fields before and the ways they affect people or aircraft or space ships. If not, you have probably at least seen sci-fi movies like *Star Trek* where the characters are concerned about some dangerous, menacing field out there. You have probably also heard references to magnetic fields.

In this book, we study three fields that are part of basic physics—gravitational fields, electric fields, and magnetic fields. When discussing fundamental forces, we always treat electrical and magnetic forces together as one basic force. But when we discuss fields, there are some interesting differences between electric and magnetic fields, so we treat them separately.

The term *field* refers to a mathematical concept: a geometric description of the strength, range, and direction associated with an object's ability to exert a force on another object. As Figure 5.5 illustrates, we know there is *something* present around a magnet because when we bring an iron or steel object near to it, the object is attracted to the magnet. This *something* that is present in the region around the magnet is the magnetic field.

With the term field, *we are talking about a geometrical description of the way one object exerts a force on another.*

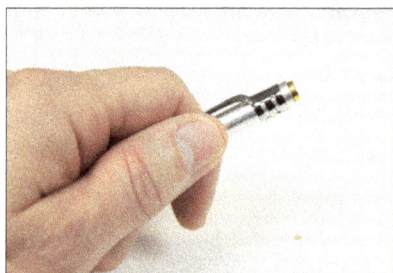

Figure 5.5. The magnetic field around this tool is not visible, but it makes itself known when some small steel parts are placed near the magnet.

On paper, we represent fields by drawing *field lines* around an object that is producing a field. At any place in a field, the direction the lines point indicates the direction of the force that would be present on an object placed there, assuming it was the kind of object that could be affected by that kind of field. The spacing between field lines represents the strength of the field.

The Gravitational Field   Figure 5.6 is a diagram of the *gravitational field* around a spherical mass, such as a planet or the sun. All the arrows point to-

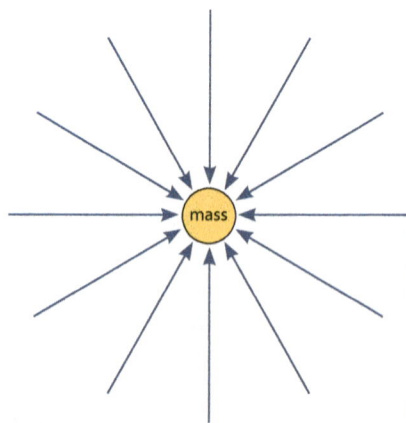

Figure 5.6. The shape of the gravitational field around a mass, such as the sun.

ward the sphere, which means that if another object is placed in the field, it feels a force pulling it toward the center of the sphere. This is why you always feel gravity pulling you toward the center of the earth, no matter where you are.

Notice that the field lines are farther apart at locations farther from the sphere, symbolizing the reduced strength of the field out there. We have seen that the gravitational force is an inverse-square law, so the farther you get from a planet, the weaker the gravitational force is.

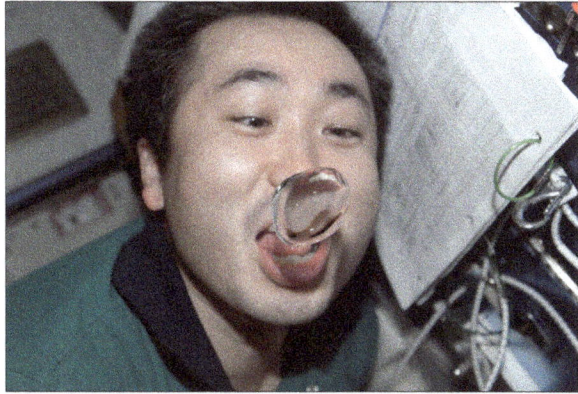

Figure 5.7. Strange things happen in a zero-gravity environment.

To give you an example, consider an astronaut weighing 170 lb going on a trip to the moon. The distance to the moon is about 240,000 miles, but by the time the astronaut is 50,000 miles from earth, he only weighs 1 lb.

Astronauts in the Space Shuttle or the International Space Station are only 225 miles or so above the earth's surface. At this altitude, the astronauts' weights have decreased some, but not really all that much. The reason the astronauts experience weightlessness is because they are orbiting the earth. Physically, this orbiting is the same as continuously *falling* around the earth. The effect is weightlessness, the same as if they were weightless deep in space. In Figure 5.7, a Japanese astronaut in the International Space Station is about to try to drink a floating blob of water, an amusing example of how strange things can be in a zero-gravity environment.

In summary, gravitational fields affect all matter and all electromagnetic radiation. Around any mass, a gravitational field pulls objects toward the center of the mass. These key points are summarized in Table 5.1.

Gravitational Fields	
What causes them?	Anything with mass (matter)
What is affected by them?	Anything with mass (matter), and electromagnetic radiation
How are they shaped?	The direction of the force of the field is toward the center of the object causing the field.

Table 5.1. Summary of key points about gravitational fields.

The Electric Field The shape of the *electric field* around an electrical charge (such as a proton or electron) is shown in Figure 5.8. As you know, there are two kinds of electrical charge, which we call positive and negative. Like charges (positive and positive, or negative and negative) repel each other, and opposite charges (positive and negative) attract, so in order to draw field lines we need to have a convention about what kinds of charges are involved. Our con-

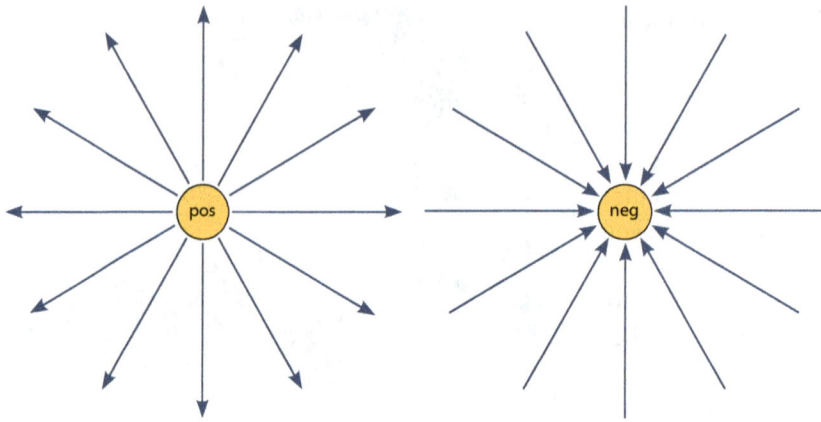

Figure 5.8. The shape of the electric field around a positive charge (left) and a negative charge (right).

vention is that the lines are drawn to show the direction of the force on a positively-charged object placed in the electric field. A positively-charged object is repelled by another positive charge, so the field lines around the positive charge point away from the charge causing the field. A positive charge is attracted to a negative charge, so the lines point toward the negative charge.

Compared to the gravitational force, the strength of the electromagnetic force is 1,000,000,000,000,000,000,000,000,000,000,000,000 (10^{36}) times greater. Although it is nearly impossible to wrap your brain around a number this large, it is still pretty easy to see the difference. In Figure 5.5 the little dinky magnet on that tool easily holds up the steel parts, even though the gravitational field of the entire earth is pulling them down. It is easy to see the effects of magnetism and electric attraction on things around us. But we never notice the gravitational attraction between objects unless we are dealing with something the size of the moon or the earth.

The strength of the electromagnetic force is 10^{36} times that of the gravitational force.

Since the electromagnetic force is so strong, it is interesting to think about what happens to the shape of an electric field if the charges are not in a spherical arrangement. For example, if we cover a flat metal plate with a uniform layer of negatively-charged electrons, the electric field produced is uniform, with parallel field lines as shown in Figure 5.9.[1] Near the edges of the sheet of charge

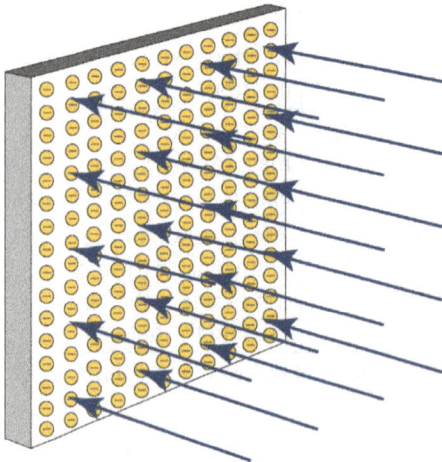

Figure 5.9. A sheet of charge produces a uniform electric field.

---

1   The same thing would happen with the gravitational field near a flat sheet of matter. But the spherical gravitational field from the earth would completely dominate it, and no uniform gravitational field would be detectable. So it's not really worth talking about.

Scientists, Experiments, and Technology

The fact that a sheet of charge produces a uniform electric field led to the invention of the capacitor, one of the simplest and yet most fundamentally important devices in all of modern technology. A *parallel-plate capacitor* is formed when two metal plates are placed close together, face to face, as shown in the sketch to the left. When connected to a battery, electrons are pumped onto the plate connected to the battery's negative terminal, forming a negatively-charged plate. Electrons on the plate connected to the battery's positive terminal are attracted to the battery, leaving more positive charges behind on the plate than negative charges, and thus forming a positively-charged plate. The two plates of charge produce a uniform electric field in between them.

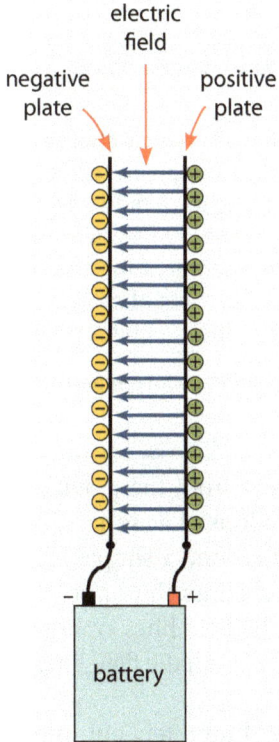

electric field

negative plate

positive plate

battery

When configured this way, the two plates of the capacitor are able to *store charge*, which is just another way of saying they can *store the energy* required for the battery to put the charges on the plates. It turns out that storing electrical energy this way has thousands of applications in electronics, electrical motors, and electrical power distribution systems. Every radio, television, computer, and electronic telephone ever made makes use of capacitors in its electronic circuitry.

A practical parallel-plate capacitor can be made from two long strips of aluminum foil separated by a thin layer of wax paper or plastic. This foil sandwich is then rolled up and placed in a protective container, such as the canisters shown in the image to the right. There are many other capacitor designs as well, represented by the various smaller capacitors shown.

the field lines become distorted, but near the center they are nice and uniform. This phenomenon led to the invention of the capacitor, one of the most important devices in modern technology (see the accompanying box).

Now let's summarize the key features of electric fields. An electric field is produced by any charged object and affects any charged object. A charged object is something like a proton, a cloud of protons, an electron, or an object with more electrons than protons (or vice versa). Around a spherical arrangement of charge, an electric field has a spherical geometry like that of the gravitational field around a planet. But charges can be arranged other ways to produce other field arrangements, such as the uniform electric field produced by a sheet of charge. These key points are summarized in Table 5.2.

Electric Fields	
What causes them?	Any charged object
What is affected by them?	Any charged object
How are they shaped?	For a spherical charge distribution, the direction of the force of the field is toward or away from the center of the charge. A flat sheet of charge produces a uniform field.

Table 5.2. Summary of key points about electric fields.

The Magnetic Field

The *magnetic field* is a bit different from the gravitational and electric fields we have looked at so far. To understand why, first look at the shape of the magnetic field around a simple bar magnet, shown in Figure 5.10. As you can see, the magnetic field lines show a completely different pattern than what we have seen before. All the field lines emerge from the "north end" of the magnet, which is called the *north pole*. All the field lines end at the "south end" of the magnet, the *south pole*.

With gravitational and electric fields, the field lines extend infinitely out into space from the object causing the field. But the field lines surrounding a magnet always start on the magnet and end on the magnet. If you cut the magnet in half, you won't have a north half and a south half. Instead, you will have two complete magnets, each with a north pole and a south pole. Scientists describe this unique behavior with the phrase "there is no magnetic monopole." All this means is that you can't have a magnetic pole by itself. Every

Magnetic poles always come in pairs; there is no magnetic monopole.

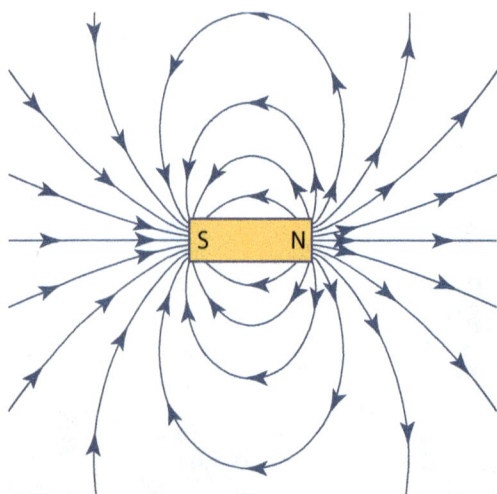

Figure 5.10. The magnetic field around a bar magnet.

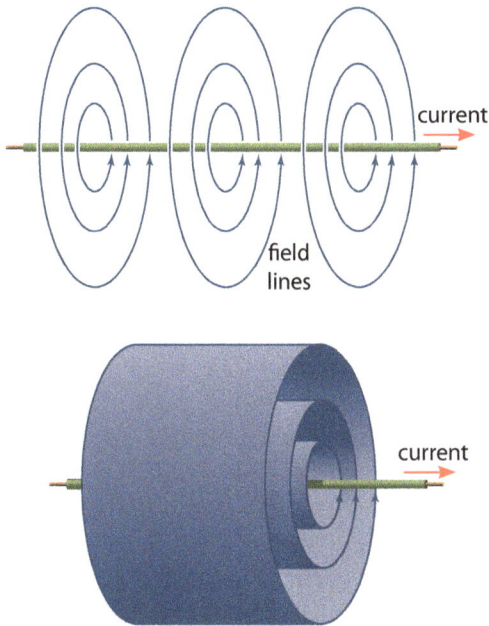

Figure 5.11. Two ways of visualizing the magnetic field surrounding a current-carrying wire.

magnet has a pair of them, north and south.

As with the electric field, there are some interesting geometrical features associated with magnetic fields. The first two I will mention arise from the very important fact that an electric current flowing in a wire creates a magnetic field around the wire. This principle is known as *Ampère's law*. We will study the nature of electric current in more detail later. Here we focus on the magnetic fields that can be produced by electric current.

According to Ampère's law, a magnetic field is produced around a current-carrying wire.

So first, the shape of the magnetic field that exists around a current-carrying wire is shown in Figure 5.11. As you see, the field lines wrap around the wire, all along the wire's length. Recall from the previous section that the strength of the magnetic force follows an inverse-square law, so the field gets weaker as the distance to the magnet increases. Sometimes this field geometry is represented with circles, as in the upper part of Figure 5.11, but another way to represent the field is to show the field lines as concentric cylinders. The term *concentric* refers to circles drawn around one another, with their centers in the same place. By showing the field lines as cylinders the drawing emphasizes the fact that the strength and direction of the field is constant all along the wire, at any given radial distance from the wire.

A second important feature of magnetic fields also relates to fields produced by current-carrying wires. As shown in Figure 5.12, when the wire is wrapped into a coil, the

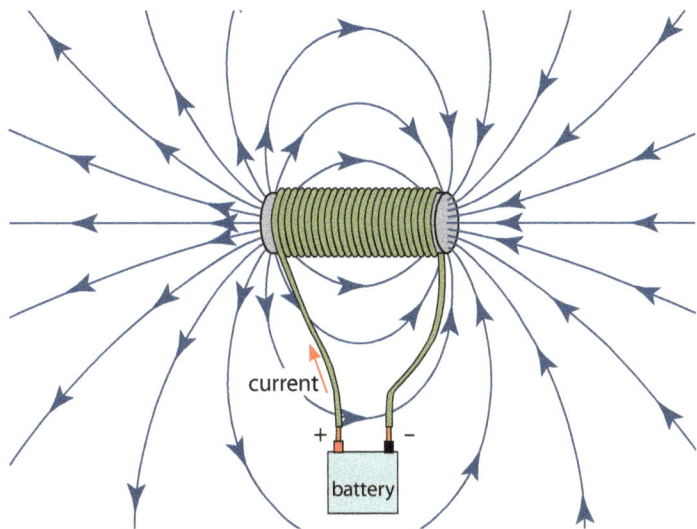

Figure 5.12. Magnetic field produced by an electromagnet.

field around the coil is shaped exactly like the field around a bar magnet. In fact, a coil of wire like this is called an *electromagnet* because it is a magnet produced by electricity. Electromagnets have north and south poles, just like other magnets. In devices making use of electromagnets, the coil is nearly always wound around a steel core. Steel is composed primarily of iron, and the iron core in an electromagnet increases the strength of the magnetic field considerably. (This is because iron is *ferromagnetic*, as we will see in Chapter 14.)

A third interesting feature of magnetic field geometry occurs when the north and south poles of the magnet are parallel and face each other, such as with the C-shaped magnet shown in Figure 5.13. The interesting feature here is that these parallel, facing poles produce a uniform magnetic field between them. As we saw with the uniform electric field produced between charged plates, the ability to produce a uniform magnetic field is enormously useful. I will mention one example in a moment, and we will look at others later in the book.

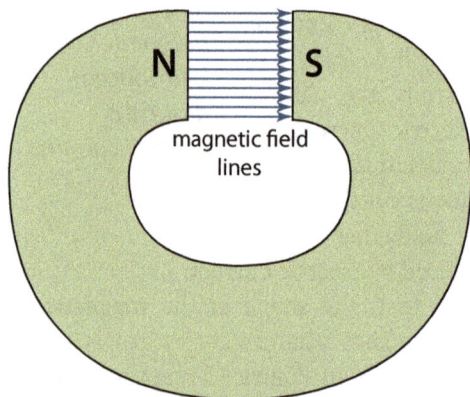

Figure 5.13. A C-shaped magnet with parallel pole faces produces a uniform magnetic field between the poles.

We have seen that magnetic fields are produced by magnets and by current-carrying wires. Let's now look briefly at the kinds of things that are affected by magnetic fields.

First, you already know that magnets are affected by magnetic fields. (Pardon me if this is obvious; I'm just trying to be thorough!) The poles of two magnets always attract or repel one another.

Second, you know that some metals are affected by magnets. Iron is the metal most readily affected by magnetic fields. Any metal that has iron in it is called a *ferrous metal*. (The term *ferrous* comes from the Latin name for iron, *ferrum*.) As I mentioned above, steel contains iron, and is a common example of a ferrous metal. Ferrous metals are affected by magnetic fields. This is why you can pick up steel objects such as nails and paper clips with a magnet. Other types of metal—copper, aluminum, silver, gold, and others—are not affected by magnetic fields. So, you cannot pick up coins, silver jewelry, gold jewelry, or aluminum objects with a magnet.

Third, moving charges (protons and electrons) are also affected by magnetic fields. A moving electric charge is always pulled *sideways* by a magnetic field. Positive particles like protons are pulled one direction, and negative electrons are pulled in the opposite direction. A sketch depicting what this looks like is shown in Figure 5.14. In this figure, the magnetic field is uniform, just like the field we have between the parallel facing poles of Figure 5.13. In this figure, a proton fired into this magnetic field. The figure shows the proton just as it enters the magnetic field. The sideways force on the positively-charged proton by the magnetic field causes the

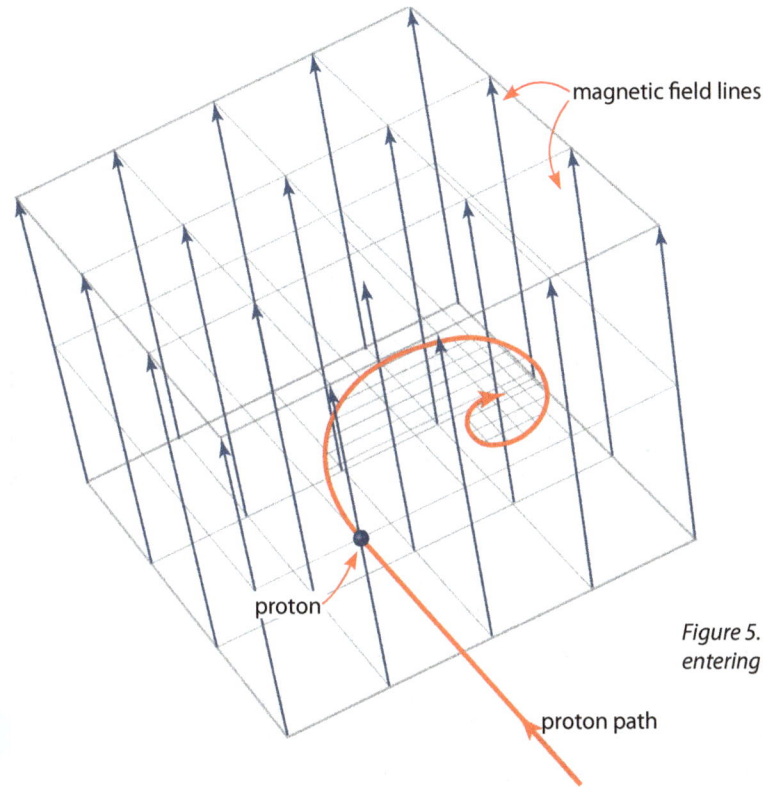

Figure 5.14. The path of a proton entering a uniform magnetic field.

proton to turn in a curved pathway. An electron fired into the field from the same direction curves in the opposite direction.[2]

This curving phenomenon has been enormously useful in physics for studying the world of subatomic particles. With an invention from the early 1950s called a bubble chamber, scientists can actually see tracks left by particles as they fly through the chamber. By studying the direction and diameter of the spiral as particles pass through a magnetic field, scientists determine the mass and charge of the particle causing the track and thus identify the particle. An actual image

Figure 5.15. Particle tracks in a bubble chamber allow physicists to determine the mass and charge of unknown particles.

---

2   The reason the particle moves in a spiral instead of a circle is that when a charged particle moves in a curve it gives off energy and slows down. This causes it to spiral inward.

Scientists, Experiments, and Technology

New technologies using magnetism have been under constant development for 150 years. Human life has been permanently altered as a result. The system shown in the first photo to the right is used in *magnetic resonance imaging* (MRI). MRI is now a widely used medical technology because of its ability to generate images of tissues inside the body. This allows doctors to diagnose problems involving muscles, ligaments, brain tissue, and tumors without invasive surgery. The large round object is a magnetic coil which generates a magnetic field around the patient. The magnetic field causes water molecules in tissues to rotate and line up. As they move, they radiate energy that is detected and decoded for the information about the tissue.

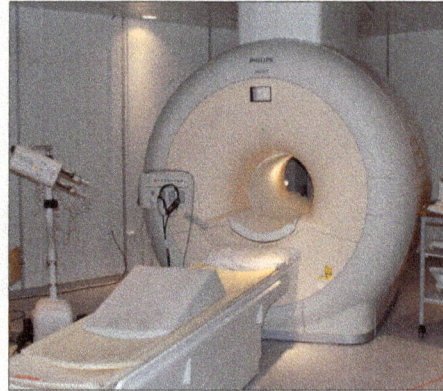

Magnetic levitation is now being explored as a transportation technology. A maglev train like the Japanese train shown in the photo below uses magnetism for both levitation and propulsion. Since the train moves without touching any rails, high speeds are readily achieved. The world record at present is 361 miles per hour for a test train in Japan. Maglev technology is expensive, so debates are ongoing about how practical it can be for wide use.

The third image shows the inside of a hard disk drive in a computer. Hard disk technology is amazing. At the tip of the arm is a C-shaped electromagnetic head, which records and reads data on the disk. The data are recorded on the disk by magnetizing a microscopic region on the disk surface to make it a north or south pole. With modern drives, the head floats on a cushion of air as the disk spins. The air cushion flies past the head at 80 miles per hour, allowing the head to glide along only 10 nanometers (100 atomic diameters!) above the disk surface. The closer the head gets to the surface, the more data the disk can store.

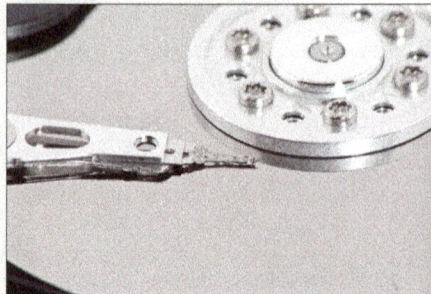

Magnetic Fields	
What causes them?	Magnets Current-carrying wires
What is affected by them?	Magnets Current-carrying wires Moving charged particles
How are they shaped?	Field lines around a magnet go from north pole to south pole. Field lines around a current-carrying wire are arranged in concentric circles.

Table 5.3. Summary of key points about magnetic fields.

from a particle physics lab in Switzerland is shown in Figure 5.15. Particles leaving straight tracks have no charge.

The key points about magnetic fields are summarized in Table 5.3.

Learning Check 5.2

1. Make a chart listing the three types of fields and their distinctive features.
2. For each of the three types of fields, identify five objects that are affected by the field, and five that are not affected by the field (30 items total).
3. What is an electromagnet, and how can one be assembled from common materials?
4. What is a uniform field, and how can uniform electric and magnetic fields be produced?

Chapter 5 Exercises

Answer each of the questions below as completely as you can. Write your responses in complete sentences.

1. Compare the three types of fields we have studied by identifying and writing out the similarities and differences between.
2. Use the information in the chapter to compare the strengths of the four fundamental forces.
3. What happens to a moving charged particle, such as an electron, when it enters a uniform magnetic field?
4. State Ampère's law and describe the type and shape of field that the law describes.
5. Explain why a small magnet can hold up a steel object, even though the gravity of the entire earth is pulling the object down.
6. What is a magnetic monopole and what is interesting about them?
7. Describe several examples of electric and magnetic fields in technology.

Experimental Investigation 3: Electrostatic Forces

Overview

- *The goal of this investigation is to observe the effects of forces caused by accumulations of electric charge. These accumulations are called "static electricity."*
- Make simple "pith ball" *electroscopes* out of pieces of Styrofoam and fishing line. Apply electric charges by rubbing on hair or fur, and observe how they interact.
- Observe how the Styrofoam "pith balls" behave when close to metal objects.
- Make a simple aluminum-foil electroscope and observe how the leaves interact under the influence of charge accumulations.

Basic Materials List

- polystyrene foam (Styrofoam) cup
- fishing line (8 inches)
- copper wire (8 inches)
- aluminum foil (1 inch × 2 inch strip)
- quick clamps with plastic clamping surfaces, small (2)

Many of the negatively-charged electrons in metals (*conduction electrons*) are free to move around, while the positively-charged protons are bound in place. Other substances, such as hair and cat fur, will readily donate electrons to objects made of substances such as rubber or polystyrene foam (Styrofoam). In this experiment, we will make use of these two properties to gather up electrons and observe what they do when they accumulate.

Our investigation will take place in three stages. During each stage, carefully document all your observations in your lab journal.

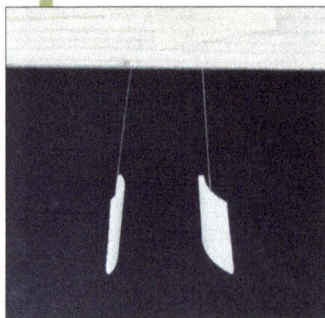

In the first stage, use small pieces of polystyrene foam glued to fishing line. The polystyrene foam easily accumulates electrons from hair or cat fur when you rub the polystyrene on the hair. (Clean dry hair works pretty well. Cat fur works even better.) In experiments like this back in the 17th century, scientists used a spongy material called *pith*, found in the core of some kinds of trees. The scientists would use little balls of pith attached to silk threads. In our day, polystyrene and fishing line are a lot easier to come by than pith and silk. (Feel free to use pith balls and silk threads if you have some.)

Cut two squares of polystyrene foam about 3/4 inch square. Glue a piece of fishing line to the back of each one. Charge up each piece by holding at the edges and rubbing it on your hair or your cat. Then dangle each piece of polystyrene by the fishing line and try to bring their charged surfaces together, as illustrated in the photo above.

Now for stage two. Charge up one of your little "pith balls" again. This time, hold it by the fishing line and watch what happens when you bring the charged surface near any metal object. Try it with as many different types of metal as you can, including painted metal surfaces.

For stage three, make a simple *electroscope* out of an 8-inch piece of solid copper wire and two strips of aluminum foil, as illustrated in the photograph. Cut the foils strips about 1/2 in × 2 in, and cut a small hole near the end of each one. Hang the foil strips from the bent legs of the copper wire. These foil strips are called *leaves*, because back in the 18th century when the electroscope was invented, scientists used light weight "gold leaf" instead of aluminum foil. (Aluminum foil wasn't around back then.)

Use two clamps with insulated (plastic) clamping surfaces to support the copper electroscope vertically. You will need to locate your setup in a calm room, with no fans or open windows to blow the foil leaves around. With the electroscope in place, we will experiment with it in three steps.

For the first step, take a polystyrene cup, or the big chunk of cup left over from the first stage of this investigation, and rub it vigorously on your clean, dry hair or your cat. This accumulates a *lot* of electrons on the cup. Now bring the charged cup very near to the top of the electroscope—without letting them touch—and observe the leaves. You should see them swing apart from each other. Withdraw the cup and observe what the leaves do.

Now for step two. Repeat step one, but this time actually touch the charged cup to the top of the electroscope and rub it around, as if you are scraping electrons from the charged cup onto the electroscope (which, in fact, you *are* doing). If you do this right, the leaves should swing apart and stay apart, even when the cup is withdrawn.

Finally, while the leaves are apart, touch the top of the electroscope with your finger and observe what happens.

Analysis

Make sure all your observations have been carefully documented in your lab journal. In your report, address the following questions.

1. Explain the behavior of the two small polystyrene squares (the "pith balls") when they were charged up and brought near one another. Don't simply describe what happened. Use the principles of electric charge, electric fields and force to explain *why* it happened.
2. Consider how your charged square of polystyrene behaved when brought near a metal. Recalling that the conduction electrons move easily in metals, explain why the charged polystyrene is *attracted* to the metal. Again, be specific, using the principles you have studied.
3. Consider the first step of investigation with the electroscope. Explain why the leaves swing apart when the charged cup is brought near the electroscope (without touching) and relax when the cup is withdrawn. If you are working in a group with other students, discuss your thoughts on this item until you have all agreed on the explanation. Figuring this out may take a bit of discussion and thought.
4. Consider the second step of the electroscope, when the cup is scraped on the copper. Explain why the foil leaves remain apart, even when the charged cup is removed.
5. Finally, explain why the leaves relax when you touch the top of the electroscope.

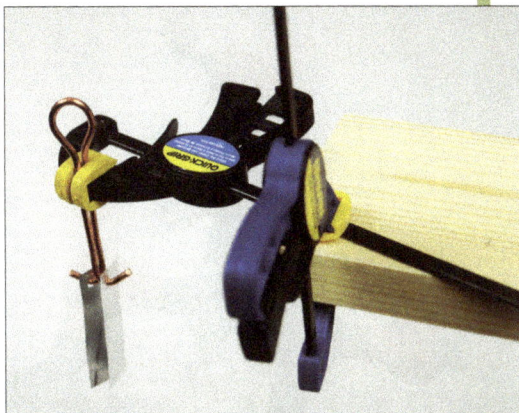

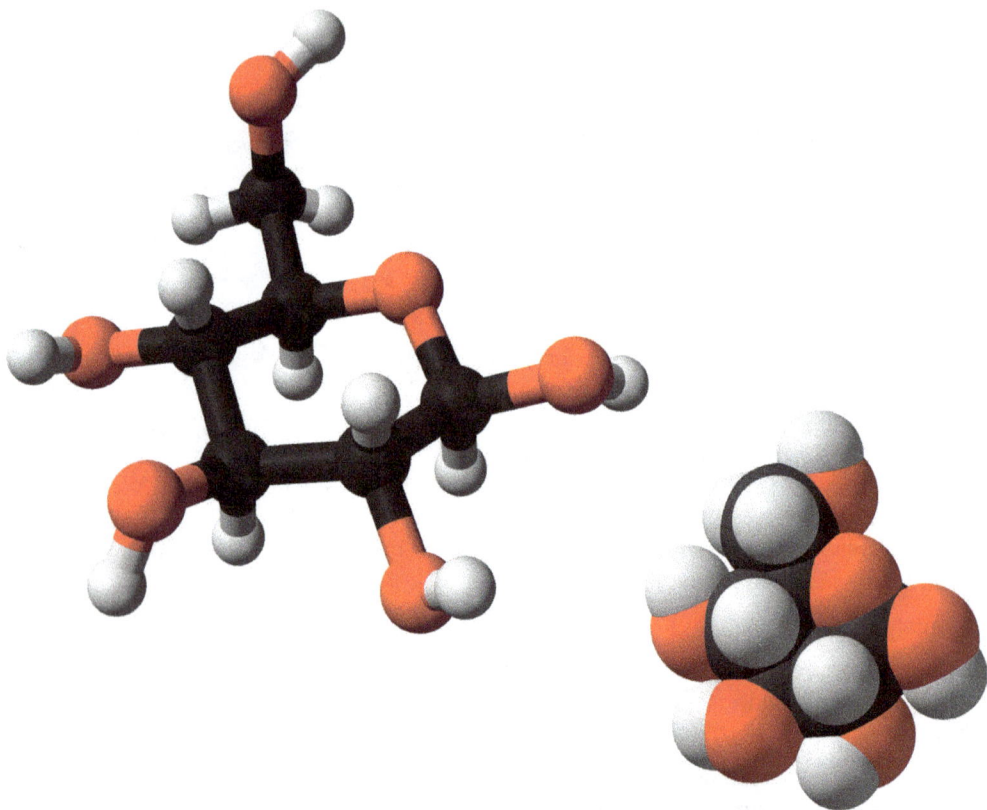

Above are models of a single molecule of glucose, one of the simple sugars found in many plants and animals. The model on the left, a ball-and-stick model, is helpful for showing how atoms are bonded to other atoms. The model on the right, a space-filling model, is better for showing how the compact molecule actually fills up space.

In humans, glucose is often called blood sugar because the glucose in the blood is tightly regulated by the body's metabolism. In the models, white balls represent hydrogen atoms, black balls represent carbon atoms, and red balls represent oxygen atoms. If you count the balls of each color, you will see why the chemical formula for glucose is $C_6H_{12}O_6$. In nature, glucose is found in several different isomers, molecules using the exact same number of each type of atom, but arranging them in slightly different ways.

OBJECTIVES

After studying this chapter and completing the exercises, you should be able to do each of the following tasks, using supporting terms and principles as necessary.

1. Name five examples of pure substances that are composed of molecules and five examples of crystalline substances.
2. Explain the difference between molecular and crystalline compounds.
3. From a chemical formula, identify the elements and their atomic ratios in a compound.
4. Describe the three main regions in the Periodic Table of the Elements.
5. Describe the origin of the Periodic Table of the Elements.
6. Define and distinguish between *pure substances* and *mixtures*.
7. Define and distinguish between *elements, compounds, homogeneous mixtures*, and *heterogeneous mixtures*.
8. Define *atomic number, atomic mass, isotope*, and *ion*.
9. Describe the process of dissolving for both molecular and crystalline solutes.

VOCABULARY TERMS

You should be able to define or describe each of these terms in a complete sentence or paragraph.

1. acid rain	11. dissolve	19. mixture
2. atom	12. element	20. molecule
3. atomic mass	13. heterogeneous	21. Periodic Table of the
4. atomic number	mixture	Elements
5. chemical formula	14. homogeneous	22. polar
6. chemical symbol	mixture	23. pure substance
7. compound	15. hydrocarbon	24. solute
8. crystal	16. ion	25. solution
9. crystal lattice	17. isotope	26. solvent
10. dissociate	18. metalloid	

6.1 Atoms, Molecules, and Crystals

As you know, all substances are composed of atoms. But on earth, it is not very common to find substances existing as single atoms. The huge majority of atoms on earth are electrically bonded to other atoms. There are two basic configurations atoms take when bonding with other atoms: molecules and crystals.

Molecules When atoms form *molecules*, they join together in groups, such as in the glucose molecule shown on the opposite page. Glucose is a

Figure 6.1a. Water molecule: H_2O

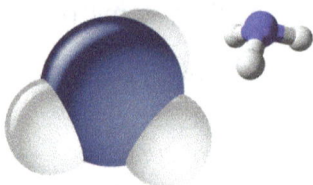

Figure 6.1b. Ammonia molecule: NH_3

Figure 6.1c. Carbon dioxide molecule: CO_2

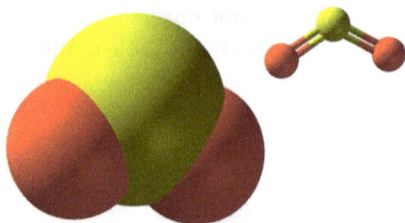

Figure 6.1d. Sulfur dioxide molecule: SO_2

Figure 6.1e. Propane molecule: C_3H_8

rather complicated molecule, composed of 24 atoms all connected together. But there are many substances familiar to us in every day life that are composed of simpler molecules. A few examples are shown in Figures 6.1a through 6.1j. The same color scheme is used in each of these models for different kinds of atoms: white for hydrogen, black for carbon, blue for nitrogen, red for oxygen, yellow for sulfur, and green for chlorine. All these models are computer renderings designed to show the relative sizes and orientations of the atoms in the molecules. The larger "space filling" models show how close the centers of the atoms are to one another in the molecule and how much space the molecule occupies. The smaller "ball and stick" models give more information about how the atoms are bonded together.

Molecules are so small and exist in such large numbers that it is difficult to wrap our brains around the numbers involved. In a *single drop* of water there are about 1,340,000,000,000,000,000,000 water molecules. Just how big is this number? This is almost five times the number of *drops* of water in Lake Superior—one of the largest freshwater lakes in the world.

In the six molecules shown in Figures 6.1a through 6.1f, each molecule includes atoms from two different elements. Ammonia (Figure 6.1b) is used in household cleaners and as an agricultural fertilizer. Carbon dioxide (Figure 6.1c) is a common gas given off by mammals as they breathe and by the combustion of fossil fuels, as we saw in Chapter 2.

Sulfur dioxide (Figure 6.1d) is produced by burning of fossil fuels, and is a major contributor to *acid rain*. In the

1970s, scientists learned that sulfur dioxide and other compounds released into the atmosphere by fossil-fuel power plants reacted chemically with moisture in the atmosphere to produce sulfuric acid and other acids. These acidic compounds mix with the moisture in the atmosphere, and appear in rain water. The acids in rain were destroying forests, contaminating bodies of water, and damaging buildings, bridges and statues. Since then, clean air standards have required the use of "scrubbers" and other technologies to remove these compounds from the waste gases at power stations, and acid rain has decreased significantly. Efforts to reduce the release of these pollutants even further continue to this day.

Propane and methane (Figures 6.1e and 6.1f) are common fuel gases derived from petroleum. The two molecules are examples of a class of compounds known as *hydrocarbons* because their molecules are composed entirely of hydrogen and carbon. Oxygen and nitrogen (Figures 6.1g and 6.1h) are the major components of earth's atmosphere (21% and 78%, respectively). Ozone (Figure 6.1i) is produced naturally in the earth's atmosphere and helps protect life on earth by blocking much of the sun's ultraviolet radiation. In its pure form, chlorine (Figure 6.1j) is a very poisonous substance. But chlorine gas is a significant part of the public water treatment processes that make our drinking water safe, and chlorine tablets are commonly used to prevent algae growth in swimming pools.

This short list of molecules is only a drop in the bucket compared to the huge number of different molecular compounds scientists have identified.

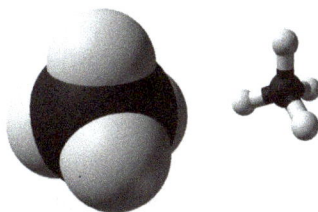

Figure 6.1f. Methane molecule: CH_4

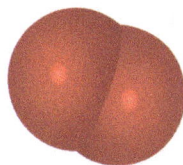

Figure 6.1g. Oxygen molecule: O_2

Figure 6.1h. Nitrogen molecule: N_2

Figure 6.1i. Ozone molecule: O_3

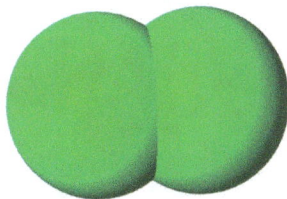

Figure 6.1j. Chlorine molecule: Cl_2

Crystals The second major way atoms bond together is in *crystals*. A crystal is formed when atoms organize themselves into a repeating grid structure, or *array*. This array is called a *crystal lattice*. As with molecules, there are thousands and thousands of different crystal forms found in nature. The images in Figures 6.2a through 6.2f depict a few examples of this vast spectrum of crystal forms.

Calcium oxide (Figure 6.2a) is a simple alternating arrangement of calcium and oxygen atoms. (In Figures 6.2a and 6.2b the white spheres are calcium atoms.) Sodium chloride, ordinary table salt, has a crystal structure exactly like that of calcium oxide. A model of sodium chloride looks the same as Figure 6.2a.

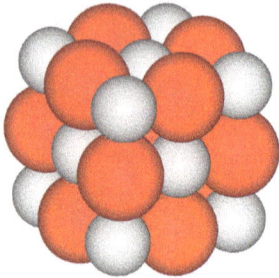

Figure 6.2a. Calcium oxide crystal: CaO

The more interesting geometry of calcium chloride (Figure 6.2b) incorporates two chlorine atoms for each calcium atom. The figure shows a single *unit cell* of the crystal. The crystal lattice of a crystal is formed by chaining and stacking of multiple units of the unit cell, with the corner atoms held in common between the cells. For a brief tutorial on how the 1:2 ratio of calcium to chlorine atoms is achieved in the lattice, read the box on the opposite page.

The even more interesting (and complex) structure of a calcite unit cell is depicted in Figure 6.2c. Calcite is composed of the compound calcium carbonate, which is itself composed of calcium, carbon, and oxygen. A photograph of some natural calcite crystals is shown in Figure

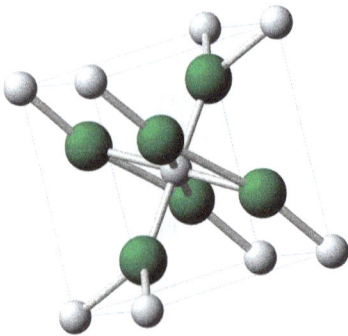

Figure 6.2b. Calcium chloride crystal: $CaCl_2$

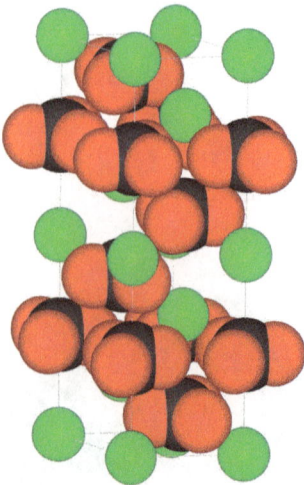

Figure 6.2c. Calcite crystal: $CaCO_3$

Figure 6.2d. Calcite crystal: $CaCO_3$

We Pause Here to Talk About **Counting Atoms in Crystals**

The formula for calcium chloride, $CaCl_2$, indicates that there is a ratio of one calcium atom for every two chlorine atoms in the calcium chloride crystal. We should be able to arrive at this same conclusion by looking at the geometry of the calcium chloride unit cell (Figure 6.2b).

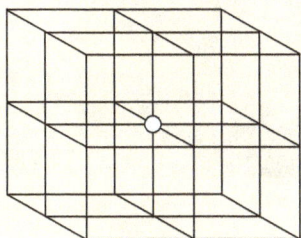

We begin by considering the calcium atoms at the eight corners of the unit cell. The sketch to the left depicts one of these corner atoms, surrounded by eight neighboring unit cells. This sketch shows that each corner atom is shared by eight unit cells. Since there are eight corners for each unit cell, the eight corner atoms taken together contribute one calcium atom per cell in the crystal lattice. Adding this to the calcium atom in the center of the cell gives us two calcium atoms per unit cell in the crystal lattice.

For the chlorine atoms, the two chlorine atoms on the front face of the cell are shared by that cell and the one in front of it, as shown in the sketch to the right. This means that the pair of them contributes one atom per cell in the lattice. The same thing goes for the two chlorine atoms on the back face of the cell. Together then, these four atoms contribute two chlorine atoms per cell in the lattice. Adding these to the two other chlorine atoms that are completely within the cell gives us a total of four chlorine atoms per unit cell in the crystal lattice.

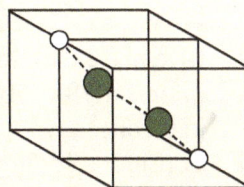

Two calcium atoms and four chlorine atoms per cell gives us a ratio of one calcium atom for every two chlorine atoms in the crystal lattice.

6.2d. Calcium carbonate appears in many different forms in nature. Limestone, chalk, and sea shells are all composed of calcium carbonate.

Two final examples of crystals are shown in Figures 6.2e and 6.2f. The cerussite (orange) and malachite (green) crystals in Figure 6.2e are a wonderful example of the pleasing beauty often found in natural crystals. Cerussite is lead carbonate, with the formula $PbCO_3$. Malachite is also a carbonate material, but with copper (and some oxygen and hydrogen) instead of the lead. As you can see from our examples so far, carbonate (CO_3) is quite common in natural crystals.

The crystal in Figure 6.2f is made of pure gold. Gold is a metal, as you know, and all pure metals are crystals. This gold crystal is synthetic, meaning that it was grown artificially by allowing gold vapor to deposit on the crystal lattice very slowly, one layer of atoms at a time. When allowed to form a crystal structure in this gentle and careful fashion, many different metals exhibit stunning crystal structures.

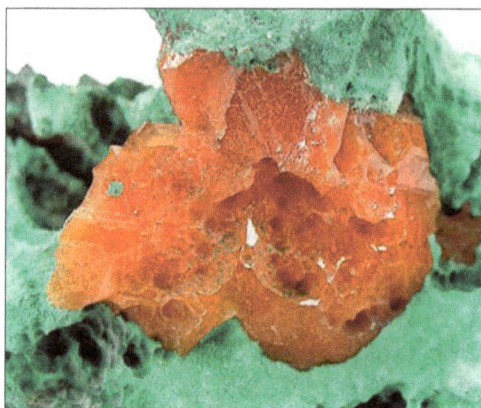

Figure 6.2e. Cerussite (orange) and malachite (green) crystals.

Figure 6.2f. Gold crystal, synthetically grown from gold vapor.

Learning Check 6.1

1. Describe the difference between molecules and crystals.
2. What is a crystal lattice?
3. Give five examples of molecular substances. For each one, state one important fact about the way it appears in nature or the way it is used in technology.
4. Consider a sodium atom in a crystal of table salt. If you consider all directions (front to back, side to side, vertical and all angles) how many atoms are there surrounding this sodium atom in the crystal lattice?
5. Imagine a lump of pure nickel and a lump of pure copper, both of which are metals. Are the atoms in these two lumps arranged in molecules or in a crystal lattice? How do you know?

6.2   The Substances Family Tree

To help us in our discussion of different types of substances, it is helpful to classify them into different categories according to the purity of their composition (what they are made of). In general, there are two types of substances: *pure substances* and *mixtures*. There are two kinds of pure substances: *elements* and *compounds*. There are also two kinds of mixtures: *homogeneous mixtures* and *heterogeneous mixtures*. To remember how these different kinds of substances are related, think of them in a sort of "family tree," as shown in Figure 6.3.

The two basic kinds of substances are pure substances and mixtures.

In the next few sections, we look at each of these four types of substances. As we go, keep their positions in the family tree in mind.

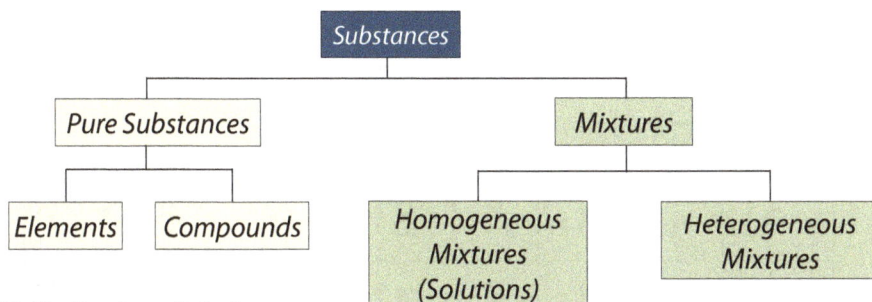

Figure 6.3. The "family tree" of substances.

6.3   Elements

Since ancient times, scientists have wondered about the basic, or elemental, stuff everything is made of. A popular theory back in ancient Greece was that everything was composed of earth, air, fire, and water. Since these were believed to be the most elemental substances, they were regarded as the four *elements*. It took many centuries for experimental science to develop to the point where scientists knew that this theory was not correct. But by the time John Dalton came out with his atomic theory in 1803 (Chapter 1), many scientists were thinking that the term element was an *atomic* term. There was something about different kinds of atoms that distinguished one element from another.

The identity of elements is determined by the number of protons in the atoms' nuclei.

Recall from Chapter 1 that part of Dalton's atomic theory was that every atom of a given element is identical. In stating this principle, Dalton was partially correct. Every atom of a given element does have the same number of protons in the nucleus. And this is how we distinguish elements from one another today—by the number of protons they have in their nuclei.

In 1869, Russian scientist Dmitri Mendeleev (Figure 6.4) discovered that when the elements were arranged in order according to their atomic weight, many of their properties were *periodic* or *cyclic*. That is, a given property of the elements, such as the density, increases from one element to the next up to a point and then drops down and starts increasing over again. When Mendeleev arranged the elements in rows according to the cycle of their properties, he had made a major discovery: he had designed the first *Periodic Table of the Elements*.

The contemporary Periodic Table of the Elements with the 118 elements discovered so far is shown in Figure 6.5. This table is the first and most important chart for the study of chemistry. Each box in this table represents one element. In today's pe-

Figure 6.4. Russian scientist Dmitri Mendeleev.

liquid at room temperature

radioactive

1	2		3	4	5	6	7	8	9	10	11	12	13	14	15	16	17	18
1 **H** Hydrogen 1.0079																		2 **He** Helium 4.0026
3 **Li** Lithium 6.941	4 **Be** Beryllium 9.0122												5 **B** Boron 10.811	6 **C** Carbon 12.011	7 **N** Nitrogen 14.0067	8 **O** Oxygen 15.9994	9 **F** Fluorine 18.9984	10 **Ne** Neon 20.1797
11 **Na** Sodium 22.9898	12 **Mg** Magnesium 24.3050												13 **Al** Aluminum 26.9815	14 **Si** Silicon 28.0855	15 **P** Phosphorus 30.9738	16 **S** Sulfur 32.066	17 **Cl** Chlorine 35.4527	18 **Ar** Argon 39.948
19 **K** Potassium 39.098	20 **Ca** Calcium 40.078		21 **Sc** Scandium 44.9559	22 **Ti** Titanium 47.88	23 **V** Vanadium 50.9415	24 **Cr** Chromium 51.9961	25 **Mn** Manganese 54.9380	26 **Fe** Iron 55.847	27 **Co** Cobalt 58.9332	28 **Ni** Nickel 58.6934	29 **Cu** Copper 63.546	30 **Zn** Zinc 65.39	31 **Ga** Gallium 69.723	32 **Ge** Germanium 72.61	33 **As** Arsenic 74.9216	34 **Se** Selenium 78.96	35 **Br** Bromine 79.904	36 **Kr** Krypton 83.80
37 **Rb** Rubidium 85.468	38 **Sr** Strontium 87.62		39 **Y** Yttrium 88.9059	40 **Zr** Zirconium 91.224	41 **Nb** Niobium 92.9064	42 **Mo** Molybdenum 95.94	43 **Tc** Technetium 98.9072	44 **Ru** Ruthenium 101.07	45 **Rh** Rhodium 102.9055	46 **Pd** Palladium 106.42	47 **Ag** Silver 107.8682	48 **Cd** Cadmium 112.411	49 **In** Indium 114.82	50 **Sn** Tin 118.710	51 **Sb** Antimony 121.76	52 **Te** Tellurium 127.60	53 **I** Iodine 126.9045	54 **Xe** Xenon 131.29
55 **Cs** Cesium 132.905	56 **Ba** Barium 137.327		71 **Lu** Lutetium 174.967	72 **Hf** Hafnium 178.49	73 **Ta** Tantalum 180.9479	74 **W** Tungsten 183.85	75 **Re** Rhenium 186.207	76 **Os** Osmium 190.2	77 **Ir** Iridium 192.22	78 **Pt** Platinum 195.08	79 **Au** Gold 196.9665	80 **Hg** Mercury 200.59	81 **Tl** Thallium 204.3833	82 **Pb** Lead 207.2	83 **Bi** Bismuth 208.9804	84 **Po** Polonium 208.9824	85 **At** Astatine 209.9871	86 **Rn** Radon 222.0176
87 **Fr** Francium 223.0197	88 **Ra** Radium 226.0254		103 **Lr** Lawrencium 262.11	104 **Rf** Rutherfordium 261.11	105 **Db** Dubnium 262.114	106 **Sg** Seaborgium 263.118	107 **Bh** Bohrium 262.12	108 **Hs** Hassium (265)	109 **Mt** Meitnerium (266)	110 **Ds** Darmstadtium (281)	111 **Rg** Roentgenium (281)	112 **Cn** Copernicium (285)	113 **Nh** Nihonium (284)	114 **Fl** Flerovium (289)	115 **Mc** Moscovium (288)	116 **Lv** Livermorium (293)	117 **Ts** Tennessine (294)	118 **Og** Oganesson (294)

57 **La** Lanthanum 138.9055	58 **Ce** Cerium 140.115	59 **Pr** Praseodymium 140.9077	60 **Nd** Neodymium 144.24	61 **Pm** Promethium 144.9127	62 **Sm** Samarium 150.36	63 **Eu** Europium 151.965	64 **Gd** Gadolinium 157.25	65 **Tb** Terbium 158.9253	66 **Dy** Dysprosium 162.50	67 **Ho** Holmium 164.9303	68 **Er** Erbium 167.26	69 **Tm** Thulium 168.9342	70 **Yb** Ytterbium 173.04
89 **Ac** Actinium 227.0278	90 **Th** Thorium 232.0381	91 **Pa** Protactinium 231.0359	92 **U** Uranium 238.0289	93 **Np** Neptunium 237.0482	94 **Pu** Plutonium 244.0642	95 **Am** Americium 243.0614	96 **Cm** Curium 247.0703	97 **Bk** Berkelium 247.0703	98 **Cf** Californium 251.0796	99 **Es** Einsteinium 252.083	100 **Fm** Fermium 257.0951	101 **Md** Mendelevium 258.10	102 **No** Nobelium 259.1009

Figure 6.5. The Periodic Table of the Elements.

riodic table, the elements are ordered—1 through 118—according to their *atomic number*. This number is usually displayed at the top of the box for each element. The atomic number is the number of protons an atom of that element has in its nucleus.

17	← atomic number
Cl	← chemical symbol
Chlorine	← element name
35.4527	← atomic mass

Figure 6.6. The basic information in each cell of the periodic table.

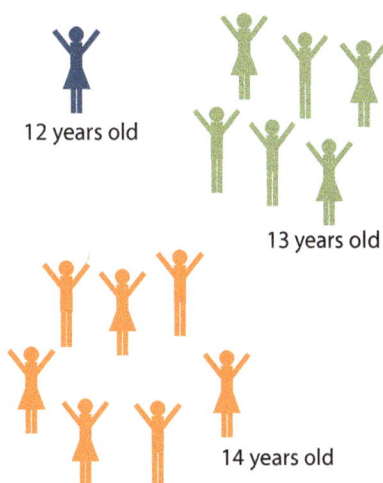

Find element 17, chlorine, in the periodic table. We will use chlorine as an example to learn about the other information found in the box for each of the elements. As shown in Figure 6.6, the atomic number for chlorine is 17. This means every atom of chlorine has 17 protons in its nucleus. We can also say this the other way around: if an atom has 17 protons in its nucleus, it is definitely a chlorine atom. The one- or two-letter abbreviation in the center of each box is the *chemical symbol* of the element. All chemical equations and chemical formulas use the chemical symbols of the elements.

Below the chemical symbol is the element's name. At the bottom of the box is the element's *atomic mass*. To explain the atomic mass, we need to go back to John Dalton for a moment. Remember that I wrote that Dalton was *partially* correct that every atom of a given element was identical. Dalton's statement was correct when applied to protons (even though the proton was not discovered for another 103 years after Dalton published his atomic theory). But when it comes to the neutrons in the nucleus, the atoms of an element are not necessarily identical. For every element, the neutron count in the nucleus of an atom can take on several different values.

For example, the number of neutrons in a chlorine atom can be 18, 19, or 20. This means there are three different types of chlorine atoms. These different types of atoms are called *isotopes*. The three isotopes of chlorine each have 17 protons, but the number of neutrons in each isotope is different.

Now, the isotopes of an element are not all equally abundant in nature—far from it. Some are common; some are scarce. For chlorine, about 76% of atoms have 18 neutrons, and most of the rest have 20. Nuclei with 19 neutrons are very scarce. But if you calculate the average number of nucleons (protons and neutrons, giving totals of 35, 36, and 37) in chlorine atoms, taking into account their relative abundance in nature, you find the average to be 35.4527. That value is the atomic mass.

Here's an analogy as an illustration. Figure 6.7 shows a classroom of students grouped by age. In this classroom, the average age, which I am sure

12 years old

13 years old

14 years old

Figure 6.7. The average student age in this class is 13.429 years.

you can easily verify, is 13.429 years. The fourteen-year-old students in the class are the most abundant. The 13-year olds are also quite common, and the 12-year olds are somewhat scarce. The different age groupings are like different isotopes of an element. The number of students in each group is like the abundance of an isotope in the chlorine atoms found in nature. The atomic mass is the average number of nucleons in the nucleus for a given element.

Different isotopes of an element have different numbers of neutrons in their nuclei.

An element's position in the periodic table is our major tool for understanding that element's properties and behavior. There are three distinct regions in the periodic table that you should know about, shown in Figure 6.8. As you see from the figure, the huge majority of elements are metals! Isn't that amazing? Eighteen of the elements fall into the class of nonmetals.

Between the metals and the nonmetals, there are seven elements called *metalloids*. As you might guess, the metalloids have properties that are not cleanly metallic or nonmetallic, so they are grouped separately. One interesting thing to know about the metalloids is that they are commonly used in the manufacturing of computer chips called *semiconductors*. You know that metals conduct electricity. Semiconductors conduct as well, but only under certain conditions. Once scientists discovered this strange behavior, the computer age was born. Element 14, silicon, is a metalloid. The use of silicon in the manufacturing of computer chips is what gave the area around the San Francisco Bay the nickname "Silicon Valley."

By the way, you may have noticed the two rows of elements separated out at the bottom of the periodic table. The only reason those two rows are commonly shown that way is that when we put them in their place (shown by the arrow) the periodic table becomes too wide to show conveniently. It's not that they're more special than other elements or anything.

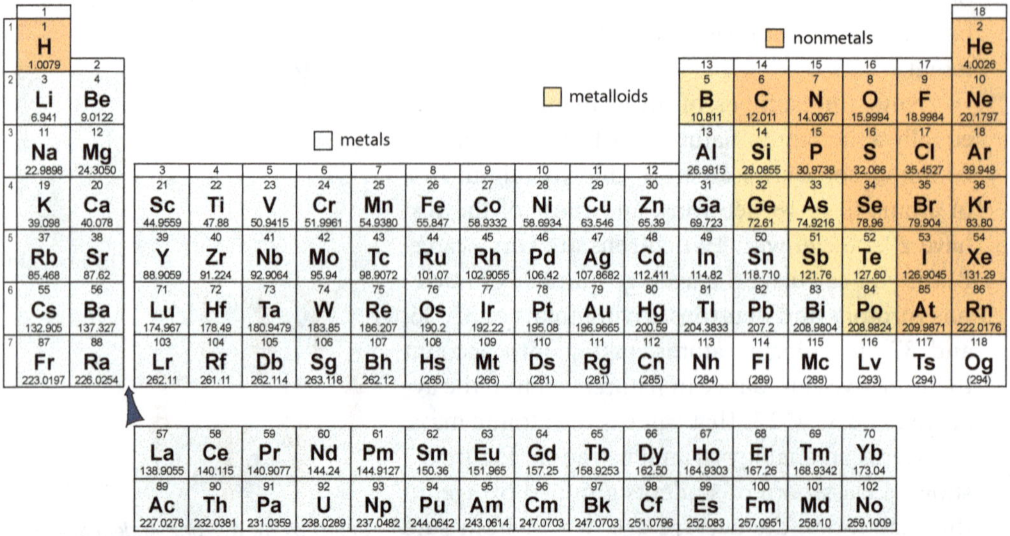

Figure 6.8. The major regions in the periodic table.

There is much we could say about different elements. Many books have been written describing their properties and the history behind their discovery. I will close this section with Table 6.1, showing some images and a few notes about some of the common elements we have encountered in this chapter.

Calcium, like many metals, is shiny with a silvery color. We never see it this way in nature because it reacts readily with other substances to form compounds. One of the most common is calcium carbonate, the compound chalk and sea shells are made of.	
Oxygen makes up 21% of the earth's atmosphere, and is one of the major elements involved in fires (combustion) and corrosion. Liquid oxygen, shown here, has a pleasant blue tint, and is strongly *paramagnetic* (see Chapter 14). Oxygen condenses to a liquid at −183°C.	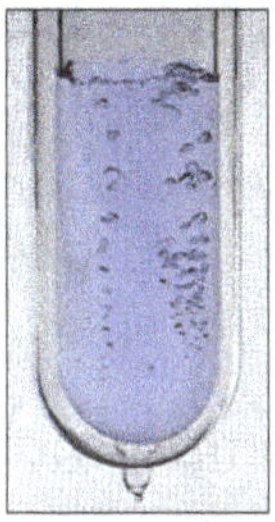
Nitrogen makes up 78% of the earth's atmosphere. When zapped with high-voltage electricity, nitrogen gas emits a purple glow, as shown here. If you ever witness a lightning strike (as I have), you may see the surrounding air glowing violet like this for a few seconds. That's the nitrogen atoms in the air emitting light.	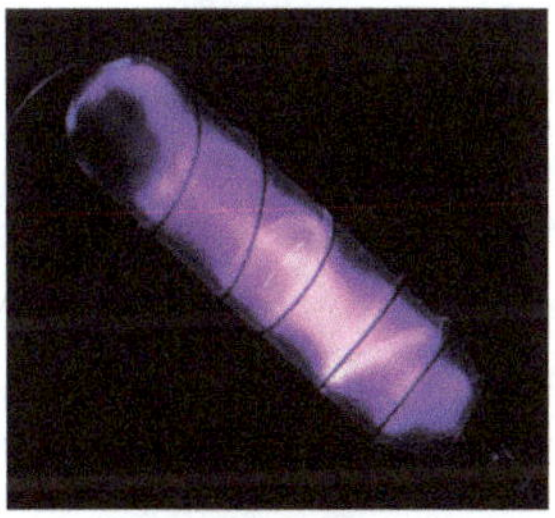
Sulfur is famous for its yellow color, as well as for the stinky, rotten-egg smell it produces in certain compounds. Sulfur is one of the primary ingredients in gunpowder.	
Sodium is an extremely dangerous metal. It slices like cheddar cheese, as shown here, and when freshly cut has a bright silvery-peach color.	
Chlorine, a poisonous gas, is yellow-green in color. Chlorine gas was first used by Germany as a chemical weapon in World War I. Production of other chemical weapons by Germany Britain, France, and the U.S. followed.	

Table 6.1. Notes and images for a few common elements.

1. What determines the element identity of an atom?
2. What are isotopes?
3. What is the difference between an element's atomic mass and its atomic number?
4. If a substance is neither an element nor a compound, what kind of substance is it?
5. Describe the origin of the Periodic Table of the Elements.
6. Name five elements from each of the three major groups in the Periodic Table of the Elements.
7. Where did Silicon Valley get its name?
8. Find these important elements in the periodic table. Write the chemical symbol, atomic number, and atomic mass for each one: a) iron, b) nickel, c) magnesium, d) uranium, e) mercury, f) iodine, and g) neon.

6.4   Compounds

A *compound* is formed any time atoms of two or more different elements are chemically bonded together. As we have seen, when atoms bond together like this, they either form molecules or crystal lattices. Regardless of the form the atoms in the compound take, atoms bond together through chemical reactions. In fact, that's what a chemical reaction is—the making, breaking, or rearranging of the chemical bonds between atoms.

Compounds are formed by chemical reactions.

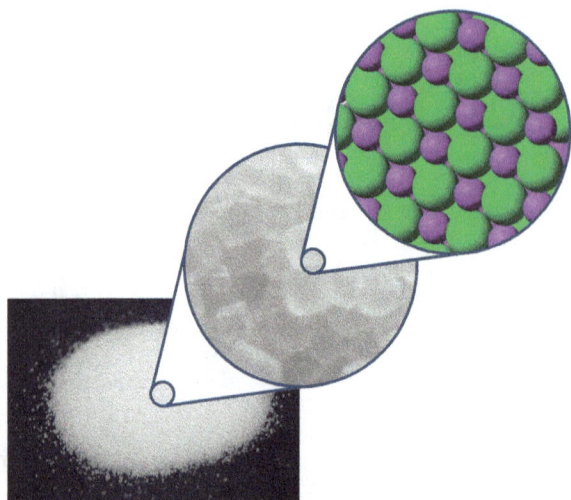

Figure 6.9. In a pure substance such as a compound, the composition is uniform all the way down to the atomic level.

There are very many different compounds; we don't even know how many are possible. The first six molecules in Figure 6.1 are compounds, and there are thousands more molecular compounds as well. The first five images in Figure 6.2 represent crystalline compounds, and again, there are untold numbers of others.

In every compound, the atoms are held together by electrical attractions between the protons and electrons in the atoms. Recall from Chapter 5 that the electromagnetic force is the main interaction that governs the way things behave in the every day world around us. This holds for chemistry, too! Electrical forces between protons and electrons are holding all those atoms together in the molecules and crystals of every substance.

No matter how closely you look, the chemical composition of a pure substance is the same—right down to the molecular level.

Compounds are pure substances. By this we mean that however closely you look, the composition of a compound is the same—right down to the level of the atoms or molecules the substance is made of. Figure 6.9 illustrates this idea. The compound depicted is ordinary table salt—sodium chloride (NaCl). The composition of this crystalline substance is uniform all the way down to the crystal lattice of sodium and chlorine atoms.

John Dalton was the first to propose that atoms always combine together in whole number ratios to form compounds. This was part of his 1803 atomic model.

The chemical formula for a compound shows the ratio of atoms in the compound.

The *chemical formula* for a compound is a way of writing the atoms and their ratios in a given compound. In the formula, an element's chemical symbol indicates its presence in the compound, and a subscript indicates the number of atoms of the element in one basic unit of the compound (if that number is greater than one). For example, the familiar chemical formula for water, H_2O, indicates that in water there are two hydrogen atoms for every oxygen atom. We can write this as 2:1, hydrogen : oxygen. This ratio should make perfect sense. Water, as you know, is composed of molecules, and every molecule of water has two hydrogen atoms and one oxygen atom in it. When parentheses are included in a chemical formula, the subscript after the parentheses is a multiplier that applies to every atom or group of atoms inside. Thus, the formula $(NH_4)_2SO_4$ (ammonium sulfate) indicates an atomic ratio of 2:8:1:4, nitrogen : hydrogen : sulfur : oxygen. More examples of chemical formulas and the atomic ratios they represent are shown in Table 6.2.

We return to the study of compounds later when we consider chemical reactions. For now, it's time to look at the other major category of substances—mixtures.

Formula	Name	Type	Atomic Ratio in Compound
NH_3	ammonia	molecular	1:3, nitrogen : hydrogen
SO_2	sulfur dioxide	molecular	1:2, sulfur : oxygen
C_3H_8	propane	molecular	3:8, carbon : hydrogen
CH_4	methane	molecular	1:4, carbon : hydrogen
$Mg(NO_3)_2$	magnesium nitrate	crystalline	1:2:6, magnesium : nitrogen : oxygen
$CaCO_3$	calcium carbonate	crystalline	1:1:3, calcium : carbon : oxygen

Table 6.2. Examples of formulas and atomic ratios in compounds.

Learning Check 6.4

1. Imagine that you are to explain what a compound is to someone who has never studied science at all. This person has never heard of atoms, elements, or compounds. Write a paragraph explaining what a compound is to this person, making your explanation as brief as possible.

2. For each compound listed below, determine the name of each element in the compound (from the formula and the periodic table) and the ratio of atoms of each element in the compound (from the formula).

a. $NaOH$	d. NH_4OH	g. $CaSO_3$	j. $BaCO_3$
b. H_2SO_4	e. KI	h. KNO_2	k. BeF_2
c. $KMnO_4$	f. UF_6	i. $LiBr$	l. CCl_4

6.5   Mixtures and Solutions

As you have probably guessed by now, *mixtures* are not pure substances. They are composed of a bit of this and a bit of that, all mixed together. Another way to say this is that when substances are combined and no chemical reaction occurs to change the composition of the substances, a mixture is formed. When a chemical reaction occurs, the original substances are changed into different substances, with different properties. This distinction bears repeating, so I have illustrated it in Figure 6.10.

In the world of mixtures, there are two basic kinds: *heterogeneous mixtures* and *homogeneous mixtures.* (These two terms are pronounced het-ur-oh-JEEN-ee-us

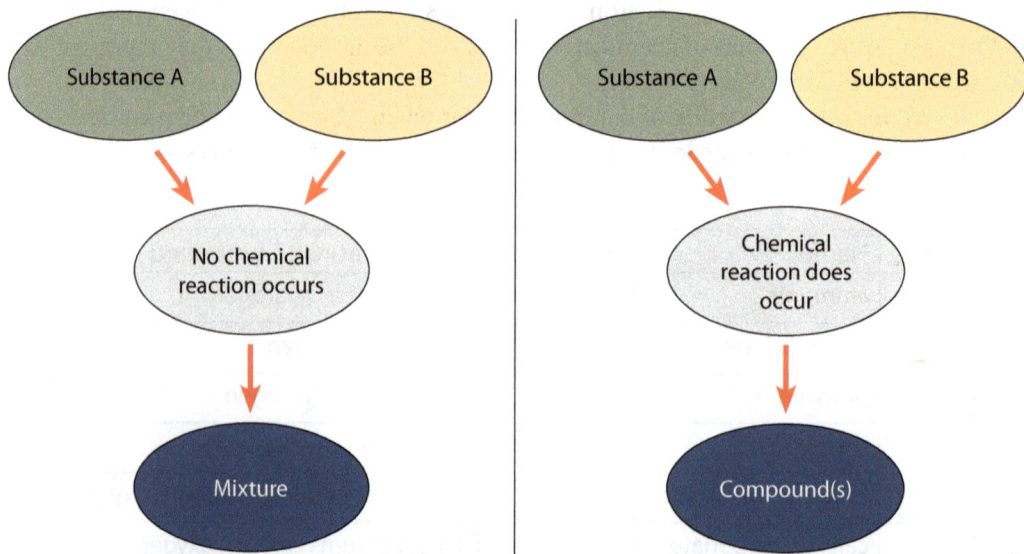

Figure 6.10. Distinguishing between mixtures and compounds.

and home-oh-JEEN-ee-us.) In heterogeneous mixtures, you can see with your
eyes or a microscope that at least two different substances are in the mixture. The
cooking seasoning mix shown in Figure
6.11 is an example of a heterogeneous
mixture. You can see at least three or four
different substances in there, just from
the different colors of the grains. Other
examples of heterogeneous mixtures are
pizza, salads, and garden soil.

Figure 6.11. Seasoning mix is a heterogeneous mixture.

In a homogeneous mixture, the sub-
stance is completely uniform, as with the glass of Gato-
rade shown in Figure 6.12. Even if you look at this drink
under a microscope, it is completely uniform with no
lumps, a yellow-green liquid. However, the sports drink
is a mixture, not a compound. If we could examine the
drink at the molecular level, we would see water molecules, sugar molecules, and other particles all jumbled
together. Other examples of homogeneous mixtures are
salt water, coffee with sugar, vinegar, ammonia cleaning
solution, and Kool-Aid.

Another name for homogeneous mixtures is *solutions*. Solutions appear everywhere in both everyday life
and chemistry. A solution is made when one substance,
called the *solute*, is *dissolved* into another substance,
called the *solvent*. This process is illustrated in Figure
6.13. The solute can be a gas, like the CO_2 dissolved in a
soft drink, a liquid, such as flavoring or syrup, or a solid,
such as sugar or salt. (In fact,
the solvent can be a solid too,

Figure 6.12. A sports drink is a homogeneous mixture.

but we have to melt it first to make a solution with it.)
Water is by far the most common liquid solvent, and
water solutions are found everywhere in nature, from
oceans to lakes to the human body.

The process of dissolving does not involve a chemical reaction. Thus, solutions are mixtures, not compounds. What's going on in the process of dissolving is
that the particles in the solute are separating and mixing
in with the particles in the solvent.

The way this happens depends on whether the sol-
ute is a molecular substance or a crystalline substance.
To explain, let's assume we have a molecular solvent.
(Most common liquid solvents, such as water, alcohol,

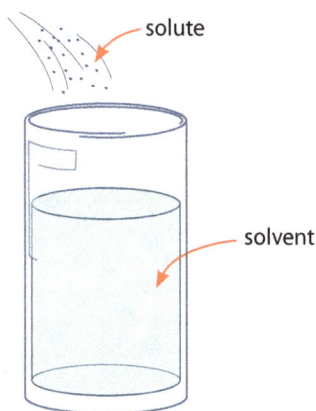

Figure 6.13. The two components of a solution.

and turpentine, are molecular.) Figure 6.14 illustrates what happens when a molecular solute is added to our molecular solvent. The molecules in the solute simply separate from each other and distribute themselves among the molecules of the solvent.

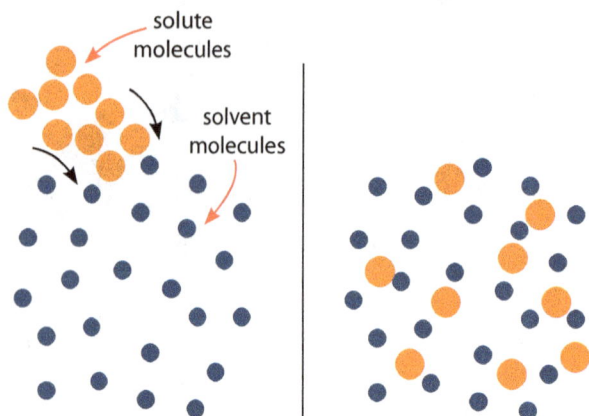

Figure 6.14. Molecules of the solute mixing with the molecules in the solvent.

When a crystalline solid, such as sodium chloride (table salt) is added to our molecular solvent, the process is slightly different. Figure 6.15 illustrates what happens.

Before we launch into this explanation, I need to mention that the atoms in a crystal lattice are not electrically neutral; they have a net electric charge. A charged atom is called an *ion*. Now, the reason the atoms in the lattice are ions has to do with some chemical details that we will save for Chapter 11. But suffice it to say that when the crystal was originally formed, the atoms traded one or more electrons with one another. Those that lost electrons ended up with more protons than electrons in the atom, becoming positive ions. Those that gained electrons gained some negative charge, becoming negative ions. These ionic charges are the source of the electrical forces that hold the crystal together.

Now, back to the process of dissolving crystals. Electrical attractions between the solvent molecules and the ions in the crystal cause the crystal to *dissociate*, which literally means to come apart. One by one, the ions in the crystal leave the crystal lattice and float off among the molecules in the solvent. There are a few more details involved. These are presented in the box on the opposite page. Here's a teaser: Why does salt dissolve in water, but not in vegetable oil?

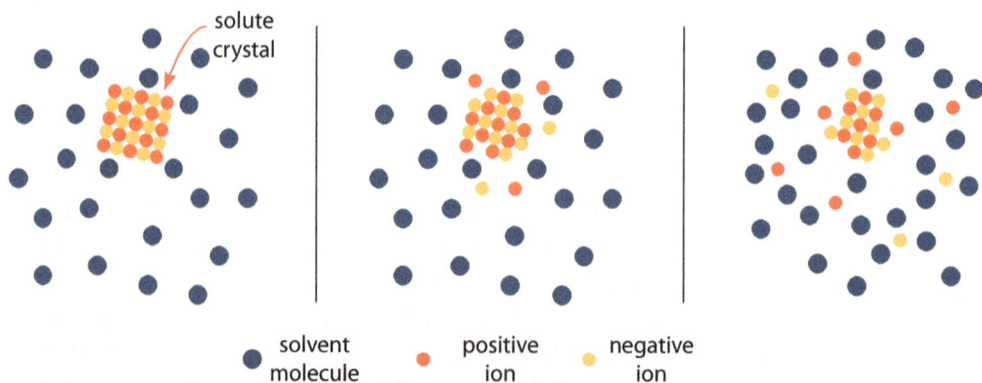

Figure 6.15. Dissociation of crystalline solute in a molecular solvent.

*We Pause Here to Talk More About **Solutions***

I assume you already know that salt dissolves in water. But if you didn't know, go into your kitchen and pour yourself a small glass of water. Now take the salt shaker, add some salt to the water, and stir. The salt dissolves in the water and the crystals vanish. Now repeat this little experiment using a few tablespoons of vegetable oil or olive oil in place of the water. Even after extended and vigorous stirring, those salt crystals will still be there. They won't dissolve!

The explanation for this has to do with the fact that some molecules are *polar*, and some are not. Water molecules are polar; oil molecules are not. When a molecule is polar, this means there are one or more places on the molecule that are more positively charged and one or more places that are more negatively charged. This happens because the atoms of different elements differ in the amount of attraction they have for electrons. For example, in the water molecule, represented in the sketch to the left, the oxygen atom attracts the electrons in the atoms of the molecule more strongly than the hydrogen atoms do. This attraction causes the electrons to shift slightly toward the oxygen end of the molecule. As a result, the elbow of the molecule is more negative and the hydrogen ends of the molecule are more positive. This is what it means for a molecule to be polar.

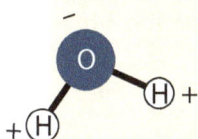

Molecules of alcohol and soap also have polar areas on them, and this is why they dissolve in water. The positively charged areas are attracted to the oxygen end of the water molecule, and negatively charged spots are attracted to the hydrogen end of the water molecule.

Now you might be starting to see how this works for crystals. When a crystal dissociates, the atoms coming out of the crystal are ions; they have a charge. These ions are attracted to the polar ends of the solvent molecules, and this attraction pulls them out of the crystal lattice and into solution. As shown in the sketch to the right, positive ions end up surrounded by the oxygen end of several water molecules, and negative ions end up being surrounded by the hydrogen ends of several water molecules.

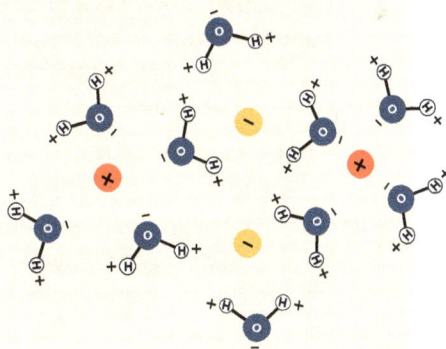

Without polar molecules in the solvent, there is nothing to pull the ions out of the crystal. And that is why the salt won't dissolve in the oil!

Scientists, Experiments, and Technology

Before the 1960s, every electronic circuit was put together with wires, solder, and electronic parts. Such is no longer the case.

The high-tech world you are growing up in is entirely dependent on crystals. This is because all the computers and high-tech devices we use every day—mobile phones, laptops, tablets, and so on—are operated and controlled by crystalline semiconductors. More specifically, the enormously complicated circuits that run our digital devices are all built in to the crystal structure of metalloid crystals. Microscopic wiring is still involved to connect these devices to one another, but the devices themselves are in the crystals.

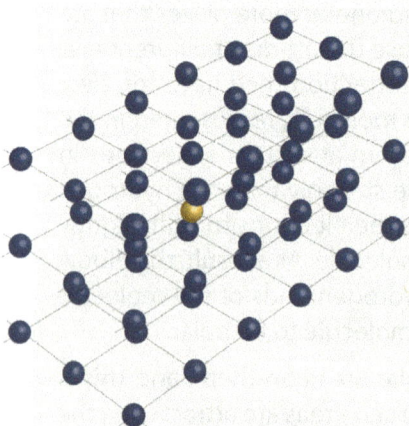

As you learned in this chapter, the metalloids are a small group of elements in the periodic table between the metals and the nonmetals. They conduct electricity, but only under certain conditions. As it turns out, these conditions are perfectly suited to the needs of electronic circuits. However, there is one catch. When a metalloid is conducting electricity, it doesn't conduct nearly well enough.

The scientists and engineers who developed semiconductor technology solved this problem through a process called *doping*. When a crystal is doped, about one out of every 10,000 atoms in the crystal is an atom of a different element. For example, assume the crystal structure shown above is a lattice of germanium atoms, and the single yellow atom is an atom of arsenic. If only 0.001% of the atoms in the crystal are arsenic, the ability of this crystal to conduct electricity is 10,000 times greater.

The whole process is much more complicated, of course. But to make it work, the crystals in semiconductors must be grown in extremely pure conditions. Ordinary air in a city can have around 35 million particles of size 0.5 micrometers in diameter and larger in a single cubic meter of air. But in a high grade clean room, the largest particles are 0.3 μm in diameter, and there is a maximum of only 12 of them per cubic meter of air. Workers in such clean rooms must wear clean room garments such as the outfit you see to the right (fondly referred to as a "bunny suit" by those in the industry).

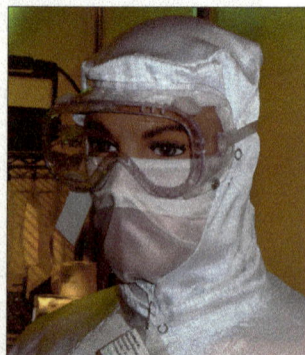

Learning Check 6.5

1. Explain the difference between a mixture and a compound.
2. Explain the difference between heterogeneous and homogeneous mixtures.
3. What is a solution?
4. What is an ion?
5. Describe, in your own words, the process by which a crystalline substance dissolves.

Chapter 6 Exercises

Answer each of the questions below as completely as you can. Write your responses in complete sentences.

1. At the top of a page, draw the substances family tree. Underneath it, write a thorough description of each of the six major categories of substances we have studied.
2. Write a paragraph or two explaining what isotopes are and relating them to the atomic mass of an element.
3. Distinguish between molecules and crystals, and give several examples of each type of substance.
4. Define acid rain and briefly explain its cause.
5. Explain why compounds are regarded as pure substances and mixtures are not.
6. Describe the origin of the Periodic Table of the Elements.
7. Explain what we mean when we say that water molecules are polar.
8. Describe the difference between the way a molecular solute dissolves in water and the way a crystalline solute dissolves in water.
9. Explain why salt dissolves in water but not in olive oil.
10. Explain the difference between a metal and a metalloid.
11. Explain how the elements are ordered in the Periodic Table of the Elements.
12. What are hydrocarbons, and what is one of their primary uses in modern society?
13. For each compound listed below, determine the names of each element in the compound (by finding them in the periodic table) and the ratio of atoms of each element in the compound (from the formula).

a. CO_2	d. $(NH_4)_2SO_4$	g. $BeBr_2$	j. Li_2CO_3
b. $CaCl_2$	e. K_2NO_3	h. K_3PO_4	k. H_2O_2
c. $Ca(NO_2)_2$	f. HF	i. $NaHCO_3$	l. $Mg(ClO_3)_2$

Experimental Investigation 4: Growing Crystals

Overview

- *The goal for this investigation is to grow crystals, and to observe and describe the crystal formations you see.*
- Grow crystals of alum, sodium chloride, and Epsom salt. If possible, describe the crystal structures you see developing over the course of several hours.
- Harvest the crystals from each solution. Describe their size and shape.

Basic Materials List

- alum, table salt, Epsom salt
- glass jars and bowls
- measuring cups and spoons
- stirring spoon
- magnifying glass
- nylon thread (fishing line), lead weight, and pencil
- beakers, borosilicate, 250 mL and 600 mL
- burner or hot plate
- mass scale
- distilled water

Safety Precautions

Use care when working with a burner or hot plate. Keep clothing and hair away from flames and hot surfaces. Handle hot liquids with caution. Use only borosilicate glassware (e.g., Pyrex) for heating.

Growing crystals is fascinating, but sometimes, when things don't work out the way you hope, it can be frustrating as well. In this investigation you will try to grow crystals from three different salts. If the process does not seem to work well the first time, try it again.

In each case, the process consists of making a *saturated* solution of the salt in water and then letting the solution cool. A saturated solution is formed when the maximum possible amount of solute is dissolved in a solvent. For many salts, more salt dissolves in water when the temperature is high than when the temperature is low. If you heat your solvent (water), dissolve as much salt in the solvent as possible, and then let the water cool, you have a *supersaturated* solution. When the solution cools, it has more solute dissolved into the solvent than the solvent can hold at that temperature. This condition is unstable. Crystals begin to grow as the excess solute "precipitates" out of the solution. In some cases, it helps to give the crystals a "seed" crystal to start growing on. Once crystal growth starts, the crystal keeps on growing until the excess solute finishes precipitating out of the solution.

Begin with alum, which is the easiest crystal to grow. *Alum* is the common name for a compound called hydrated potassium aluminum sulfate. It has all kinds of uses, including water purification in ancient times, underarm deodorant, pickling (to keep vegetables fresh and crisp), and styptic pencils (to stop bleeding when you cut yourself shaving).

To grow alum crystals, place a small pan or 250-mL beaker with 120 mL of distilled water on a burner or hot plate. Stir in 30 g of alum. Heat the water slowly over medium heat until the alum has all dissolved. Pour the solution into a small, clear glass bowl and let it cool. You should observe some sizeable crystals growing on the bottom of the bowl over the next few hours.

With the next solution you will have crystals of sodium chloride, ordinary table salt. To give the sodium chloride crystals something to grow on, attach a small lead fishing sinker to a 6-inch length of nylon thread (fishing line). Tape the fishing line to a pencil so that when placed across the top of a glass the weight does not touch the bottom.

Place a 600-mL beaker with 475 mL of distilled water on a burner or hot plate. Stir in 165 g of salt. Heat the water slowly over medium heat until the salt has all dissolved, but don't allow the water to boil. Pour the solution into a tall, clear drinking glass. Dip your fishing line and weight down into the hot solution, then remove it and set the pencil across another glass while the solution cools. When the solution has cooled to room temperature, gently lower the nylon thread into the solution so it is held by the pencil. Then cover the glass with a cloth. Some crystal growth will occur within the next couple of hours, but for the growth to run its course you will need to let the solution stand undisturbed overnight. The bottom of the glass will fill with crystals, and you should have tiny crystals formed all along the length of nylon thread.

In the third solution, we will use Epsom salt as the solute. Epsom salt (hydrated magnesium sulfate) gets its name from a saline spring in Epsom, England, where the salt was originally produced. Epsom salt is commonly used as a bath salt to relieve aching muscles or sore feet.

To prepare the Epsom salt solution, place a 250-mL beaker with 120 mL of distilled water on a burner or hot plate. Stir in 170 g of Epsom salt. Heat the water slowly over medium heat until the salt has nearly all dissolved, but don't let the water boil. Pour the solution into a small, clear glass bowl and place it in the refrigerator. Over the next three hours you should see the bottom of the bowl fill with needle-like crystals.

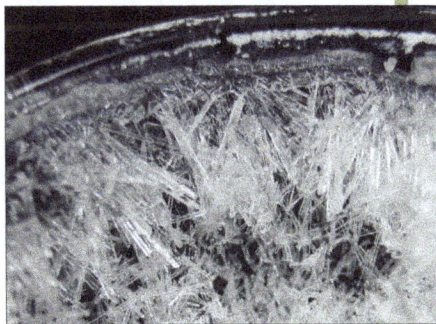

Analysis

Harvest the largest and most perfectly formed crystals from each of your solutions. Handle them as carefully as you can, although the table salt and Epsom salt crystals may break anyway. Place the crystals on a dark cloth under bright light. Measure your crystals and sketch them or photograph them. (If you have access to a DSLR camera with a macro lens, it will really help with the photographs. Inexpensive macro lenses are also available now for use with smart phones.)

In your report, include these comments:

1. Describe your observations as the crystals were growing.
2. Describe the size and shape of the largest or most well-formed crystals from each of the solutions.

Chapter 7
Science, Theories, and Truth

In the photograph above, the man pointing is Mike Mullen. Admiral Mullen was the Chairman of the Joint Chiefs of Staff in the U.S. military from 2007 to 2011. In the photo, Admiral Mullen is reviewing a model of the demilitarized zone between North Korea and South Korea with other military personnel.

Models are often used to represent complex realities where absolute knowledge of the situation is unattainable. Models of battles have always been used by military personnel to plan or track the progress of a military conflict. Models are also commonly used in architecture, medicine, urban planning, and many other disciplines.

One of the central messages in this chapter is that science is the process of mental model building.

OBJECTIVES

After studying this chapter and completing the exercises, you should be able to do each of the following tasks, using supporting terms and principles as necessary.

1. Define *science, theory, hypothesis,* and *scientific fact.*
2. Describe the Cycle of Scientific Enterprise by describing each major step in the cycle and explaining their relationships to each other.
3. Distinguish between theories and hypotheses and give actual examples of each.
4. Distinguish between facts, theories, and truth.
5. Define *truth,* and give examples of true statements.
6. Explain why science can be described as "mental model building."
7. Describe three ways we know truth.
8. Explain why scientific facts and theories are regarded as provisional.

VOCABULARY TERMS

You should be able to define or describe each of these terms in a complete sentence or paragraph.

1. caloric theory
2. Cycle of Scientific Enterprise
3. experiment
4. hypothesis
5. mental model
6. science
7. scientific fact
8. theory
9. truth

7.1   Science is Mental Model Building

People tend to think of science as a giant mountain of facts: facts about atoms, clouds, electricity, whales, chemical reactions, vitamins, rivers, and on and on. As with every other field of study, science does involve a lot of facts. (Hopefully, you have learned many interesting scientific facts so far in this text!) Not only is it common to think of science in terms of facts, it is also common to think of science textbooks as proclamations of "the way it is," assuming scientific knowledge is settled.

There is a problem with thinking this way about science. The problem is that the natural world is extraordinarily complex, and what we know about it changes all the time. Scientific knowledge is not settled; it is dynamic, which means it is always changing. Today's fact can be tomorrow's discarded idea. For example, people used to think the planets orbited the earth. Incorrect—they orbit the sun. People used to think there were exactly seven "heavenly bodies." Incorrect—there are scores. People used to think atoms were indivisible, indestructible particles. Incorrect—there is an entire world of structure inside atoms, as we have already seen. "Junk DNA" got its name because scientists formerly thought those sections of DNA serve no purpose. Incorrect—now it appears that "junk DNA" performs important functions after all. Scientific knowledge is changing all the time. And how could it be otherwise, when the world is such a complex place?

Two centuries ago, scientists generally thought that science was a process of discovering the truth about nature. But during the 20th century, it became clear that we never do know the actual truth about nature. Even basic scientific facts that we think are settled are sometimes called into question because of new discoveries. All scientific knowledge is subject to change in this way. So we do not know the truth about atoms and planets and DNA. When we do science, we are striving to know the truth but we never get there. We just keep refining our knowledge more and more.

So then, what kind of knowledge is scientific knowledge? And if we are not learning the truth in our science classes, *what are we learning*? It is our task in this section to answer this question. To put it another way, what is science?

To address this question, let's consider the colorful model of an animal cell shown in Figure 7.1. This model would be very useful to have in a biology class. It shows the interior parts of the cell and how they are arranged. As students examine the model, the teacher can explain what the parts are for and what they do, at least as far as we understand such things.

Does this model tell everything there is to know about animal cells? Of course not. Might it be possible to improve the model as we learn more about cells? Probably so. Is the model useful? Yes it is. Even though we know the model does not tell the whole story about cells, and even though the model might need to be corrected or improved, the model is useful. And of course, the model is correct so far as we know, but there may be much more to animal cells than we know at present.

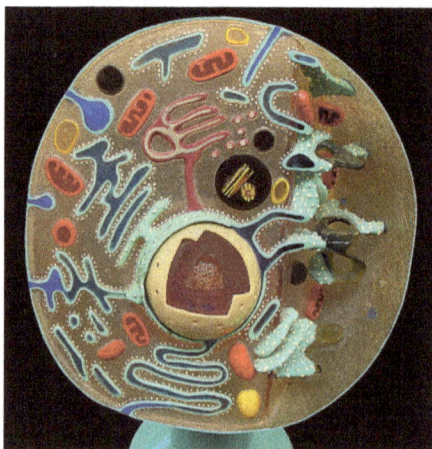

Figure 7.1. A model of an animal cell.

There is a reason I have written all this about models. We can describe science as *mental model building*. This is the best way to think about the nature of scientific knowledge. We do not presently have the ability to ask some all-knowing person how everything works. We must do our best to develop the best possible explanations about how things in nature work. And the explanations we develop can be thought of as *mental models*.

Let's dig in a bit more to this idea of mental models. When I say that science is mental model building, of course I do not mean that scientists are sitting around all the time with glue and bits of plastic building things. What I mean is that just as the model in the figure above is a representation of a cell, the theories we develop in scientific study are *representations* in our conceptual understanding of how the

Science is mental model building.

natural world works. A scientific theory is a mental model, a conceptual representation of how part of the world works.

Now, this definition of science as mental model building is a bit skimpy on detail, so let's expand on it a bit. A fuller definition of science is this:

> *A scientific theory is a mental model—a conceptual representation of how part of the world works.*

> **Science is the process of using experiment, observation, and reasoning to develop mental models of the natural world. The mental models scientists develop are called theories.**

This definition says four things. First, it says that science is a process. Science is not just a fixed body of facts and information. It is a process of learning and explaining. Second, the definition highlights three tools we use in the scientific process: experiment, observation, and reasoning. If you want to know the temperature at which alcohol boils, perform an experiment; heat some alcohol and measure its temperature. If you want to know what insects do at night, observe them and find out. And if you want to know what's going on in distant galaxies where experiments and observations cannot be performed, then collect data from telescopes and use reasoning to deduce what you can about those distant locations.

Third, as we have already seen, the business of science is developing mental models to explain how the natural world works. Our goal in science is to make these models increasingly accurate. And fourth, we have a word for these mental models: theories. The four parts of our definition for science are summarized in Figure 7.2.

Defining Science

1. Science is a process.

2. The basic tools of science are experiment, observation, and reasoning.

3. The goal of science is developing mental models that accurately describe how the natural world works.

4. Our mental models are called theories.

Figure 7.2. Key points in defining science.

Learning Check 7.1

1. What is science?
2. What is the goal of science?
3. Explain what a model is, and identify your own example of a model that could be used to explain something.
4. What is a theory?
5. What do we mean by the phrase "mental model"?

7.2   The Cycle of Scientific Enterprise

To help students understand how science works, I developed a scheme I call the *Cycle of Scientific Enterprise*. This scheme is shown in Figure 7.3. This diagram illustrates the scientific *process* we discussed in the previous section. Scientific learning is not "linear." By this I mean that science is not a process of simply learning one new fact after another until we have them all. Instead, science is a *cyclic* process—it relies on feedback. Let's review the diagram in detail so that you understand this process and all the terms involved. As we go, I will illustrate with an example—an actual case study from scientific history.

Theory We begin on the left side of the diagram with *theory*. As mentioned before, the goal of science is to develop theories—mental models—that accurately explain how the natural world works. So the process of scientific inquiry begins with a theory that explains most of the known facts in some scientific field of study. (I say *most* because a theory never explains *all* of the related facts. There are almost always a few pesky facts we know that our best theory cannot yet explain.) Scientists construct a theory about these facts

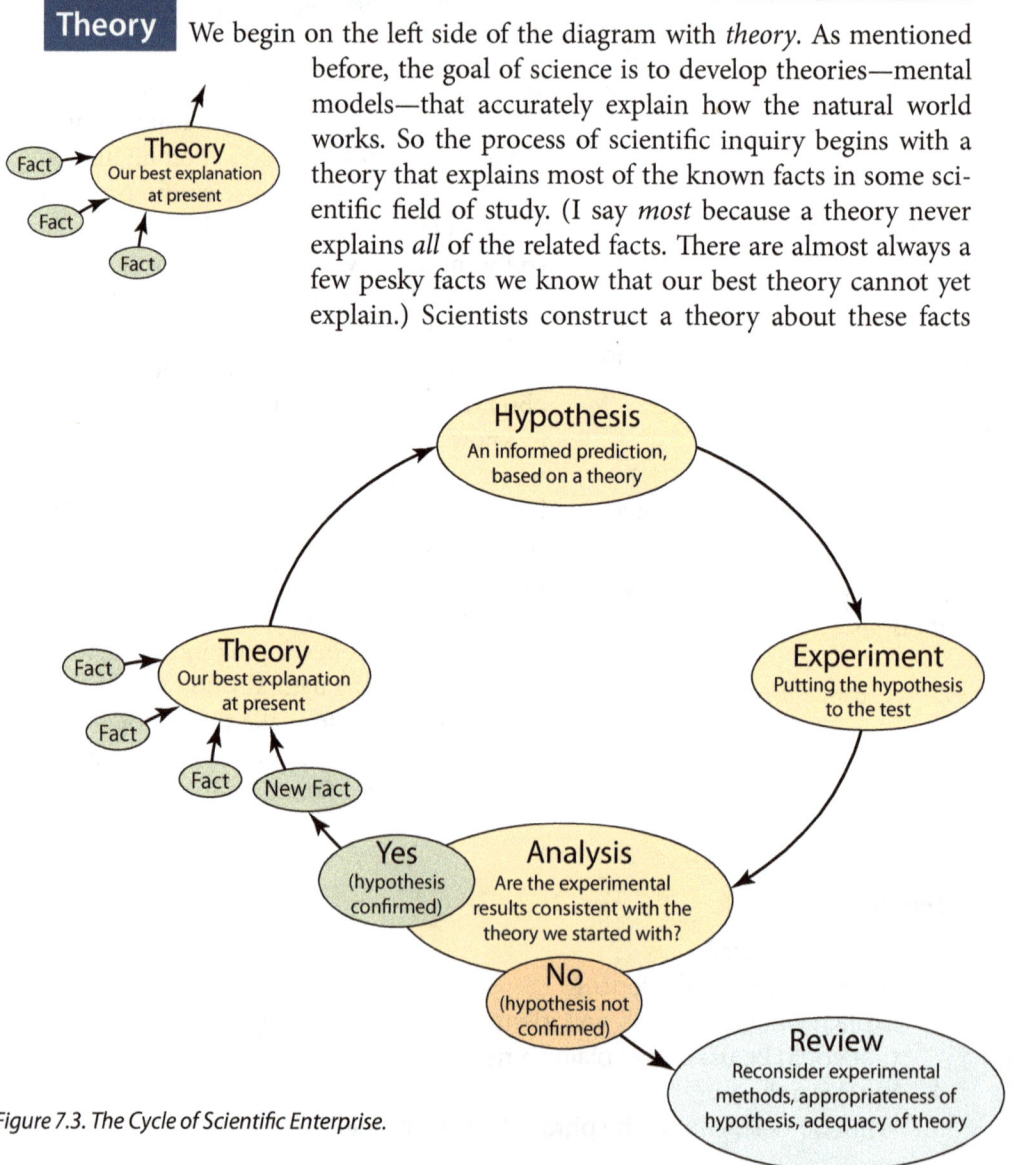

Figure 7.3. The Cycle of Scientific Enterprise.

that explains them and relates them together. This theory is our model—our best explanation for how nature works.

| Case Study: Part 1 | In the 18th century, many facts about the behavior of hot objects were known. No one knew what fire and |

heat were, but everyone knew that fire is hot. It was also well known that heat flows upward out of boiling water. A third well-known fact was that when a hot object and a cool object are put together, the hot object cools down and the cool object warms up.

In 1783, French chemist Antoine Lavoisier proposed that heat was caused by a substance objects contain, a substance he called *caloric*. Caloric was understood to be a weightless gas that flows through pores in substances. The caloric theory held that caloric repels itself, so it expands out of hot substances like fire and boiling water, and flows from hot objects to cool objects, as illustrated in Figure 7.4. Hot objects have more caloric in them, and cool objects have less, so the caloric flows to where there is more room to spread out. Caloric theory explained many facts about heat very well and became a popular theory.

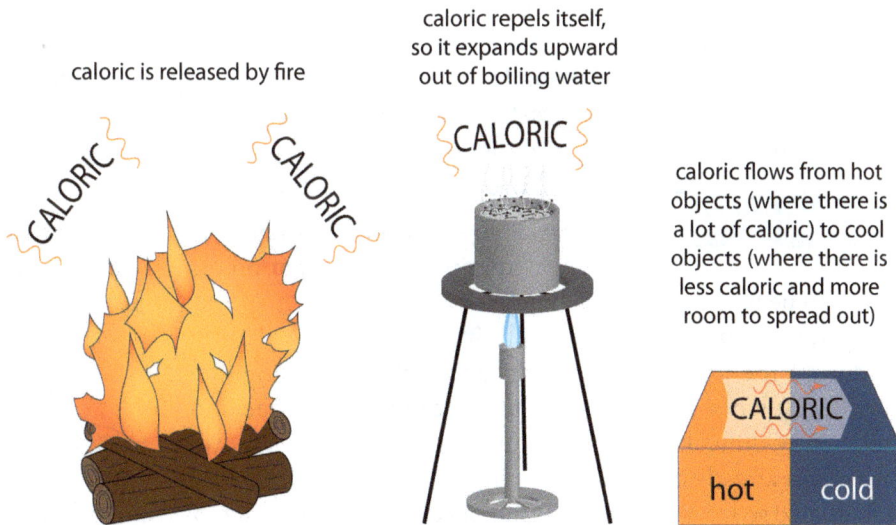

caloric is released by fire

caloric repels itself, so it expands upward out of boiling water

CALORIC

caloric flows from hot objects (where there is a lot of caloric) to cool objects (where there is less caloric and more room to spread out)

CALORIC

hot   cold

Figure 7.4. Some of the facts behind the caloric theory.

| Hypothesis | The first job of a theory is to explain the known facts as accurately as possible. But if this were all a theory did, science could go no- |

where. Anyone can make up any crazy theory they like, but if there is no way to test it, there is no way to find out whether the theory is an accurate description of nature. So the second thing a useful theory must do for us is enable scientists to formulate new *testable* hypotheses.

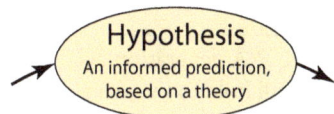

Hypothesis
An informed prediction, based on a theory

A successful theory accounts for the known facts and enables new hypotheses to be developed and tested.

A *scientific hypothesis*[1] is a testable, informed prediction, based on a theory, of what will happen in certain circumstances. It is an *informed* prediction because a hypothesis is always based on a certain theoretical model. To make a prediction, the prediction has to be based on something. A scientific hypothesis is based on a particular theory, and different theories lead to different hypotheses—different predictions of what will happen.

Scientific hypotheses must also be *testable*. If a prediction is not testable, it is not a valid scientific hypothesis. The horoscopes make predictions all the time. But those predictions are so vague that no one can ever test them, so they do not qualify as scientific predictions. For example, horoscopes make predictions such as, "You will meet someone important soon." Give me a break. Who's to say what people or what meetings are important, or might prove to be so 30 years later? And how soon is "soon"? A day? A year? You can see the problem with such vague predictions.

Case Study: Part 2 Caloric theory produced a number of different testable hypotheses. One of the most important was to enable scientists to revise the theoretical calculation of the speed of sound worked out by the great scientist Isaac Newton. This is certainly a testable hypothesis. Measuring the speed of sound is simple in principle; simply have someone fire a pistol half a mile away. Start a timer when you see the smoke of the gun, and time how long it takes the sound to arrive.

Experiment The purpose of an experiment is to put a hypothesis to the test. A successful experiment is one that produces definitive results and can be replicated by other scientific teams. (To *replicate* an experiment means to set up the same experiment with similar equipment and procedures.) The scientific community does not base new thinking on the result of a single experiment. Successful experiments are often very difficult to perform, and a lot can go wrong. (You may have already discovered this in the experiments you have conducted so far in this course!) But once results have been replicated by other scientists, they can be considered reliable and the results subjected to analysis in the next stage of the cycle.

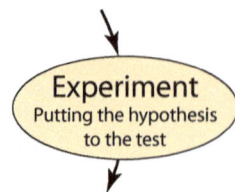

Experiment
Putting the hypothesis to the test

Analysis In the analysis of the experimental results, we seek to determine whether the experimental results turn out the way the hypothesis predicts. If they do, then we say *the hypothesis was confirmed*. When a hypothesis is

---

1   A generation ago, correct usage required us to write *an* hypothesis, and some writers still hold to this usage. I am using the more common contemporary usage by writing *a* hypothesis. Note also that the plural of *hypothesis* (a term derived from Greek) is *hypotheses*.

confirmed, the credit goes all the way back to the theory the process started with. The theory has scored a victory! If a theory enables scientists to make correct predictions about nature, it is a useful theory. A successful theory repeatedly leads to correct predictions that are confirmed by experimental results.

Yes (hypothesis confirmed)

Analysis Are the experimental results consistent with the theory we started with?

No (hypothesis not confirmed)

A confirmed hypothesis means the experimental results are consistent with the theoretical model the scientists are working with. This is because the hypothesis is based on the theory in the first place. When the theory leads to successful predictions like this, the experimental results take their place among the facts the theory successfully explains, as shown by the completing of the Cycle of Scientific Enterprise in Figure 7.3. At this point we say that the theory has been strengthened by the new experimental results.

A successful theory repeatedly leads to correct predictions.

It is very important to notice some things here that we do *not* say when a theory is supported by the results of an experiment. We do not say the experimental result prove the theory. Theories are models. Models are not proven or disproven, they are simply useful or not useful, more accurate or less accurate. Also, just because the theory produces a correct prediction does not mean the theory is the truth. Again, a scientific theory is a model; it is not a truth claim. We return to these important matters in the next section. For now, let's continue with our case study.

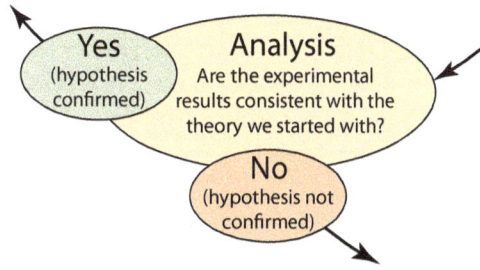

Case Study: Part 3   The new prediction of the speed of sound according to caloric theory was very successful. Not only were experimental results more in line with the prediction, but new experiments over the next century continued to agree closely with the new prediction, even as equipment and instruments became more precise. Caloric theory was strengthened by these experimental successes and became the leading theory about heat.

Review   A critical part of the Cycle of Scientific Enterprise pertains to what happens if the experimental result does not turn out as the hypothesis predicts. The question here is, What do these unexpected results mean?

An unexpected result means that some part of our understanding is incorrect somewhere and we need to find out what it is. The process of doing this is similar to what you would do if you got home from school one day and found you did not have your

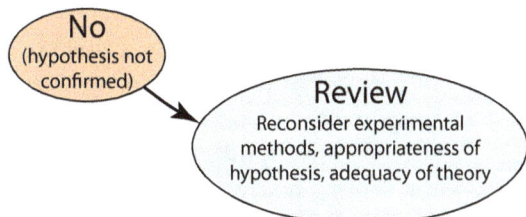

No (hypothesis not confirmed)

Review Reconsider experimental methods, appropriateness of hypothesis, adequacy of theory

purse or wallet. What would you do? You would retrace your steps, going back-wards from where you are. In science, we do the same thing.

First, scientists must revisit the experiment. All results are reviewed and com-pared to the results of other experiments. As mentioned before, experiments are difficult to perform and a lot can go wrong. So this is the first place to look in the re-view process. A big part of this review is to look at the measurements that produced the experimental data. Often slightly different measurement techniques produce data that tell a different story.

If the experimental results appear to be valid, the next stage in retracing our steps is to revisit the hypothesis. Is the hypothesis a correct consequence of the theory? Are the calculations correct? Is the theory correctly understood?

Finally, if the experiment and hypothesis both check out, scientists are forced to take another look at the theory. This is a big deal. A theory is not just cooked up over night. It takes a long time for a theory to be accepted as our best explanation. Scientists are not going to toss a theory out the window just because of one negative experimental result. Nevertheless, if the theory does not explain the experimental result, something is amiss. If enough of these problems turn up, scientists must be-gin looking for a way to modify the theory. And if this can't be done, a completely new theory might be needed.

Case Study: Part 4 Back on the opening page of Chapter 3, you were intro-duced to Benjamin Thompson, better known as Count Rumford. Rumford's interest in science started when he was a young teenager, and by the time he was 16 he was already conducting experiments on heat. As an adult, Thompson conducted a lot of ex-periments on explosives and artillery. To make a cannon, molten metal is cast in a mold, and then a hole is bored with a drill down the center of the cannon. While boring cannon (the word is both singular and plural), Rumford conducted experiments on the heat produced by the drill-ing process. Rumford noticed that even when bored re-peatedly, cannon never seemed to run out of heat (Figure 7.5). According to caloric theory, the caloric fluid inside a cannon should eventually equalize with the caloric in the air outside the cannon and stop flowing. Thus, here were experimental data that caloric theory could not account for.

The continuous release of heat while boring led Rumford to propose a new theory about heat. In 1798, Rumford proposed that the heat released during the boring process was originating with the motion—the kinetic energy—of the drilling machine. Rumford theorized that the kinetic

Figure 7.5. When boring a cannon repeatedly, it never seems to run out of heat.

energy of the drill's motion was being converted into energy released as heat during the drilling.

Other scientists continued working on heat theory over the next half century. One of these was James Prescott Joule, the English physicist whose name is now used in the metric unit for energy, the joule. Out of this work emerged the principle we now refer to as the *mechanical equivalent of heat* and the law of conservation of energy. Caloric theory hung around throughout the 19th century, but by the end of the century the mechanical equivalent of heat and the law of conservation of energy were well established. This new theoretical model is today the universally accepted theory of heat.

Learning Check 7.2

1. What is the difference between a theory and a hypothesis?
2. What are two major characteristics of a successful scientific theory?
3. Describe the caloric theory of heat.
4. What is the purpose of an experiment?
5. If an experimental result fails to confirm the hypothesis, how do scientists respond? Describe the steps in the review process, and the order in which they are taken.

7.3   Facts and Theories

At the beginning of this chapter, I give several examples of scientific facts that had to be revised in light of new experimental data. The point of those examples is to show that scientific facts are not etched in stone; they can and sometimes do change as new information is acquired. And as we have seen in the previous section, the same thing applies to theories. Both facts and theories change as the Cycle of Scientific Enterprise continues.

We are now ready to define what a scientific fact is. A scientific fact is a statement, supported by a lot of scientific evidence, that is correct so far as we know. However, facts can and sometimes do change as new evidence becomes known. Hopefully, it is apparent at this point that since scientific facts can change, scientific facts are not regarded as "the truth." (We will get to talking about truth later on.) The best we can say about facts is that they are correct so far as we know. A summary of these defining statements about facts is in Figure 7.6.

Let's talk about theories a bit more. I have said a couple of times already that the goal of science is to develop accurate theories—successful mental

Defining Scientific Fact

1. A scientific fact is a statement, supported by a lot of scientific evidence, that is correct so far as we know.

2. Facts can change as new evidence is obtained.

Figure 7.6. Key points in defining scientific facts.

127

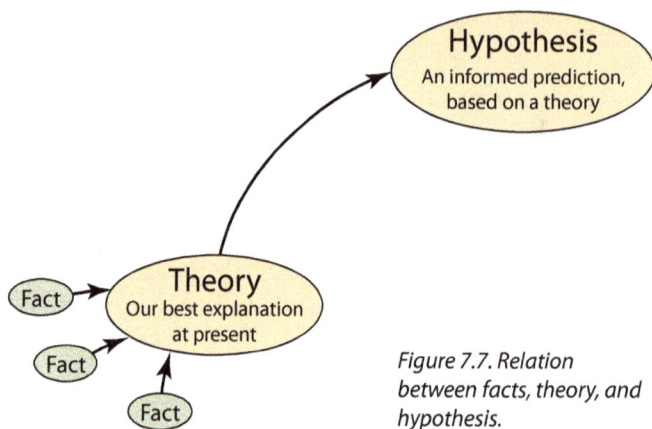

Figure 7.7. Relation between facts, theory, and hypothesis.

models of how the natural world works. So theories are at the very heart of what science is all about. Now that I have defined facts, theories, and hypotheses, I want to summarize and add to what I said about scientific theories.

First, to emphasize again the relationship between facts, theories, and hypotheses, Figure 7.7 shows a portion of the Cycle of Scientific Enterprise from Figure 7.3. You can see that there is an order—a collection of facts leads to the formulation of a theory that explains them, and the theory leads to new hypotheses that scientists can test. We have no way of understanding scientific facts apart from some kind of model or representation that explains them and ties them together. This is what the theory does. And since theories are at the heart of scientific knowledge like this, we can make this very strong claim: *all scientific knowledge is theoretically based.* There isn't any scientific knowledge apart from the theoretical models we use to explain the facts we know. The main points I have presented in defining theories and relating them to facts and hypotheses are summarized in Figure 7.8.

All scientific knowledge is theoretically based.

Second, in common speech today people often use the terms theory and hypothesis interchangeably. This is illustrated for you with all the humor I can muster in Figure 7.9. As you have seen from our review of the Cycle of Scientific Enterprise, these two important terms mean quite different things. Perhaps one more example will help clarify the distinction. This example involved a young man who was a freshman in high school. (He also happens to be my nephew.) Somehow we got onto the subject of Big Bang theory. I was explaining that according to the Big Bang theory, the universe began 13.8 billion years ago with a mighty explosion of energy. I further explained that in the early period of the universe, the universe was not transparent, and photons could not travel freely. As things cooled and expanded, photons from the energy of the original explosion were able to travel freely throughout the universe,

Defining Theory

1. A theory is a representation of how part of the natural world works.

2. Theories are "mental models."

3. Theories should account for the known facts (so far as possible).

4. Scientific theories must enable scientists to formulate new hypotheses.

Figure 7.8. Key points in defining theories.

and so it has been to this day. This is a rough and general sketch of part of the Big Bang theory.

Upon hearing this, the teenager I was talking to said, "If that is the way things started, then shouldn't we still be able to see remnants of all that original photon energy? If all that light has been propagating through the universe since the beginning, shouldn't we be able to see it today?" His question was a hypothesis. He considered the theory, and made a prediction based on that theory: that we should be able to see the remnants of the light in the universe from the Big Bang.

I've got a new theory about what happened to my lost homework paper. I think my pesky brother swiped it!

That's not a theory—that's a hypothesis **based** on a theory. The theory is the idea that your brother is a joker!

Figure 7.9. Distinguishing between theory and hypothesis.

Well, as it turns out, this was a brilliant hypothesis, and I told him so. In fact, the remnant radiation he was asking about was discovered, quite by accident, in 1964! At that time, scientists discovered that the universe is uniformly full of microwave radiation in every direction. This radiation is the Cosmic Microwave Background (CMB) (see page 18), and current theory holds that the CMB is indeed the remnant photon energy left over from the Big Bang. The experimental discovery of the CMB is now considered strong supporting evidence for the Big Bang theory.

Third, it is painfully common to hear people today speaking incorrectly about theories. So Figure 7.10 provides a short tutorial on correct ways to refer to theories. Based on everything I have written so far in this chapter, the comments on the "Appropriate" side of the figure should all make sense. Here are a few remarks about the comments on the "Inappropriate" side.

- The first statement is inappropriate because, as I mentioned previously, theories are never proven. Theories are strengthened gradually until we have high confidence in them (like general relativity), or are weakened gradually to the point of becoming obsolete (like caloric theory).
- It is inappropriate to speak scornfully of theories just because they are theories. In science, theories are all we've got, and developing good ones is the goal of science. Moreover, widely accepted theories take a long time to develop (typically decades). They are not just untested speculations. They have also been successful—repeatedly, for decades—at producing confirmed hypotheses. Theories are *not* hunches, and they are not ideas one can quickly write off or dismiss.
- As I discuss further later, scientific claims are not truth claims. Scientists do not claim their theories are true. Theories are simply models; they are either accurate models of nature (well-supported theories) or less so.

Theories are not just hunches or wild, untested ideas.

Speaking About Theories —Inappropriate—	Speaking About Theories —Appropriate—
• The new data prove the theory. • I don't believe that; it's just a theory. • Theories aren't true until they are proven. • That's merely a theory; you don't know that for sure. • A recent discovery destroyed that theory.	• The new data support the theory. • Supporters of that theory are disappointed, because several recent experiments have failed to confirm hypotheses based on it. • The theory has gained strength because it has led to many successful predictions. • That theory has led to successful predictions for so long that it is now almost universally accepted.

Figure 7.10. Speaking appropriately about theories.

- The fourth statement is a combination of the second and third statements. Similar comments apply.
- No theory is ever established or destroyed by the result of a single experiment or observation. The failure of an important experiment could deal a serious blow to a theory, but a theory is not abandoned by the scientific community until mounting evidence from several experiments forces scientists to look for a better explanatory model.

One final word about theories, just in case you've missed my point so far. *Successful theories are the glory of science*. Developing theories that accurately model nature is what science is all about.

Successful theories are the glory of science.

Learning Check 7.3

1. What is a scientific fact?
2. Are scientific facts true? Explain.
3. Imagine that you hear someone say, "Oh, we don't need to believe that; it's just a theory." Write a paragraph as if you were speaking (respectfully) to this person, explaining why such statements are inappropriate.
4. How does a theory become widely accepted? Is it because of scientists' personal tastes, or for some other reason? Explain your answer.
5. Can a theory be disproven by a single experiment? Explain.

7.4   Experiments and the Scientific Method

You have probably heard of the *scientific method* before. Most students first begin learning about the scientific method in about 4th grade. But I don't like to emphasize the scientific method too much. The fact is, in scientific research there is not just one scientific method. There are many different scientific methods, as the box on the page 133 suggests. And ironically, if you were to walk up to a PhD scientist and ask him or her about the scientific method, it's quite possible the scientist won't even know what you are talking about.

So, in my opinion, the scientific method kids are always taught in schools is really overrated. Here's another thing: the so-called scientific method isn't some kind of grand program. It's really just a description of how to conduct valid experiments.

Okay, so do you need to learn the scientific method? Yes. Even after all I said, you need to know it. One reason is that schools and scientific literacy tests everywhere typically expect kids to know it, so there's no way out. But more importantly, the scientific method does describe a solid procedure for conducting valid experimental research. This is the real reason it is worth learning. So perhaps we should call it *Steps for Conducting Valid Experimental Research*. The following chart shows the steps in the scientific method and explains what each one means.

Step	Task	Remarks
1	State the problem.	All experimental research is targeted at filling in gaps in our knowledge. I think this step should be called, "Identify exactly what it is you want to find out." You can call this a "problem" if you want to.
2	Research the problem.	If a scientist can perform step 1, she has probably already done step 2. I think what this step is trying to say is, "Thoroughly understand your theory." You need to understand your theory in order to perform step 3.
3	Form a hypothesis.	Yes. This is what the experiment is designed to test, so you need to have one.
4	Conduct an experiment.	Roger that. I think this step is here to emphasize that conducting experiments is the *acid test* in science. (Science pun!) Theories that do not lead to hypotheses and experiments get us nowhere. However, I repeat that designing valid experiments is very challenging. It's not like pouring Kool-Aid and saltwater on plants like you might have done a few years ago for your science "research project."
5	Collect data.	Actually, collecting data (along with many other tasks) is part of the experiment in step 4. One doesn't do experiments without collecting data.
6	Analyze the data.	This is a hard step. Sometimes it is difficult to determine what the data are saying. There can be more than one way to interpret a set of data. So analysis of experimental data is just as hard as designing a good experiment in the first place.
7	Form a conclusion.	The conclusion in mind here is whether the hypothesis was confirmed by the experiment. So this step should be called, "State whether the hypothesis was confirmed."

Step	Task	Remarks
8	Repeat the work.	This silly step has confused many a student. This step does *not* mean that every time scientists do an experiment they have to turn around and do it again. But it does mean two things. First, when collecting data it is necessary to make measurements multiple times. It is easy for a single measurement to be inaccurate. Second, as I write in Section 7.2, new research has to be replicated by other scientists before the results are considered valid. If one group of scientists publishes a new experimental result, it is not they who will repeat the work, it is other scientists who will try to replicate the work.

There is still much more that goes into good experimental research. But we will leave those things for future science courses. This is plenty for now.

Learning Check 7.4

1. How does the scientific method relate to doing experimental research?
2. Describe several other methods for doing science besides the list of steps described in the scientific method.
3. At what points does a scientist's hypothesis come into play in the scientific method?
4. What is really meant by the phrase "repeat the work" in the scientific method?
5. What makes analysis of experimental results challenging?

7.5   Truth

Understanding truth is important, so the next three sections are devoted to it.

You will avoid a lot of confusion if you remember this simple principle: *science is not in the business of making truth claims*. As we have seen in this chapter, science is in the business of developing successful theories to model the natural world. When scientists speak carefully, they avoid using the term *truth* with respect to scientific knowledge. Of course, like everyone else, scientists talk about truth when they are discussing their personal opinions about the meaning of life and so on. But when referring to scientific facts, scientists say that a fact is correct so far as we know. When referring to an accepted theory, a scientist says that the theory is a successful or promising model of how nature works.

All our scientific knowledge about the way the natural world works is provisional or temporary—it is our best understanding for now, but it is subject to change as we learn more. This goes for facts as well as theories.

Science is not in the business of making truth claims.

Is there *truth* about nature? Of course there is. But to know the truth of the technical details

Scientists, Experiments, and Technology

There is not one scientific method; there are hundreds of scientific methods. In the past few centuries, thousands of discoveries have been made by scientists using every different method under the sun. Here's just a taste.

"Try to see the world from a different point of view." Richard Feynman used unconventional mathematical methods, invented Feynman diagrams, and won the Nobel Prize in Physics for work in quantum electrodynamics.

Process one ton of raw material using slow, delicate methods in order to get 1/10 of a gram of what you are looking for. Marie Curie discovered radium and polonium, and won Nobel Prizes in both Physics and Chemistry.

Think a lot until the answer just pops into your head. Hans Oersted discovered that electric current produces a magnetic field, an effect now known as electromagnetism.

Decode the genetic information embedded in billions of base pairs in DNA. Francis Collins discovered the genes for cystic fibrosis and Huntington's disease, and headed the Human Genome Project.

Apply mathematics in unjustifiable ways and see what happens. Max Planck solved the black body radiation problem by first originating quantum theory, and won the Nobel Prize in Physics.

Develop film that you haven't even used yet. Wilhelm Roentgen exercised exhaustive attention to detail in the lab, discovered X-Rays as a result, and won the first Nobel Prize in Physics.

Look for patterns in diffracted light. Francis Crick and James Watson discovered the double-helix structure of the DNA molecule and won the Nobel Prize in Medicine.

Play with mathematics for 20 years until you finally find equations that work. Albert Einstein conceived the special and general theories of relativity, and won the Nobel Prize in Physics for correctly explaining the photo-electric effect.

All scientific knowledge is regarded as provisional, or temporary.

about nature, we would have to be all-knowing about nature—which we aren't. Scientists all hope that our scientific theories are getting closer and closer to the truth, but saying that is as far as we can go.

So now you may be wondering: what is truth, and how do we know it? That's an important question, and we need to address it. Here's a definition for truth:

> **Truth is the way things really are.**

There is a reality about the way nature works. Whatever that reality is, it is the truth. But logically, since facts and theories change and develop, they cannot be regarded as *the truth*, and scientists don't regard them that way. Matter may be made of atoms, as present atomic theory holds, or it may be constructed of strings, or multi-dimensional membranes, or something else. Again, as our theories develop, we hope they are getting closer to the truth. But hoping that our theories are getting closer is all we can say. We cannot say that a scientific theory has arrived at the truth.

7.6   Ways of Knowing Truth

To help illustrate the distinction between scientific knowledge and truth, let's address the question of how we know truth. According to the traditional teachings of philosophy, there are two ways we know truth. Many faith traditions accept a third way of knowing truth.

Direct Observation   The first way to know truth is for it to be obvious to us from direct observation, that is, our own experience. Where are you sitting right now as you read these words? In your classroom? At home? For the sake of example, let's say you are reading this in your bedroom. If you are, then you can make the following truth claim: I am reading this book in my bedroom. How do you know that's true? Because it is obvious to you from your own direct experience. There are many, many things that you know are true because of this kind of direct, first-hand knowledge. You know the truth about where you go to school, what you look like, the grade you got on your last science paper, and the name of your cat.

One way to know truth is from your own direct experience.

Valid Logic   According to traditional philosophy, there is a second source of knowing truth: valid reasoning (logic) based on true premises. We are not going to go into this very deeply here, just enough so that you get the basic idea. The study of formal logic tells us that valid reasoning from true premises results in knowledge of truth. To illustrate this statement, let's consider a form of logical reasoning called a *syllogism*. A syllogism consists of two or more statements (called *premises*) followed by a conclusion. Here is an example:

Premise 1:   All dogs have four legs.
Premise 2:   Buster is a dog.
Conclusion: Buster has four legs.

A second way to know truth is by the use of valid logic, starting from true premises.

Formal logic says that if the two premises are true, and if the logic is *valid*, then the conclusion must be true. In the syllogism above, you have probably noticed that the first premise is not true—there are dogs that do not have four legs, either because they were born that way or because of some disease or injury. In fact, it is difficult to come up with premises that we know with certainty to be true. Here is a classic example, this time using premises that are true:

Premise 1:   All men are mortal.
Premise 2:   Socrates is a man.
Conclusion: Socrates is mortal.

For logic to give us truth, two things are required—the premises must be true and the logic must be valid. In the two syllogisms above, the logic is valid. Here is an example of logic that is not valid:

Premise 1:   All dogs have four legs.
Premise 2:   Buster has four legs.
Conclusion: Buster is a dog.

Do you see the problem? Buster might be a cat or some other animal. Just because Buster has four legs doesn't mean he is a dog. Thus, the logic in this syllogism is not valid.

Most of us don't use formal logic very much, but if we do, then if we reason with valid logic from premises that are certainly true we know our conclusion is true. But that's enough on logic for our purposes here.

Divine Revelation

Many faith traditions hold that there is a third way we can know the truth—revelation from a divine being such as God or angels. These faith traditions teach that God reveals truth to people in several ways. The first is that God speaks through prophets or angels. The second is that God reveals truth through sacred writings such as the Hebrew Scriptures, the Christian Bible, or the Koran of Islam. The third is that God reveals truth in nature. Note, however, that if there is a God who speaks to people through nature, this is different from the scientific facts scientists obtain from the study nature. As discussed previously, truth is the way things are and thus does not change, whereas scientific facts are un-

According to many faith traditions, a third way to know truth is by divine revelation.

derstood to be provisional—subject to change as we learn more. I address this in more detail in the next section.

In summary, there are two ways we are able to know truth according to traditional philosophy, and a third way taught by many faith traditions. From philosophy we understand that we can know truth from our own direct observations and from valid logic based on true premises. Various faith traditions teach that we can also know truth by revelation from God, through prophets, holy writings, or in nature.

Learning Check 7.5/6

1. Explain why it is correct to say that science is not in the business of making truth claims.
2. Define truth.
3. List five examples of truths that you know from the obviousness of your own direct experience.
4. Make two syllogisms, one based on premises you know to be true and one based on premises that are clearly not true.
5. Describe the three ways of knowing truth.

7.7 Relating Scientific Knowledge and Truth

There are two ways in which our discussion of scientific knowledge relates to our discussion of truth. First, we have seen that one way to know truth is by direct observation. We have also seen that scientists use observation as a way of discovering scientific facts. How do these two uses of observation relate to each other?

Imagine a scientist studying tigers who observes a tiger eating the meat of another animal. The scientist can say, "It is true that this tiger eats meat." The scientist might observe 75 other tigers exhibiting this same behavior. She can then say, "These 75 tigers all eat meat." So far, all the scientist has done is to say things that she has found to be true by direct observation.

But now suppose the scientist takes this information and makes a general claim about tigers: "It is a scientific fact that tigers are carnivores." The scientist has now made a leap from tigers she has directly observed to many other tigers she has not directly observed. Who knows whether there might be a species of vegetarian tiger somewhere out there? We have no way of knowing the eating habits of every single tiger. This is why we cannot say that meat eating is a truth about all tigers. We can only say that it is a scientific fact about tigers. The scientific fact about tigers is a statement based on a lot of evidence that is correct so far as we know, but it may need to be changed if further research shows that there are species of tigers that do not eat meat.

Second, we have studied tigers for a long time and are pretty sure that the statement, "all tigers are carnivores" is true. We are so sure that most of us probably do regard this statement as true. This is fine, but we need to keep in mind that it is always possible that a scientific claim may turn out to be false.

7.8   Summary: The Nature of Scientific Knowledge

In recent years, many scientists have realized that the scientific community has not done an adequate job of educating the public about what theories are and the role they play in scientific research. It is common to hear people dismissing theories because they are not "proven," even though theories never are. It is also common to hear people making grand claims about theories, as if they were the final word on our understanding of nature, even though theories never are this either.

We need to keep in mind the distinction between two kinds of knowledge. On one hand, we have knowledge we recognize as truths or truth claims. The truths you know are the way things really are. These truths are permanent, so long as the conditions giving rise to them exist. If you live in Texas, then it is true to say "I live in Texas." You might move elsewhere, but so long as you live there you know it is true that you do live there. On the other hand, we have the facts and theories of science. Scientific knowledge is always provisional—it is always subject to change if new contradictory evidence is discovered. Facts are correct so far as we know. Theories are our best explanations of the facts for the present.

Scientists hope that scientific research brings us closer and closer to the truth about nature. If we do discover the complete truth, we won't know we have. We will simply know that our theories explain every question we can think up and always lead to predictions that stand up to every test. That day is a long way off, and may never come at all. This means we have plenty to learn about as scientific research continues. The adventure of exploring the wonders of nature never ends!

Chapter 7 Exercises

Answer each of the questions below as completely as you can. Write your responses in complete sentences.

1. Describe each of the main steps in the Cycle of Scientific Enterprise. Define each term, and explain how they relate to one another.
2. Choose an actual example of a scientific theory, and use it to illustrate how theories change. In this book, we have looked at atomic theory, caloric theory, gravitational theory, and Big Bang theory.
3. Write a paragraph or two explaining the difference between scientific theories and truth.
4. Explain the real purpose for the sequence of steps in the scientific method.
5. Explain why it is inappropriate to ridicule a scientific model by calling it "just a theory." In your explanation, describe appropriate ways to speak of scientific models.

Chapter 8
Measurement and Units

The image above shows a small cylinder made of 90% platinum (element 78). This image is computer-generated, but it is an accurate representation of the official one-kilogram mass maintained in a vault in Sèvres, France by the International Bureau of Weights and Measures. The mass of the platinum cylinder in that vault is the official definition of the kilogram.

The kilogram is the only one of the seven base units in the metric system that is still defined by a man-made physical object (an artifact). The others are now defined in terms of various constants found in nature. Officials are looking to change the definition of the kilogram so that it, too, is defined in terms of a constant instead of a physical object. Research is now underway to determine the best definition.

The kilogram is also the only base unit in the metric system that includes a metric prefix in its name.

OBJECTIVES

After studying this chapter and completing the exercises, you should be able to do each of the following tasks, using supporting terms and principles as necessary.

1. Define and use the five most common metric prefixes: *kilo–*, *mega–*, *centi–*, *milli–*, and *micro–*.
2. Perform unit conversions involving common English units for length, volume, and time; metric units; and the metric prefixes.
3. Use the conversion factors in Table 8.6 from memory (except for factors shown in orange cells at the bottom).
4. Calculate the volume of right rectangular solids and right circular cylindrical solids.
5. Describe the origin of the SI System of units and explain why it is universally used in scientific work.
6. Explain why making accurate measurements is important in scientific study.

VOCABULARY TERMS

You should be able to define or describe each of these terms in a complete sentence or paragraph.

1. base unit
2. derived unit
3. International System of Units
4. kilogram
5. liter
6. meter
7. metric prefix
8. metric system
9. right circular cylinder
10. right rectangular solid
11. second
12. SI System
13. unit conversion factor

8.1   Science and Measurement

Physical science is the study of the physics and chemistry of nature. Making measurements is central to this work. So far in this course, we have focused on descriptive information. I have mentioned measurements a few times, and two of the experiments you have conducted so far have involved taking measurements. But in coming sections, we begin looking at a few important types of calculations. Once we begin using equations to perform calculations, we will be dealing with measurements frequently.

As I said, making measurements is of central importance in science. Accordingly, two of the most basic and important skills in scientific study are making accurate measurements and correctly using units of measure. Using units of measure correctly requires you to know the common units of measure, the common prefixes used in the metric system, and some common "conversion factors" for converting from one unit of measure to another. These matters are the subject of this chapter.

The ability to make measurements and use different units of measure is not only important in science; it is important in just about every area of life. Everyone uses measurements. Carpenters, plumbers, cooks, merchants, farmers, and people in many other trades and professions all use measurements nearly every day. Figure 8.1 is a photograph of a baby bottle marked with three different measurement scales. The message here is that from mothers to carpenters, from farmers to engineers, from surveyors to chemists, measurements are used by everyone.

If you are interested in the history of science, you will find that reading about the history behind measurement systems and units of measure is fascinating. Since ancient times, measurements have been of great importance for architecture, commerce (weights of grain, etc.), travel, and geography.

Since I love to read, I have always been interested in the units of length that come up when reading the great works of literature. Let's look at a few of these. An ancient unit of length is the *cubit*, illustrated in Figure 8.2. The cubit was a measure of length based on the length of a man's arm from the

Figure 8.1. A baby bottle, marked with three different measurement scales for three different units of measure.

elbow to the tip of the middle finger, about 18 inches or 46 centimeters. The cubit was the preferred unit for small lengths among the Egyptians and Mesopotamians. Among the Greeks and Romans, the foot was preferred. We'll come back to the foot in a moment.

Figure 8.2. A measuring cubit from about 3,000 BC in Mesopotamia, currently on display in a museum in Istanbul, Turkey.

When reading English literature, the *furlong* comes up a lot when writers describe distances horses travel, and this unit is still in use today in the sport of horse racing. A furlong is equal to 1/8 of a mile, or 220 yards. In the metric system, this is about 201 meters.

In English sailing literature, one often comes across the *fathom*. A fathom is two yards (six feet) long, or 1.83 meters. When sailing in shallow waters, it is important for a ship's captain to keep track of the water depth so the ship does not run aground. In the days before modern electronics, crews on ships would measure the depth of the water by a process called *sounding*, illustrated in Figure 8.3. To take a sounding, the crew would lower a *sounding lead* (pronounced led), which was a rope with a lead weight on the end. The sounding lead was marked off in fathoms with pieces of leather. On the Mississippi River in the 19th century, when a crew on the river took a sounding and found the water to be two fathoms deep, they would call out "Mark—twain!" This is the origin of Samuel Clemens' pen name.

One of these days, I hope you will fall in love with the great Russian writers such as Fyodor Dostoyevsky and Leo Tolstoy. If you read Russian literature, you will frequently come across the *verst*, a length unit used to describe travel distances for horses and distances between stations or inns. It is handy to know that a verst is just a shade over one kilometer (1.07 km), which makes it right at 2/3 of a mile. (You heard it here first!)

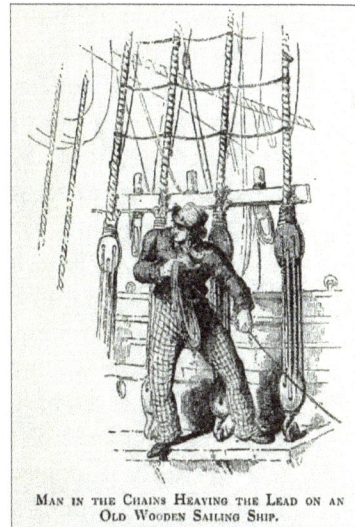

MAN IN THE CHAINS HEAVING THE LEAD ON AN OLD WOODEN SAILING SHIP.

Figure 8.3. Heaving the lead.

Moving on now to the modern length units, the Greeks and Romans preferred the *foot* as the common unit for short lengths. The foot was based on the length of an average man's foot. The Romans divided the foot into 12 *uncia*, which is the origin of our term *inch*. The inch was very close to being 2.54 centimeters. In the 1950s, momentum was building for the publication of the International System of Units (next section). In 1959, the inch was redefined to be *exactly* 2.54 centimeters. This redefinition was very important for allowing precise conversions of common American units such as inches, feet, and miles into metric units.

The *mile* was originated by Roman soldiers marching in the days of the Roman Empire. The Romans defined the mile as 5,000 feet, but in 1593 an English Act of Parliament—called a statute—defined the mile as 5,280 feet, our definition today. This is why our mile is sometimes called a *statute mile*.

Learning Check 8.1

1. List several reasons why it is important for everyone to learn about measurement and units of measure.
2. Briefly describe the origin of the English units inch, foot, and mile.
3. What is a sounding lead?
4. How is the inch defined today?

8.2   The International System of Units

The International System of Units, called the *SI System*, is used universally today in scientific work. This system, which we Americans often refer to as the *metric system*, was published in 1960. The metric system originated in France during the French Revolution. The original system included only the meter and the kilogram. Over the years, as measurement treaties were signed and scientific learning advanced, the system grew into the formal SI System that has now been in use since 1960. The SI System is administered by an organization in Sèvres, France (near Paris) known as the International Bureau of Weights and Measures. The SI System has been adopted almost globally. There are only three nations in the world that have not accepted the SI System as their official system of measurement: Myanmar, Liberia, and the United States. But even though our road sign markers still give distances in miles, in scientific work the SI System is the one we use.

There are seven *base units* in the SI System, listed in Table 8.1. We will be using the first five of these units in this and later chapters. All other SI units of measure, such as the joule (J) for measuring energy and the newton (N) for measuring force, are based on these seven base units. Units based on combinations of the seven base units are called *derived units*. A few common derived units are listed in Table 8.2.

There are seven base units in the SI System.

Of the five base units we use in this text, you are already familiar with the SI unit for time—the second. You may or may not be familiar with the other units, so here are some facts and photos to familiarize you with these. A meter is just a few inches longer than a yard (3 feet). Figure 8.4 shows a wooden measuring rule one meter long, commonly called a meter stick, along with a metal yardstick for comparison.

Unit	Symbol	Quantity
meter	m	distance
kilogram	kg	mass
second	s	time
ampere	A	electric current
kelvin	K	temperature
candela	Cd	luminous intensity
mole	mol	amount of substance

Table 8.1. The seven base units in the SI System.

Unit	Symbol	Quantity
joule	J	energy
newton	N	force
cubic meter	m³	volume
watt	W	power
pascal	Pa	pressure

Table 8.2. Some SI System derived units.

On earth, a mass of one kilogram weighs about 2.2 pounds. The six-volt lantern battery shown in Figure 8.5 weighs just under 2.2 pounds, so the mass of the battery is just about one kilogram.

I can't show you a picture of one ampere of electric current, but it may be helpful to know that a standard electrical receptacle (or "outlet") such as the one shown in Figure 8.6 is rated to carry 15 amperes of current. However, the largest continuous current that the receptacle is allowed to supply is 80% of its rating, or 12 amperes. This is why vacuum cleaners are often advertised as having 12-amp motors. That's the upper limit of the current available to run them.

Finally, we will look at temperature scales in more detail in the next chapter. However, here I will note that a temperature change of one kelvin is the same as a temperature change of one degree Celsius, and both of these are almost double the change represented by one degree Fahrenheit. Room temperature on the three

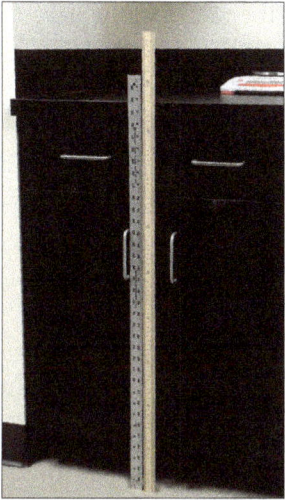

Figure 8.4. A wooden meter stick (right), with a metal yardstick for comparison.

Figure 8.5. The mass of this battery is about one kilogram.

Figure 8.6. A standard receptacle in the home is rated for a current of 15 amperes.

scales is 72°F, 22.2°C, and 295.4 K. If the temperature in the room increases by 6°F to 78°F, the other new temperatures are 25.6°C and 298.8 K.

As indicated on the opening page of this chapter, the kilogram is the only base unit in the SI system that is defined by a man-made physical object (an artifact). Formerly, the meter was also defined that way; the meter used to be defined as the length of a metal bar, such as depicted in Figure 8.7. The bar shown in the figure was the standard in the U.S. for the meter from 1893 to 1960. But this method of definition was not at all convenient.

The kilogram is the only base unit still defined by a man-made object.

For starters, there was no way to tell if the standard meter bar length had changed except by comparing it to other meter bars. There was no outside standard to compare it to. Copies of the official meter bar were very expensive to make and maintain. One or more copies had to be made for many different countries to use, and copies of these copies had to be made for various states to use. The

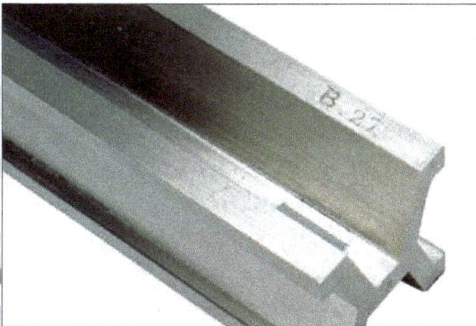

Figure 8.7. Standard meter bar number 27, owned by the U.S. and used as our standard meter from 1893 to 1960.

lengths of these copies had to be checked regularly against the length of the official meter bar, which required a lot of complicated travel with the precious meter bar copies. And all these standard meter bars had to be kept in vaults under very tightly controlled temperature and humidity conditions because the slightest change in temperature causes the bar's length to change as the metal expands or contracts. All this was necessary in a world with rapidly growing technology and global commerce, but it was a hassle.

In 1960, the definition of the meter was changed so that the meter was equal to a certain number of wavelengths of a certain color of light emitted by a certain isotope of the element krypton. In 1983, the definition of the meter was changed again, and now is defined as equal to the distance light travels in a vacuum in exactly 1/299,792,458 seconds. Quite an interesting history. Hopefully, soon the definition for the kilogram will change as well. Then all the kilogram standard masses can be put in museums.

We conclude this section with a caution you should be aware of as you enter the world of higher-level science studies. Unfortunately, the kilogram is often used in common speech outside the U.S. as a unit of weight. The kilogram is not a unit of weight, so this usage is confusing to say the least! Weight is a force, and in the SI System forces are measured in newtons. (In the common system used by Americans, forces are measured in pounds.) At sea level on earth, a certain mass is pulled on by the earth with a certain force, and that force is the object's weight. But to avoid confusion, you should always keep weight and mass—and their units of measure—separate in your mind. The concept of mass is difficult enough to understand all by itself, without the additional complication of mixing up the units.

> ### Learning Check 8.2
>
> 1. Explain what a base unit is in the SI System.
> 2. Describe the three different standards that have been used in the U.S. since 1893 to define the meter.
> 3. Why is it important to define base units of measure in terms of a natural standard, instead of an artifact?
> 4. The three most commonly used base units are the meter, the kilogram, and the second. For each of these three units, make up two examples of your own that help illustrate their size.
> 5. Explain what a derived unit is, and give some examples of derived units.

8.3   Metric Prefixes

In the system of units we commonly use in the U.S., we have different units to use for different sizes of objects. For example, for short lengths we might use the inch or the foot, whereas for longer distances we switch to the mile. For the small volumes used in cooking, we use the fluid ounce (or pint, quart, teaspoon, tablespoon, etc.), but for larger volumes like the gasoline in the gas tank of a car, we switch to the gallon. (That's six different volume units I just listed!) Having so many different units around is really unnecessary. Also unnecessary are all the numbers used in relating these units to one another, such as 12 inches in a foot, 128 ounces in a gallon, 5,280 feet in a mile, and so on. The only reason our system seems easy to use is that we grow up using it. And even when you grow up with it, it's still challenging to remember how many ounces are in a gallon! Quick—How many teaspoons are in a quart? I don't know either, and who cares anyway?

The SI System is much simpler. (The only reason it seems difficult to students is that they didn't grow up with it!) Each type of quantity—such as length or volume—has one main unit of measure. Instead of using several different units for different sizes of quantities, the SI System uses multipliers on the units to multiply them for large quantities, or to scale them down for smaller quantities. We call these multipliers the *metric prefixes*. The complete list of the 20 metric prefixes is in Table 8.3. I'm showing the full list here just for fun. I know some of you will be very interested to see them all. But I am also aware that some of you may not have spent much time with scientific notation yet, which is used to express the factors. So, if this large table seems intimidating, don't sweat it. There are several prefixes most of us will probably never use, such as hecto– and yocto–. Others, such as giga– and nano–, are important because they are used frequently, and you will learn them over the next couple of years in school.

Multiples	Prefix	deca–	hecto–	kilo–	mega–	giga–	tera–	peta–	exa–	zetta–	yotta–
	Symbol	da	h	k	M	G	T	P	E	Z	Y
	Factor	10	10^2	10^3	10^6	10^9	10^{12}	10^{15}	10^{18}	10^{21}	10^{24}
Fractions	Prefix	deci–	centi–	milli–	micro–	nano–	pico–	femto–	atto–	zetto–	yocto–
	Symbol	d	c	m	µ	n	p	f	a	z	y
	Factor	1/10	$1/10^2$	$1/10^3$	$1/10^6$	$1/10^9$	$1/10^{12}$	$1/10^{15}$	$1/10^{18}$	$1/10^{21}$	$1/10^{24}$

Table 8.3. SI System prefixes—the complete list.

For the purposes of this course, there is a much shorter list of common prefixes that you should memorize and learn how to use. These are shown in Table 8.4. In the table, I have shown examples of how to use these prefixes. It is important for

	Prefix	Symbol	Meaning	Examples of usage
Multiples	kilo–	k	1,000	One kilojoule is 1,000 joules. There are 1,000 joules in one kilojoule, so 1,000 J = 1 kJ.
	mega–	M	1,000,000	One megawatt is 1,000,000 watts. There are 1,000,000 watts in one megawatt, so 1,000,000 W = 1 MW.
Fractions	centi–	c	1/100	One centimeter is 1/100 of a meter. There are 100 centimeters in one meter, so 100 cm = 1 m.
	milli–	m	1/1,000	One milligram is 1/1,000 of a gram. There are 1,000 milligrams in one gram, so 1,000 mg = 1 g.
	micro–	µ	1/1,000,000	One microliter is 1/1,000,000 of a liter. There are 1,000,000 microliters in one liter, so 1,000,000 µL = 1 L.

Table 8.4. Essential metric prefixes for now.

you to study these examples carefully. We use these five prefixes repeatedly in the coming pages and chapters.

To help you get a feel for what these prefixes mean and the sizes of the quantities involved, we now look at a number of photos and examples. We begin with the fractional prefixes.

You are probably already familiar with the centi– prefix from making measurements in centimeters. This prefix means "one hundredth," or 1/100. One centimeter (1 cm) is 1/100 of a meter, so there are 100 cm in 1 m.

The centimeter is, by far, the most common use for the centi– prefix. A comparison between the centimeter and the inch is shown in Figure 8.8. In this figure, these two length units are shown at actual size. As you can see, the centimeter is a bit less than half an inch.

The milli– prefix means "one thousandth," or 1/1,000. One millimeter (1 mm) is 1/1,000 of a meter, roughly twice the thickness of the lead in a mechanical pencil. The red lines shown in Figure 8.9 are 1 mm wide on the page of this book. They are also each 2 cm long. You should get out your centimeter/millimeter rule and measure them for yourself. (The green bars in Figure 8.8 are 2 mm wide.)

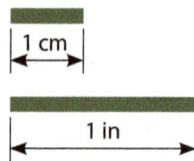

Figure 8.8. Comparison between the centimeter and the inch.

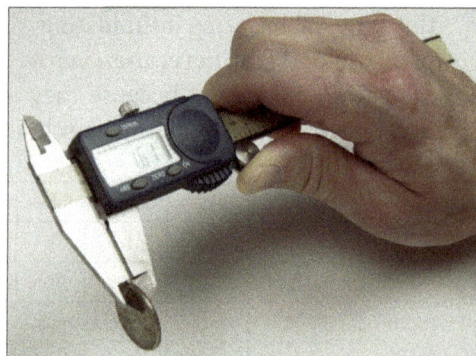

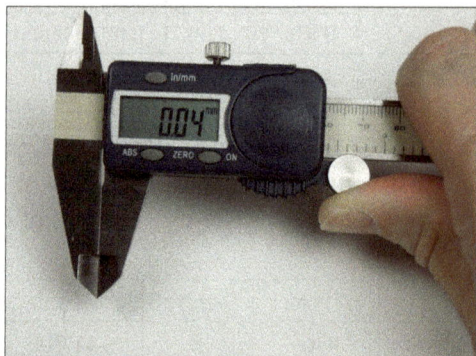

Figure 8.9. These red lines are each 1 mm wide and 2 cm long.

Figure 8.10 shows a caliper with a digital display reading in millimeters. This instrument is being used to measure several small things as examples. In the upper photo, the caliper is measuring the thickness of the rim on a quarter, and the reading is 1.61 mm.

The micro– prefix means "one millionth," or 1/1,000,000. One micrometer

Figure 8.10. Small quantities in SI System units: thickness of the rim of a quarter, 1.61 mm; diameter of a human hair, 40 μm; thickness of a sheet of printer paper, 100 μm.

(1 μm) is 1/1,000,000 of a meter. This is a very small length. The middle image in Figure 8.10 shows the caliper measuring the width of a human hair, which is barely visible in bottom left of the photo. The caliper shows the hair to be 0.04 mm thick, which is equal to 40 μm. So 1 μm is about 1/40 the width of a human hair.

As another example, the lower image in Figure 8.10 shows the caliper measuring the thickness of an ordinary piece of printer paper. The paper is 0.10 mm thick, which is equal to 100 μm. So 1 μm is 1/100 the thickness of a sheet of paper.

The milli– prefix is often used for volume measurements with the liter. We discuss volume in some detail in Section 8.5, but here we use the liter as another example illustrating the size implications of the prefix milli–. On the left of the upper photo in Figure 8.11 is a beaker with a capacity of 1,000 milliliters (1,000 mL), equal to one liter (1 L). In the center of the upper photo is a graduated cylinder containing 100 mL of water. The lower image in the figure shows a close-up of the graduated cylinder. Each line on the cylinder represents one milliliter (1 mL) of liquid. Thus, between two lines on the graduated cylinder is 1/1,000 the volume of the beaker.

An example illustrating the kilo– prefix is shown in Figure 8.12. The prefix kilo– means "one thousand." The photo shows one kilogram (1 kg) of salt in the beaker on the left and one gram (1 g) of salt in the blue tray in the center. Another common use of the kilo– prefix is in distances measured in kilometers. The easiest way to understand the kilometer is to remember that one kilometer (1 km) is close to being 2/3 of a mile. A drive of 100 km is about 62 miles.

The mega– prefix is difficult to illustrate because it is difficult to imagine one million of anything, but consider this. The flow rate of the Mississippi River at New Orleans is 17 million liters per second, or 17 ML/s!

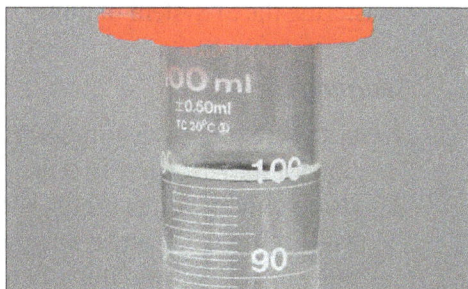

Figure 8.11. In the upper photo, 1,000 mL of water (1 L) in a 1-L beaker (left), 100 mL of water in a 100-mL graduated cylinder (center), and a coffee cup for comparison. In the lower photo, each line on the graduated cylinder is 1 mL.

Figure 8.12. 1,000 grams (1 kg) of salt (left), and one gram (1 g) of salt (center).

We conclude this section with a few notes about the metric prefixes and their symbols. First, when using the prefixes for quantities of mass, we do not add prefixes to the base unit, the kilogram. Prefixes are only added to the term gram, even though the kilogram is the base unit, not the gram. Second, note here that the capital M is used as a symbol for mega–. A lower case m is used for milli–, which is 1/1,000. You will have to practice using these prefixes to keep from getting them mixed up. Third, students often struggle to remember how to use the prefix μ ("mu," the lower-case *m* of the Greek alphabet) so take special care to learn it. Finally, pay close attention to the difference between multiplier prefixes and fraction prefixes. Learning to use the fraction prefixes properly is the most challenging part of mastering the SI System of units.

Learning Check 8.3

1. List some of the advantages the SI System of units has over the common unit system used in the U.S.
2. Using only the units meter, kilogram, second, and liter, and the five prefixes in Table 8.4, write down rough estimates with units for the quantities listed below. (Don't actually measure or research anything; just make your best guess. You may use the gram as well as the kilogram, and you may do some multiplying on paper or with a calculator.)

    a. The width of a typical classroom measuring ruler.

    b. The length of time the flash is on when you take a picture with a mobile phone.

    c. The mass of a golf ball.

    d. The volume of liquid in an ordinary cup of coffee.

    e. The distance across the Golden Gate Bridge in San Francisco.

    f. The mass of a single piece of chewing gum.

    g. The lifespan of a typical dog or cat.

    h. The thickness of a teenage girl's pinky finger.

    i. The length of time it takes a jet to travel 1 mm.

    j. The mass of a bowling ball.

    k. The wingspan of a large commercial jet.

    l. The volume of milk in a gallon jug.

    m. The length of a golf club.

    n. The distance from Los Angeles to New York City.

8.4   Unit Conversions

One of the most basic tasks people perform with units of measure is to re-express quantities using different units of measure. For example, if your family

*We Pause Here to Talk About a few **Mathematical Principles***

There are a few important mathematical principles at work behind the scenes when we perform unit conversions. We will review these here. The dot symbol (·) means the same thing as the multiplication sign (×).

1. Mathematically, we treat units of measure as multiplied by the numerical value they are attached to. We never actually write in the multiplication sign, but mentally, when you see "560 mi" you can think "560 × mi."

2. A number written by itself may be considered the same as a fraction with a denominator of 1. For example,

$$5 \text{ cm} = \frac{5 \text{ cm}}{1}$$

So, in cases when you see something that looks like this:

$$560 \text{ mi} \cdot \frac{1{,}609 \text{ m}}{1 \text{ mi}}$$

you know that it means the same thing as this:

$$\frac{560 \text{ mi}}{1} \cdot \frac{1{,}609 \text{ m}}{1 \text{ mi}}$$

3. To multiply two fractions together, multiply the numerators and leave the result in the numerator; multiply the denominators and leave that result in the denominator.

4. When we multiply quantities together, the order does not matter. In other words, $5 \times 6 = 6 \times 5 = 30$. So when you see something that looks like this:

$$\frac{560 \text{ mi}}{1} \cdot \frac{1{,}609 \text{ m}}{1 \text{ mi}}$$

you know that it means the same thing as this:

$$\frac{560 \text{ mi}}{1 \text{ mi}} \cdot \frac{1{,}609 \text{ m}}{1}$$

5. Finally, when the numerator and denominator in a fraction are equivalent, the value of the fraction is 1. For example, there are 1,609 meters in 1 mile, so

$$\frac{1{,}609 \text{ m}}{1 \text{ mi}} = 1 \text{ , and } \frac{1 \text{ mi}}{1{,}609 \text{ m}} = 1$$

drives 560 miles on your family vacation, you might want to know how far the drive was in meters. To find out, you perform a calculation called a *unit conversion* to convert 560 mi into an equivalent number of meters. Mastery of this skill is essential for every student. To get started, first read the box on page 153, then return here to continue.

As we go, we need to use standard symbols for units of measure. The common symbols for the units normally used in America (called U.S. Customary Units) and the common SI System unit symbols are shown in Table 8.5. The conversion factors we need are derived from the equations in Table 8.6. You need to memorize the conversion factors shown in white cells in the table. We will use the factors in the orange cells at the bottom for practice, but memorizing them is not as crucial for now. I explain how to use these conversion factors as we go.

Let's begin with the basic principle of how a unit conversion calculation works. First, you know that multiplying any value by one leaves its value unchanged. Second, you know that in any fraction, if the numerator

A unit conversion factor is simply a convenient way of writing "1."

and denominator are equivalent, the value of the fraction is one. *A "unit conversion factor" is simply a fractional expression in which the numerator and denominator are equivalent ways of writing the same physical quantity. This means a conversion factor is just a special way of writing "one."*

Let's now see how we can put these ideas together to perform the unit conversion I referred to in my example about the family vacation. Our goal is to convert 560 miles into an equivalent distance expressed with units of meters. We will walk through the calculation step by step. After the first run through the calculation, we will look even more closely at the "canceling out" business that occurs in the middle. So if things don't quite click for you, stay with me through both explanations.

Unit Symbols
U. S. Customary System Symbols
mi = mile
ft = foot
yd = yard
in = inch
yr = year
dy = day
hr = hour
min = minute
s = second
gal = gallon
SI System Symbols
kg = kilogram
m = meter
s = second
L = liter (not an official SI unit)

Table 8.5. Standard unit symbols.

EXAMPLE 8.1

To begin, write down the starting quantity:

560 mi

Next, select a conversion factor that includes the units you presently have, and the units you want to convert to. In our case, we have units of miles and we want to convert to meters. From Table 8.6, 1,609 m = 1 mi. Write this as a fraction with miles in the denominator so the mile units will "cancel out."

$$560 \text{ mi} \cdot \frac{1{,}609 \text{ m}}{1 \text{ mi}}$$

Written this way, the units will cancel out, like this:

$$560 \cancel{\text{mi}} \cdot \frac{1{,}609 \text{ m}}{1 \cancel{\text{mi}}}$$

Finally, multiply the numerator values and write the result with the new units.

$$560 \cancel{\text{mi}} \cdot \frac{1{,}609 \text{ m}}{1 \cancel{\text{mi}}} = 901{,}040 \text{ m}$$

Conversion Equations
U. S. Customary Unit Factors
5,280 ft = 1 mi
3 ft = 1 yd
60 s = 1 min
60 min = 1 hr
24 hr = 1 day
365 dy = 1 year
3,600 s = 1 hr
SI System Factors
1,000 cm³ = 1 L
1,000,000 cm³ = 1 m³
1,000 L = 1 m³
1 mL = 1 cm³
Inter-System Factors
2.54 cm = 1 in
0.3048 m = 1 ft
1.609 km = 1 mi
1,609 m = 1 mi
3.785 L = 1 gal

Table 8.6. Common conversion equations.

Now, let's take a closer look at that step involving the "cancelling out." This maneuver is where the math principles from page 149 come into play. I will go through the entire process, step by step. Once we get to the end, it should be clear what cancelling out is all about. It should also be clear how you know which way to write the conversion factor, since it can be written with either 1,609 m or 1 mi in the numerator.

EXAMPLE 8.2

We are starting with this expression:

$$560 \text{ mi} \cdot \frac{1{,}609 \text{ m}}{1 \text{ mi}}$$

From principle 2 (p. 149), we can write this as

$$\frac{560 \text{ mi}}{1} \cdot \frac{1{,}609 \text{ m}}{1 \text{ mi}}$$

From principles 3 and 4, we can switch things around in the denominator like this:

$$\frac{560 \text{ mi}}{1 \text{ mi}} \cdot \frac{1{,}609 \text{ m}}{1}$$

From principle 3, the above expression is equivalent to:

$$\frac{560}{1} \cdot \frac{\text{mi}}{\text{mi}} \cdot \frac{1{,}609 \text{ m}}{1}$$

But according to principle 5, mi/mi is equal to 1, because the numerator and denominator are the same. Since multiplying by 1 doesn't change anything, we can just leave it out. The act of "cancelling out" is just a short cut that occurs when you see that you have the same units somewhere in a numerator and a denominator. You should now be able to see why we wrote the conversion factor as $\frac{1{,}609 \text{ m}}{1 \text{ mi}}$ instead of $\frac{1 \text{ mi}}{1{,}609 \text{ m}}$. In general, a conversion factor can be written with either value in the numerator, because the numerator and denominator are equivalent. But you have to select the units for the denominator that you want to cancel. That is why we wrote the conversion factor with miles in the denominator.

So now we have:

$$\frac{560}{1} \cdot \frac{1{,}609 \text{ m}}{1}$$

But now there is nothing in the denominator excepts ones, so we can get rid of those, too (principle 2), giving:

$$560 \cdot 1{,}609 \text{ m}$$

Finally, principle 1 tells us that we can just multiply these together, units and all, to get

$$560 \cdot 1{,}609 \text{ m} = 901{,}040 \text{ m}$$

In our next example, the given quantity has units in both the numerator and the denominator. A very important piece of advice comes into play here, and it is this: *Never use slant bars in your unit fractions. Use only horizontal bars.* In printed materials we often see values written with a slant fraction bar in the units, as in the speed 35 meters per second, written as 35 m/s. Although writing the units this way is fine for a printed document, you should not write values this way when you are performing unit conversions. This is because it is easy to get confused and not notice that one of the units is in the denominator in such an expression (s, or seconds, in 35 m/s). Conversion factors can always be written two ways. So to make sure you write the conversion factor the correct way, write the units of measure with a horizontal bar.

When performing unit conversions, use only horizontal bars in unit fractions.

EXAMPLE 8.3

A large commercial jet typically flies at a speed of around 225 meters per second, or 225 m/s. Convert this value to kilometers per second (km/s).

To begin, write down the given value, being careful to use a horizontal fraction bar:

$$225 \, \frac{m}{s}$$

Now select the conversion factor you need. We need to convert the meter units to kilometers. (The seconds stay as they are.) This conversion factor comes from the definition for the prefix kilo– in Table 8.4. The factor is 1,000 m = 1 km. We desire the meters to cancel out, and meters are in the numerator of the given quantity. So, we write the conversion factor with the meters in the denominator.

$$225 \, \frac{m}{s} \cdot \frac{1 \, km}{1,000 \, m}$$

Now cancel out the meters. The units we have left are km in the numerator and s in the denominator. Then compute the result by dividing 225 (which is in the numerator) by 1,000 (which is in the denominator).

$$225 \, \frac{\cancel{m}}{s} \cdot \frac{1 \, km}{1,000 \, \cancel{m}} = 0.225 \, \frac{km}{s}$$

Sometimes a unit conversion requires the use of two or more conversion factors. For our third example, we will work through a unit conversion that requires several conversion factors all at once. The process is the same: we just chain the conversion factors together one at a time until we have the units we need at the end.

EXAMPLE 8.4

If your science class lasts 55 minutes, how many years of your life go by each time you are in science? (Naturally, since everyone loves science, these are the *good* years!)

We need to convert from a quantity in minutes to a quantity in years, but we don't know how many minutes are in a year. However, we do know conversion factors for converting from minutes to hours, and hours to days, and days to years.

Begin by writing down the given quantity.

55 min

Now multiply by a conversion factor that takes us from minutes to hours. Since the minutes are in the numerator of the given quantity, they must be in the denominator in the conversion factor.

$$55 \text{ min} \cdot \frac{1 \text{ hr}}{60 \text{ min}}$$

When we cancel out the units on this, we have

$$55 \cancel{\text{min}} \cdot \frac{1 \text{ hr}}{60 \cancel{\text{min}}}$$

This leaves us with units of hours. Now, one by one, add conversion factors to go from hours to days, and days to years. Each time, write the new factor in a way that cancels out the units from the previous factor.

$$55 \cancel{\text{min}} \cdot \frac{1 \cancel{\text{hr}}}{60 \cancel{\text{min}}} \cdot \frac{1 \cancel{\text{dy}}}{24 \cancel{\text{hr}}} \cdot \frac{1 \text{ yr}}{365 \cancel{\text{dy}}}$$

This leaves us this units of years. Next, multiply all the values in the numerator, and write the result in the numerator of the result. Do the same for the denominator.

$$55 \cancel{\text{min}} \cdot \frac{1 \cancel{\text{hr}}}{60 \cancel{\text{min}}} \cdot \frac{1 \cancel{\text{dy}}}{24 \cancel{\text{hr}}} \cdot \frac{1 \text{ yr}}{365 \cancel{\text{dy}}} = \frac{55}{525,600} \text{ yr}$$

The final step is to compute the value of the fraction, and write down the final result with the correct units of measure.

$$55 \text{ min} \cdot \frac{1 \text{ hr}}{60 \text{ min}} \cdot \frac{1 \text{ dy}}{24 \text{ hr}} \cdot \frac{1 \text{ yr}}{365 \text{ dy}} = \frac{55 \text{ yr}}{525{,}600} = 0.000105 \text{ yr}$$

Here is one final tip. The term "per" implies a fraction. Some units of measure are commonly written with a "p" for "per," such as mph for miles per hour, or gpm for gallons per minute. Change these expressions to fractions with horizontal bars when you work out the unit conversion. To demonstrate, we will work one final example.

This example also requires us to convert both the numerator units and the denominator units into different units of measure. To do this, simply deal with the numerator units first, chaining together as many conversion factors as you need. Then do the same thing for the denominator units. All you are doing is chaining together a bunch of conversion factors, canceling out units, and then doing the math.

The term "per" implies a fraction.

EXAMPLE 8.5

A broken water pipe is leaking 3.17 gpm into the street. Convert this value into liters per second.

The units "gpm" stand for gallons per minute, and the "per" implies a fraction. So begin by writing the given quantity with the units in a fraction with a horizontal bar.

$$3.17 \; \frac{\text{gal}}{\text{min}}$$

Now select a conversion factor to handle the numerator units. We need to convert the gallons in the numerator into liters. From Table 8.6, 1 gal = 3.785 L. We need to write this with the gallons in the denominator so they cancel out with the gallons we have in the numerator.

$$3.17 \; \frac{\text{gal}}{\text{min}} \cdot \frac{3.785 \text{ L}}{1 \text{ gal}}$$

The units in this expression are liters per minute, but we're not finished. We need to convert the minutes into seconds. Add on the conversion factor needed to do this, and cancel out the minutes.

$$3.17 \; \frac{\cancel{\text{gal}}}{\cancel{\text{min}}} \cdot \frac{3.785 \text{ L}}{1 \cancel{\text{gal}}} \cdot \frac{1 \cancel{\text{min}}}{60 \text{ s}}$$

We now have only liters in the numerator and seconds in the denominator. So, the units of this result are L/s, which is what we want. As before, multiply the numerator values and write the product in the numerator of the result. Do the same for the denominator values, writing the product in the denominator of the result.

$$3.17 \; \frac{\cancel{\text{gal}}}{\cancel{\text{min}}} \cdot \frac{3.785 \text{ L}}{1 \cancel{\text{gal}}} \cdot \frac{1 \cancel{\text{min}}}{60 \text{ s}} = \frac{12.00 \text{ L}}{60 \text{ s}}$$

Finally, compute the value of the fraction, and write the result with the correct units of measure.

$$3.17 \; \frac{\cancel{\text{gal}}}{\cancel{\text{min}}} \cdot \frac{3.785 \text{ L}}{1 \cancel{\text{gal}}} \cdot \frac{1 \cancel{\text{min}}}{60 \text{ s}} = \frac{12.00 \text{ L}}{60 \text{ s}} = 0.20 \; \frac{\text{L}}{\text{s}}$$

Learning Check 8.4

Perform each of the following unit conversions. Show all your work, writing the unit conversion factors side by side as shown in the examples. Always show the units of measure in your result.

	Convert this quantity	Into a quantity with these units
1.	12.5 in	cm
2.	358 g	kg
3.	4,500 s	hr
4.	7,700 gpm	L/s
5.	10.6 μm	m
6.	61.77 m	mm
7.	73 m³	L
8.	5.5 kg/min	g/s
9.	9.950 yr	s
10.	0.97450 m	μm
11.	0.5677 kg	mg
12.	205 MW	W

Answers

1. 31.75 cm	5. 0.0000106 m	10. 974,500 µm
2. 0.358 kg	6. 61,770 mm	11. 567,700 mg
3. 1.25 hr	7. 73,000 L	12. 205,000,000 W
4. 485.7 L/s	8. 91.7 g/s	
	9. 313,783,200 s	

8.5   Calculating Volume

Volume calculations are common in science and engineering. In this section, we review the procedure for calculating the volume of a rectangular solid and a cylinder. But first, let's look a bit more at the units of measure used for volume.

The official SI System derived unit for volume is the cubic meter (m^3), shown in Figure 8.13. A cubic meter is a rectangular volume, one meter on each side. As you can see, a cubic meter is a large volume. Figure 8.14 illustrates the cubic centimeter (cm^3), another common volume unit.

Figure 8.13. The white frame encloses a volume of one cubic meter.

The liter is not an official SI System unit, but is commonly used in science nevertheless. From Table 8.6, you can see that one liter (1 L) is equivalent to 1,000 cm^3. Figure 8.15 shows a volume of 1,000 cm^3 on the left, and 1 L on the right. These are equivalent volumes. Figure 8.16 shows a beaker containing 1 L of water. One liter is equal to 1,000 mL. It is also equal to 1,000 cm^3, and this means that the cubic centimeter and the milliliter are equivalent volumes. Compare the graduations of 1 mL on the graduated cylinder in the lower part of Figure 8.16, with the orange cube in Figure 8.17. These volumes are equivalent. And as Figure 8.17 suggests, 1 mL is equal to about 20–25 drops.

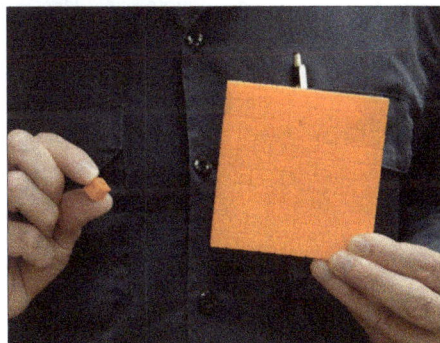

Figure 8.14. One cubic centimeter (left) and 100 cubic centimeters (right).

As one final point of comparison, placing the orange 1-L cube inside the cubic meter (Figure 8.18), illustrates the fact that 1 m^3 = 1,000 L. Since 1 L = 1,000 cm^3, one cubic meter is

One milliliter of liquid is about 20–25 drops.

Figure 8.15. Each of these objects has a volume of one liter.

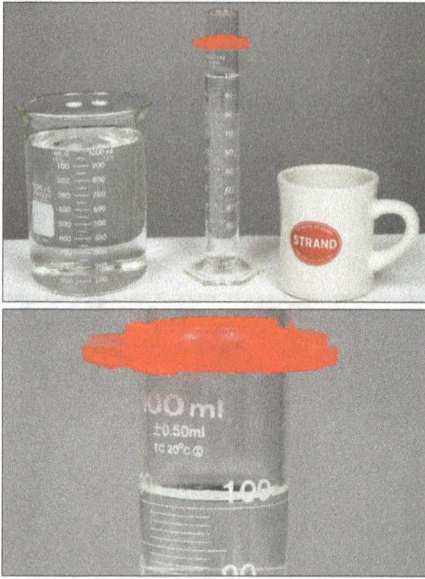

Figure 8.16. In the upper photo, 1,000 mL of water (1 L) in a 1-L beaker (left), 100 mL of water in a 100-mL graduated cylinder (center), and a coffee cup for comparison. In the lower photo, each line on the graduated cylinder is 1 mL.

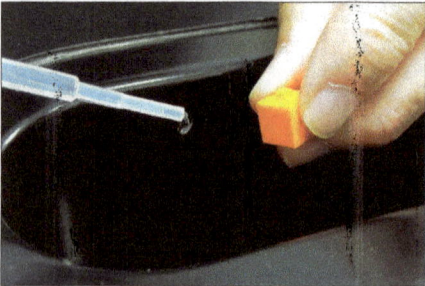

Figure 8.17. The orange cube is 1 cm³, equal to 1 mL. One mL is equal to about 20–25 drops.

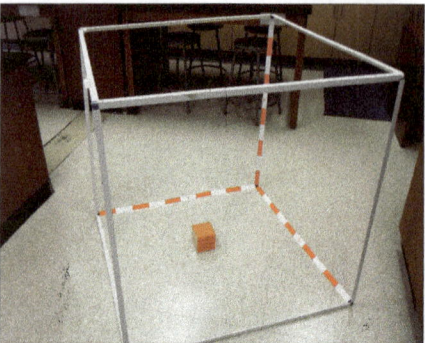

Figure 8.18. To fill the 1-m³ cube requires 1,000 1-L cubes, or 1,000,000 1-cm³ cubes.

equal to one million cubic centimeters, or $1 \text{ m}^3 = 1{,}000{,}000 \text{ cm}^3$. Now we will move on to calculating volume.

A solid made of rectangles and right angles is called a *right rectangular solid*. Calculating the volume of a right rectangular solid is easy. The object on the right in Figure 8.19 represents such a solid. On the left side of the figure, the base of the solid is shown separately. The area of this rectangular base is $A = L \cdot W$. The volume of the solid is the area of the base multiplied by the height, giving $V = L \cdot W \cdot H$. Here's an example calculation.

EXAMPLE 8.6

Calculate the volume of a right rectangular solid with dimensions $L = 25.75$ cm, $W = 15.25$ cm, and $H = 11.50$ cm.

The volume is calculated as $V = L \cdot W \cdot H$. Inserting the values into this formula gives

$$V = 25.75 \text{ cm} \cdot 15.25 \text{ cm} \cdot 11.50 \text{ cm}$$

$$V = 4{,}515.91 \text{ cm}^3$$

There is an important detail I want you to notice in this example. Recall from principle 1 on page 153 that when we handle units of measure in our mathematics, we treat them just like numbers. We can multiply them, divide them, and so on. In the example above, the computation involves multiplying the centimeter unit together three times, giving $\text{cm} \cdot \text{cm} \cdot \text{cm} = \text{cm}^3$. In case you have been wondering what "cubic" units like m^3 and cm^3 are all about, this is where these cubic units come from. Whenever a length unit such as centimeter, inch, meter, or foot is used to compute a volume, the units of measure get cubed in the calculation.

A *right circular cylinder* is a shape with a circular base, and a curving side that is at

Figure 8.19. The volume, V, of a right rectangular solid is the area of the base (A = L·W) multiplied by the height (H), giving V = L·W·H.

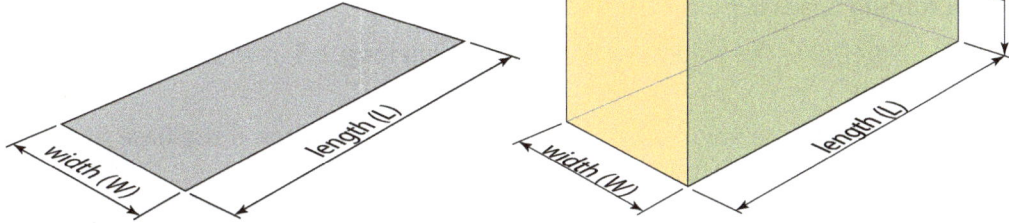

right angles with the base, as depicted in Figure 8.20. As with the rectangular solid, the volume of the cylinder is calculated by multiplying the area of the base by the height. The base of this cylinder is a circle, and the formula for the area of a circle is the number π (3.14159...) multiplied by the square of the circle's radius, or $A = \pi r^2$.

As you probably know, the number π (pi) is an irrational, non-repeating decimal. So to perform calculations with π you can either use an approximation using the first few digits of π, or if your calculator has a key for π you can use that. In science we use π so often that it is very helpful to have a scientific calculator with a π key.

Now we will work through an example calculation for the volume of a cylinder. In our previous example, the dimensions were in centimeters. This time we will use meters.

By the way, the radius of a cylinder is half the diameter. When you calculate the volume of a cylinder, always make sure you use the radius in your calculation. It is easy to use the diameter by mistake if you're not careful.

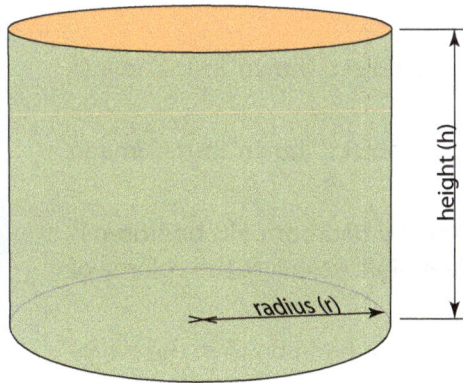

Figure 8.20. The volume, V, of a right circular cylinder is the area of the base (A = πr^2) multiplied by the height (h), giving V = $\pi r^2 h$.

EXAMPLE 8.7

Some water storage tanks are made in the shape of a right circular cylinder. Calculate the volume of such a water tank if the radius is 3.5 m and the height is 14.25 m.

We first calculate the area of the base of the cylinder, $A = \pi r^2$. Inserting the radius into this formula, we have

$A = \pi r^2 = \pi (3.5 \text{ m})^2 = 3.14159 \cdot 3.5 \text{ m} \cdot 3.5 \text{ m} = 38.48 \text{ m}^2$

Multiplying this area by the height gives us the volume.

$V = 38.48 \text{ m}^2 \cdot h = 38.48 \text{ m}^2 \cdot 14.25 \text{ m} = 548.3 \text{ m}^3$

Learning Check 8.5

1. Determine the volume of a shoe box 12 in long, 5.5 in wide, and 4.5 in high.

2. Calculate the volume of a rectangular shipping carton if the base has dimensions 3.10 ft by 3.25 ft and the height is 2.75 ft.

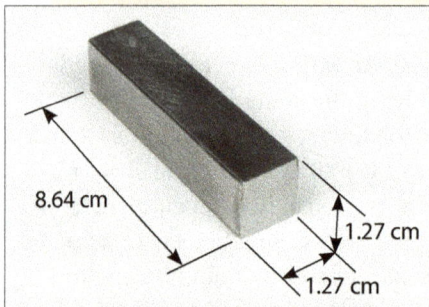

8.64 cm
1.27 cm
1.27 cm

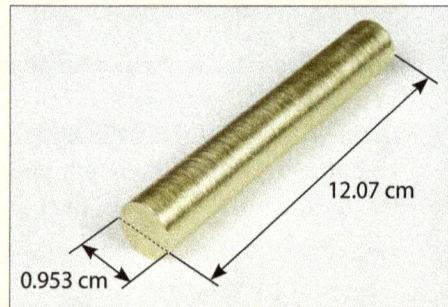

12.07 cm
0.953 cm

3. Calculate the volume of the rectangular object shown in the image above and to the left.

4. Determine the volume of the cylindrical object shown in the image above and to the right.

5. Little Rafael is fortunate to have his own tiny bedroom. His bedroom is 10.6 feet long, 8.44 feet wide, and 8.0 feet high. Calculate the volume of Little Rafael's bedroom.

6. Young Miss Toni has a cylindrical float she can hang on to in the swimming pool. Toni's float is 145 cm long and 15.24 cm in diameter. Determine the volume of Toni's float. Then convert your answer into liters.

Answers
1. 297 in³
2. 27.7 ft³
3. 13.9 cm³
4. 8.61 cm³
5. 715.7 ft³
6. 26.45 L

Chapter 8 Exercises

Answer each of the questions below as completely as you can. Write your responses in complete sentences.

1. Why is it important always to use horizontal bars in unit fractions when performing unit conversions?

2. Consider our discussion of theories and experiments in Chapter 7 and explain why making accurate measurements is important in science.
3. Briefly describe the origin of the SI System of units.
4. Explain what "base units" and "derived units" are, and give several examples of each from the SI System.
5. Perform each of the following unit conversions. Show all your work, writing the unit conversion factors side by side as shown in the examples. Always show the units of measure in your result.

	Convert this quantity	Into a quantity with these units
a.	35 mph	m/s
b.	2,300 mL	cm³
c.	345 mg	kg
d.	370,000 W	MW
e.	9.11 L	µL
f.	14.4 cm/s	m/s
g.	350 L	cm³
h.	62 g/hr	mg/s
i.	67.2 L	m³
j.	0.385 L/s	gpm
k.	745 mm	cm
l.	57,900 cm³	m³

6. A so-called "55-gallon drum" is 22.5 inches in diameter and 33.5 inches tall. Convert these dimensions into centimeters, and then determine the volume of the drum. Finally, convert your result to liters, and to gallons.

7. Consider the L-shaped rectangular figure below. Its dimensions are: A = 3.5 cm, B = 2.0 cm, C = 2.5 cm, D = 1.0 cm, and E = 1.0 cm. Determine the volume of this object. State your result in both cubic centimeters and liters.

Answers
5.

a.	15.64 m/s	g.	350,000 cm³
b.	2,300 cm³	h.	17.22 mg/s
c.	0.000345 kg	i.	0.0672 m³
d.	0.37 MW	j.	6.10 gpm
e.	9,110,000 µL	k.	74.5 cm
f.	0.144 m/s	l.	0.0579 m³

6. 218.3 L, 57.7 gal
7. 28 cm³, 0.028 L

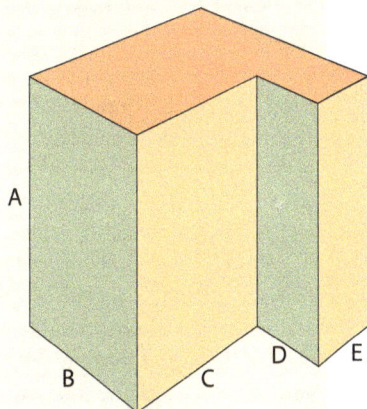

Experimental Investigation 5: Determining Volume

Overview

- Use length measurements to calculate the volumes of several different regularly-shaped objects, both cylindrical and rectangular.
- Measure the volumes of each of the same objects using a graduated cylinder and the displacement method.
- *The goal of this investigation is to develop expertise at calculating and measuring the volume of a regular solid object.*

Basic Materials List

- aluminum rod, 3/8 inch diameter × 3 inches long
- aluminum flat bar, 1/8 inch × 3/4 inch × 4 inches long
- aluminum angle, 1/8 in thick, 3/4 × 3/4 × 2.5 inches long
- brass rod, 1/2 inch diameter × 3.5 inches long
- carbon steel flat bar, 1/2 inch × 1/2 inch × 2 inches long
- graduated cylinder, 100 mL
- digital caliper (if possible)
- measuring rule, metric (if a caliper is not available)
- calculator

Safety Precautions

Use care when handling glassware. There are three ways to break glassware—carelessness, silliness, and improper procedures. These are all bad in a lab!

Scientific study often requires the calculation of an object's volume, or the measurement of volume in a lab. In this investigation, you use your calculator to calculate the volumes of five metal samples. Then you measure the volume of each one as a check on your accuracy.

As you learned in the preceding chapter, the volume of a right, regular solid is calculated by determining the area of the base, and then multiplying that by the height. For rectangles, the area of the base is just the product of the lengths of the two sides. For cylinders, the base is a circle with a certain radius (half the diameter) and an area of $A = \pi r^2$. For the sample of aluminum angle (the piece that has an L-shaped base), you can treat the base as two rectangles joined together. Just make sure you include the area of the joining corner in only one of those rectangles.

The best tool to use for making dimensional measurements on small, regular samples like these is a *caliper*. A digital caliper is easy to use, as the photograph shows, and reads out a very accurate and precise measurement directly in centimeters or inches. If you do not have access to a caliper, you can make your measurements with a measuring rule, but they will not be nearly as accurate.

First make all your measurements and record them in your lab journal in a well-organized table. Make all your measurements in centimeters, if possible. If you cannot make measurements in centimeters, then make them in decimal inches and convert each one to centimeters using the appropriate unit conversion factor. Using

dimensions in centimeters, calculate the volume of each of the six samples in cubic centimeters.

To check your calculations, we will compare them to measurements made using a graduated cylinder with the *displacement method*. First, fill the graduated cylinder about half full with water. Place the cylinder on a horizontal surface, and read the volume of the water from the scale on the glass. The measurement is in milliliters, which is equivalent to cubic centimeters. (There are some important details to attend to when reading volumes in a graduated cylinder. Please refer to Appendix A for this information.)

After taking the initial water reading, slide your sample gently down into the cylinder and take a second reading. The volume of the sample is the difference between the first and second reading. Measure and record the volume of each of the samples using this method.

When inserting a metal sample into a graduated cylinder, always tilt the cylinder over (without spilling the water), and slide the sample as gently as possible down the side of the cylinder to avoiding breaking the cylinder. When removing the sample, place your fingers loosely over the top of the cylinder to catch the sample, and tip the cylinder over a container or sink to let the water run out. Do not let the sample fall out, because it may damage the sink or become damaged itself.

To compare your calculations to your measurements with the graduated cylinder, we will compute the *percent difference*[1] for each of your six volumes. This calculation expresses the difference between a scientist's prediction and the experimental result as a percentage of the prediction. Neither of your volumes is really a prediction, since they are both based on laboratory measurements, but go ahead and use your graduated cylinder volume measurement as the predicted value and your calculated volume as the experimental value. Then calculate the percent difference as

$$\text{percent difference} = \frac{|\text{predicted value} - \text{experimental value}|}{\text{predicted value}} \times 100\%$$

The absolute value signs mean this percentage always comes out to be positive.

Analysis

Your percent difference values for each of the six samples should not be more than a few percent, maybe as high as 10%. If your difference ratio values are large, then you almost certainly have an error in your calculations or in your measurement procedures. You should find and correct this error and perform the measurements or calculations over again.

In your report, prepare a neat table listing the five samples, all your measurements (with the units of measure), your two volumes, and the percent difference for each.

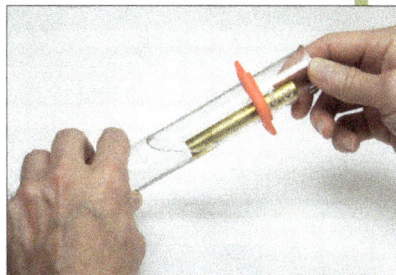

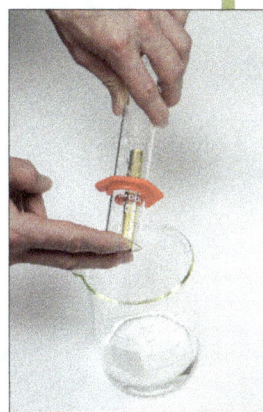

---

1   See Appendix B for a note on this terminology.

Copper's shiny luster, reddish color, and high electrical conductivity are very familiar physical properties. One of copper's chemical properties is that it reacts with the water, oxygen, and carbon dioxide in the air to form copper carbonate, an attractive green compound that is also called verdigris.

The cylindrical structure atop the Royal Observatory in Edinburgh, Scotland is made of copper. The original copper in the structure was installed in 1894, and the surface layer has turned to copper carbonate. New refurbished copper panels and trim were installed in 2010, and have yet to acquire the verdigris coating.

OBJECTIVES

After studying this chapter and completing the exercises, you should be able to do each of the following tasks, using supporting terms and principles as necessary.

1. Distinguish between physical and chemical properties.
2. Describe at least 12 different physical properties, and give examples of each.
3. State the freezing and boiling points of water on the Fahrenheit and Celsius temperature scales.
4. For gases, describe the relationship between volume and temperature, and between volume and pressure.
5. State the standard SI System units for temperature, pressure, and volume, and give examples of common reference values in standard SI System units and other common units of measure.
6. Describe the four phases of matter, including their key features.
7. Describe the two main phase transitions at the molecular level, including how the heat of fusion and heat of vaporization are involved.
8. Calculate the density of a substance.
9. State three different chemical properties, and give one or two examples of each.

VOCABULARY TERMS

You should be able to define or describe each of these terms in a complete sentence or paragraph.

1. absolute zero	13. electrical conductivity	25. phase diagram
2. barometer	14. evaporation	26. phase transition
3. barometric pressure	15. flammable	27. physical property
4. boiling point	16. hardness	28. plasma
5. brittleness	17. heat of fusion	29. pressure
6. chemical property	18. heat of vaporization	30. shear strength
7. combustion	19. incompressible	31. sublimation
8. compressive strength	20. inflammable	32. tensile strength
9. corrosion	21. luster	33. thermal conductivity
10. density	22. malleability	34. thermal properties
11. ductility	23. melting point	35. vapor
12. elasticity	24. oxidation	36. vaporization

9.1   Physical Properties

A common way to describe substances is by identifying their *physical properties*. The physical properties of a substance cover every kind of description except the substance's chemical behavior. Physical properties describe simple things like a substance's shape, color, density, and luster (shininess). But there are many physical properties that involve more technical definitions. Let's look at some examples of these.

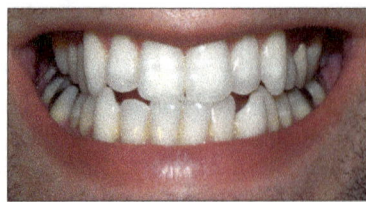

Figure 9.1. The enamel on our teeth is the hardest substance in the human body.

Hardness is an important property for many materials. The enamel on human teeth is the hardest substance in the human body (Figure 9.1). Tooth enamel is harder than steel, which is why the dentist can scrape your teeth with a sharp metal instrument without scratching your teeth. Fingernails and graphite are examples of materials with low hardness values. Glass is also harder than steel, and diamond is one of the hardest substances known. Hard substances are also sometimes brittle. *Brittleness* is a property related to how easily a substance shatters or breaks. Glass is harder than steel, but it is much more brittle.

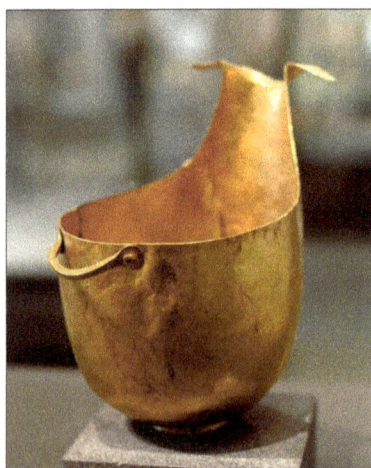

Figure 9.2 Ancient sauce boat made of hammered gold, now in the Louvre, in Paris.

Malleability and *ductility* are properties that may be described as the opposite of brittleness. Metals are well known for both malleability and ductility. If a substance is malleable, it can be hammered into flat sheets without cracking or breaking. Gold is famous for its malleability, and people have been hammering gold into objects for thousands of years. Figure 9.2 shows a vessel from 2200 BCE made out of hammered gold. A ductile substance can be formed into a thin wire by drawing it through a small hole. Many metals are ductile, and wires made of copper, aluminum, steel, and gold have been common for over a century.

Strength and *elasticity* are two physical properties that are important in construction materials. Strength may be measured in several different ways, illustrated in Figure 9.3. *Tensile strength* describes strength under stretching forces, and *compressive strength* describes strength under forces that tend to crush an object. *Shear strength* describes strength against forces that try to shear an object in two (as scissors do). Elasticity describes a material's ability to stretch without snapping or becoming permanently deformed.

Steel has very high tensile strength. And though it may not seem like it, steel is highly elastic as well. For these reasons, steel is widely used as a construction material in buildings and in bridges like

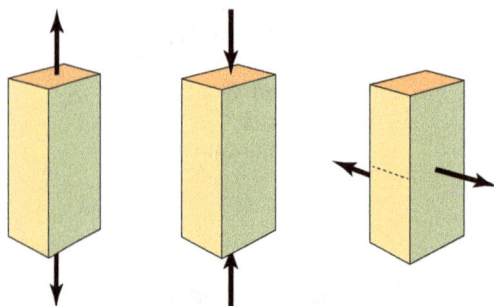

Figure 9.3. Tensile forces (left), compressive forces (center), and shear forces (right).

the one shown in Figure 9.4. The high strength allows steel to support very large loads. But because of its high elasticity, steel flexes under stress instead of breaking. Concrete does not rate very high for tensile strength, nor for elasticity, but it is king of construction for its compressive strength. This makes concrete the preferred material worldwide for building foundations.

Figure 9.4. Steel is widely used as a construction material because of its high tensile strength and elasticity.

Electrical conductivity is the ability of a substance to conduct electricity, a property usually associated with metals. Back in Chapter 3, you studied conduction of heat. The ability of a solid substance to conduct heat is the *thermal conductivity*. High thermal conductivity is also associated with metals. A comparison of electrical and thermal conductivities of some metals is shown in Figure 9.5. In this chart, the values for copper are set at 100 so that copper may be used as a reference for comparison. As you can see, the electrical conductivity generally tracks with the thermal conductivity. Copper and aluminum both have relatively high electrical conductivity and are relatively inexpensive metals. This is why these metals are universally used for electrical wiring. Gold has

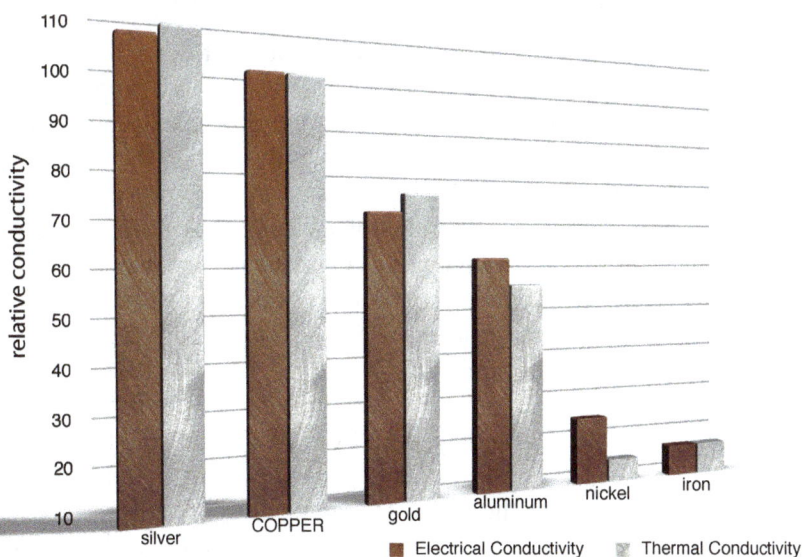

Figure 9.5. Relative electrical and thermal conductivities for some metals.

the advantage that it does not tarnish like silver, copper, and aluminum do, but its cost restricts its use to very specialized types of electrical circuits.

The list of physical properties goes on and on. But there are three more physical properties that need to be mentioned here because they come up all the time in physics and chemistry. The *melting point* of a substance is the temperature at which the substance melts or freezes, changing from solid to liquid or vice versa. Similarly, the *boiling point* is the temperature at which a substance vaporizes (boils) to change from a liquid to a vapor. The boiling point is also the temperature at which the substance condenses to change from a vapor to a liquid. Another physical property is *density*. The density of a substance is a measure of how much matter is packed into a given amount of space. We look at these three properties more closely in coming sections.

Learning Check 9.1

1. Make a chart listing each of the 16 physical properties mentioned in this section. Then write down two examples illustrating each property from your own experience of common substances.
2. Describe some physical properties that architects and engineers involved in construction of new buildings and roads should know about.
3. When concrete is used for foundations and roads, it nearly always has steel rods embedded in it. These rods are called reinforcing bars, or *rebar*, for short. See if you can use the properties of steel and concrete to explain why concrete nearly always has rebar in it.

9.2   Temperature, Pressure, and Volume

Properties such as thermal conductivity that describe the way substances behave when they are heated are often classed as *thermal properties*. Another example of a thermal property is the amount a substance expands or contracts as it is heated or cooled.

As you recall from Chapter 3, when a substance is heated, the internal energy of the substance increases. The internal energy is the total kinetic energy of all the atoms in the substance. The three variables most commonly used to keep track of the internal energy in a substance are the temperature, pressure, and volume.

Thermal properties describe the way substances behave when they are heated.

Temperature   You may recall from Section 3.5 that the temperature of a substance is related to its internal energy. The more internal energy a substance has, the higher the temperature is.

The main thing we will discuss about temperature is the different scales we use to measure it. There are many different temperature scales, but three are most important. In the U.S., we are most familiar with the Fahrenheit scale for measuring temperature because we use it every day when talking about the weather or cooking food. The rest of the world uses the Celsius scale. In scientific work, both the Celsius and Kelvin scales are commonly used.

Reference values for some temperatures are shown on all three of these scales in Figure 9.6. The important values for you to know right now are the boiling and freezing points of water on the Fahrenheit and Celsius scales. In later science courses, you will learn how to convert temperature values from one scale to another. This is often necessary in physics and chemistry. The Kelvin scale must be used for calculations because it is an absolute scale, with no negative values, but thermometers are typically scaled in °F or °C.

The kelvin (K) is one of the seven SI System base units. The other temperature scales use the term "degrees." But in 1967, this term was dropped from the Kelvin scale so that kelvin units would be spoken the same way as other units of measure. So now we say that water boils at 373.2 kelvins.

	Fahrenheit Scale	Celsius Scale	Kelvin Scale
typical baking temperature	350°F	176.7°C	450.2 K
water boils/condenses	212°F	100°C	373.2 K
alcohol boils/condenses	172°F	78°C	351.2 K
water freezes/melts	32°F	0°C	273.2 K
mercury freezes/melts	−38°F	−39°C	234.2 K
dry ice (CO_2) freezes	−109°F	−79°C	194.2 K
nitrogen boils/condenses	−320°F	−196°C	77.2 K
absolute zero	−459.7°F	−273.2°C	0 K

Figure 9.6. Reference values on three temperature scales.

As I mentioned in Chapter 8, a temperature change of one degree on the Celsius scale is the same as a one degree change on the Kelvin scale. These two scales use the same size of degree, and the only difference between them is where zero is on the scale. One degree on the Fahrenheit scale is about half (5/9) a degree on the Celsius scale.

Notice the temperature at the bottom of the scales, labeled *absolute zero*. You have learned that the temperature of a substance decreases as the internal energy decreases. You have also learned that the internal energy is based on the kinetic energies of the atoms in the substance. If you put these facts together, you can see that as a substance is cooled, the atoms or molecules in the substance move more and more slowly, and the temperature of the substance drops lower and lower. Absolute zero is the temperature at which atomic or molecular motion would cease altogether. And since nothing can go any slower than standing still, absolute zero is the lowest temperature.

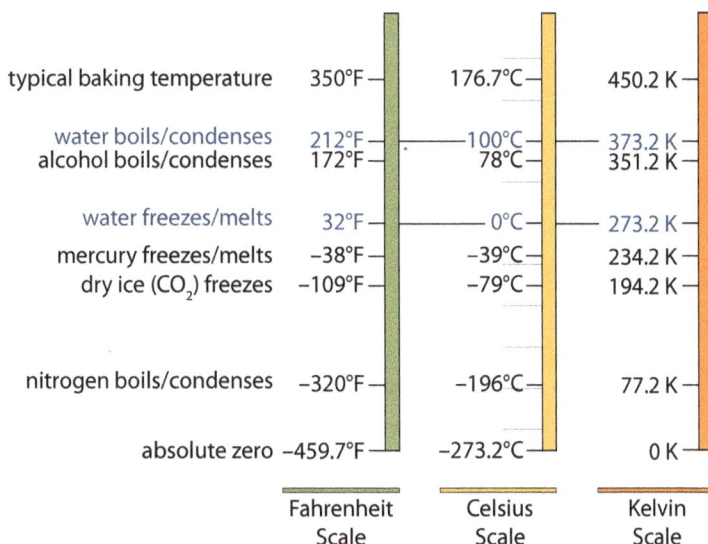

Absolute zero is understood to be the temperature at which all atomic motion would cease.

In fact, nothing in the universe can ever actually *be* at absolute zero. Even in deep outer space the temperature is not zero, although it is close. The temperature in outer space is 2.7 K, illustrated in Figure 9.7. (I threw in some other interesting information in the caption for this image, just for fun.) This temperature is because of the Cosmic Microwave Background radiation mentioned in Chapter 7.

Figure 9.7. The temperature in outer space is 2.7 K because of the Cosmic Microwave Background radiation. This Hubble Space Telescope image is of a star in the Milky Way Galaxy called V838 Mon. In 2002, the outer surface of V838 Mon suddenly expanded, and V838 Mon became the brightest star in the galaxy. Then just as suddenly, the brightness faded. Such an event had never been witnessed before, and scientists do not know what caused it.

Pressure   At the molecular level, pressure is caused by the collisions of molecules in fluids (liquids and gases). Higher internal energy means faster moving molecules, resulting in harder collisions and higher pressure.

There are several very familiar examples of pressure. The first is the pressure of the atmosphere. You can think of atmospheric pressure as being caused by the weight of the gases in the atmosphere. These gases are pulled toward the earth by gravity. As you recall, air consists mostly of molecules of oxygen and nitrogen. All these molecules are constantly zooming around at high speed, colliding with each other and with everything else exposed to the air. The combined effect of all these collisions results in the pressure in the atmosphere.

Pressure is caused by molecular collisions.

Another well-known example of pressure is pressure in gases. We use air compressors to compress air molecules together. Then we use the compressed air to inflate car tires and sports balls. Painters use compressed air to turn liquid paint into spray paint. Welders use compressed gases for welding and cutting of metal. Other gaseous substances in spray cans include spray paint, WD-40, and hair spray.

A third example of pressure is the pressure that makes liquids flow. Cities use large electric pumps to pressurize water so it will flow in pipes to all the buildings in the city. If you recall from Chapter 2, this is typically done by pumping the water up into a water tower. The weight of the water then creates the pressure in the pipes that allows water to flow to where it is needed. The weight of water is also the cause of pressure under water.

For any fluid to flow, a difference in pressure must be present. A fluid will flow toward the lower pressure.

For any liquid or gas to flow in a pipe, there must be a difference in pressure. A higher pressure is caused by more energetic collisions of molecules and forces a fluid to flow toward where the pressure is lower.

1 newton (1 N) of force

1 pascal (1 Pa) of pressure

1 m

1 m

Area = 1 m²

Figure 9.8. One pascal of pressure is one newton of force per square meter.

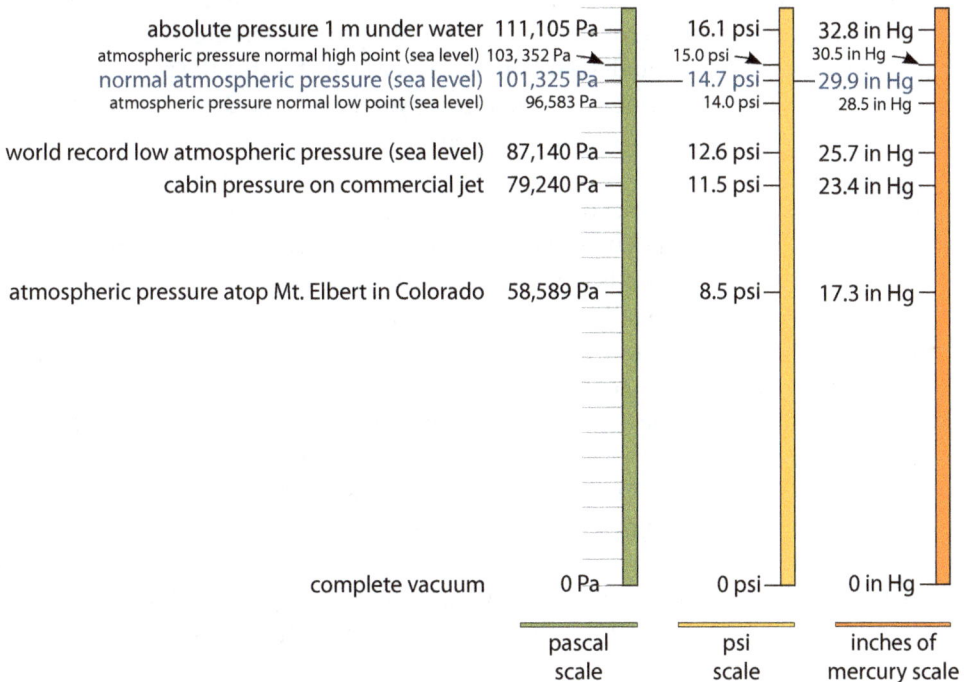

The SI System unit for pressure is the pascal (Pa). To illustrate how the pascal is defined, Figure 9.8 depicts a chamber with a freely sliding side wall. The sliding wall is one meter square, so its area is 1 m². Imagine that the chamber contains ordinary air, and then a force of one newton (1 N) is applied to the sliding wall. The wall slides in until the force of the air pushing back on the wall also equals 1 N. This force of 1 N applied to an area of 1 m² results in an air pressure inside the chamber of 1 Pa.

As with temperature, there are many different pressure scales in use. Three of these are illustrated in Figure 9.9. The pascal scale is on the left. You can see that a pressure of 1 Pa is not much pressure, because normal atmospheric pressure at sea level is 101,325 Pa. In the U.S., the pounds per square inch (psi) scale is more com-

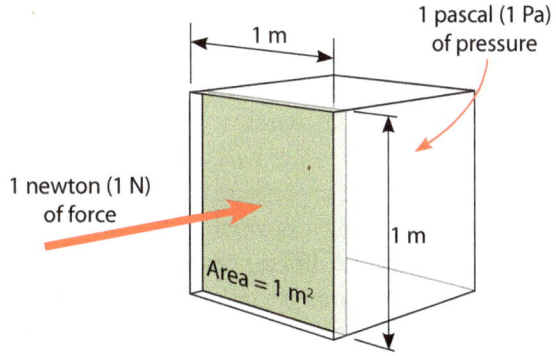

	pascal scale	psi scale	inches of mercury scale
absolute pressure 1 m under water	111,105 Pa	16.1 psi	32.8 in Hg
atmospheric pressure normal high point (sea level)	103,352 Pa	15.0 psi	30.5 in Hg
normal atmospheric pressure (sea level)	101,325 Pa	14.7 psi	29.9 in Hg
atmospheric pressure normal low point (sea level)	96,583 Pa	14.0 psi	28.5 in Hg
world record low atmospheric pressure (sea level)	87,140 Pa	12.6 psi	25.7 in Hg
cabin pressure on commercial jet	79,240 Pa	11.5 psi	23.4 in Hg
atmospheric pressure atop Mt. Elbert in Colorado	58,589 Pa	8.5 psi	17.3 in Hg
complete vacuum	0 Pa	0 psi	0 in Hg

Figure 9.9. Reference values on three pressure scales.

monly used. Weather reporters like to use a scale based on "inches of mercury." The pressure 29.9 inches beneath the surface in a column of mercury (chemical symbol, Hg) is equivalent to atmospheric pressure. (This assumes that there is a vacuum at the surface of the mercury. Otherwise, the pressure under the surface would be the atmospheric pressure *plus* the liquid pressure.)

Atmospheric pressure varies with the weather from about 2% above normal to about 5% below normal. These variations are the cause of wind, since, as I mentioned, pressure differences give rise to flowing fluid. This is what wind is—flowing air. Low pressures occur when the wind is moving rapidly, so when atmospheric pressure drops, it means stormy weather is headed your way. Atmospheric pressure is measured with a *barometer*, so weather reporters often refer to the value of atmospheric pressure as the *barometric pressure*.

A few other points are shown on the scale for reference. At the bottom is zero pressure, which is a complete *vacuum*. Remember, pressure is caused by moving molecules of liquid or gas colliding with one another or with the walls of the container they are in. If all the air molecules are removed from a container, the air pressure inside is zero. I have also shown the atmospheric pressure at a few other reference locations to help you get a feel for the range of pressures in our every day experience. You can see that the pressure at the top of Mount Elbert, the highest mountain in Colorado, is just over half the pressure at sea level. Cabin pressure in an aircraft is quite a bit below normal atmospheric pressure. This is why your ears pop during a landing. The pressure is coming back up to normal.

Finally, notice that only 1 m under the surface of water the pressure has already gone up nearly 9% from normal atmospheric pressure. We live at the bottom of a sea of air, and the weight of the air causes atmospheric pressure. But water is a lot denser than air, so as you descend under the surface of water, the pressure goes up rapidly. But as another point of comparison, you have to go 34 feet under water to get the same pressure that only 29.9 inches of mercury gives.

Volume As you know, volume is a measure of how much space an object takes up. Recall that our definition for matter is anything that has mass and takes up space, so volume is a property all matter possesses. The previous chapter addresses volume units and how to calculate volumes. Here we just need to mention a few details about how volume relates to temperature and pressure.

Solids and liquids are generally treated as *incompressible* substances. This means that they cannot be squeezed down to smaller volumes by increasing the pressure the way gases can. So when we deal with solids and liquids, we are typically dealing with fixed volumes. But when we are dealing with gases, the volume of the gas depends on both the temperature and pressure.

In a gas, the volume and the temperature are directly proportional. This means that when the temperature of a gas goes up, the volume goes up. You may already know this just from watching how balloons swell up in hot air and shrink down in cooler air.

Scientists, Experiments, and Technology

In countries where electricity is not as widely available as it is in the U.S., many communities still depend on hand water pumps to pump well water to the surface. The first image to the left shows a hand water pump in Malaysia.

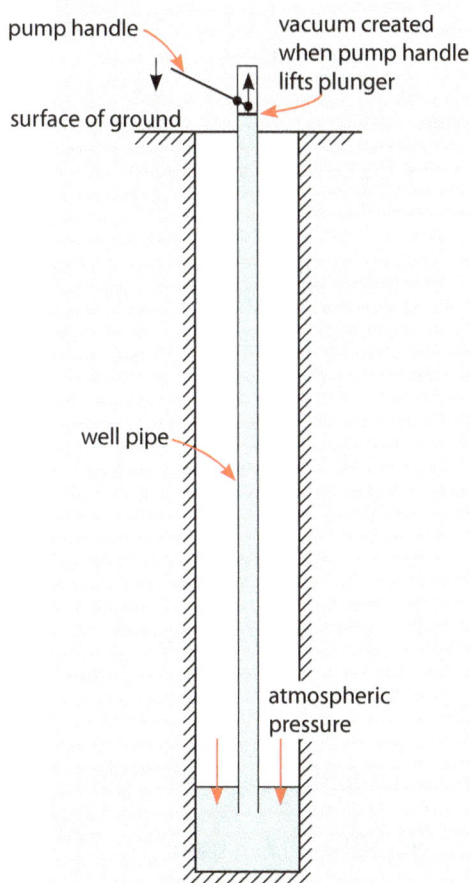

The diagram below shows how the water pump works. Atmospheric pressure down in the well is pushing on the surface of the water in the well. The pump handle is mounted on a pivot, and as the pump handle is pressed down, the other end of the handle raises a plunger inside the sealed water pipe. As the plunger is raised, it creates a partial vacuum on top of the water column, and the atmospheric pressure on the water in the well pushes the water up the pipe toward the vacuum.

The pressure at the bottom of the water column from the weight of the water is equal to the atmospheric pressure. This means that even with a total vacuum on top of the water column, atmospheric pressure can only push the water 34 feet up the pipe, because the pressure at the bottom of a 34-foot column of water (with a total vacuum on top) is the same as atmospheric pressure. If you need to get water from a well deeper than this, you either have to dig a big hole and climb down to it, or use an electric pump.

If you are interested in how the water is released from the top of the pipe where the pump is, go to commons.wikimedia.org and search for the file named water_pump.gif. The animation shows the details of the pump mechanism.

pump handle

vacuum created when pump handle lifts plunger

surface of ground

well pipe

atmospheric pressure

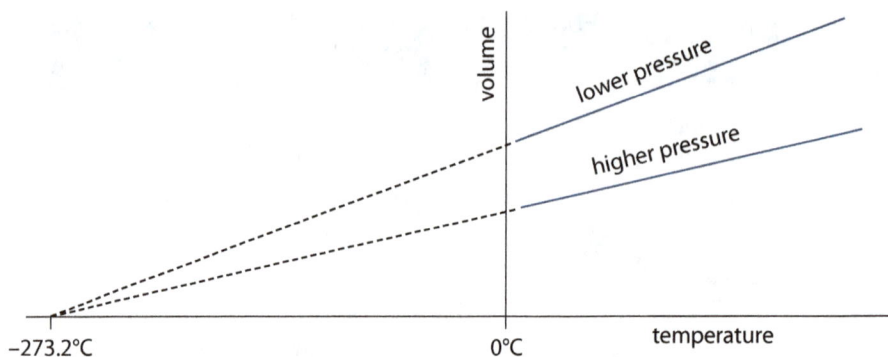

Figure 9.10. The relationship between the volume and temperature of a gas.

−273.2°C   0°C   temperature

The relationship between volume and temperature in a gas is depicted in Figure 9.10. The two blue lines in the graph represent a gas at two different pressures. In either case, as the temperature increases, the volume increases as well. If you were to collect data like this and plot it on a graph, you would find that tracing the curves backward reveals that they meet at a temperature of −273.2°C. This temperature is absolute zero.

At a given temperature, as the pressure on a gas increases, the volume of the gas decreases. This is called an *inverse* proportion, depicted in Figure 9.11. In this graph the two blue curves represent a gas at two different temperatures. In each case, as the pressure increases, the volume of the gas decreases.

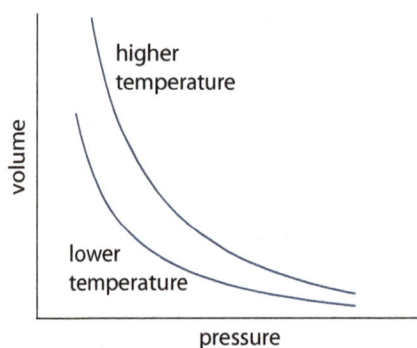

Figure 9.11. The relationship between the volume and pressure of a gas.

Learning Check 9.2

1. What is the physical meaning of the temperature known as absolute zero?
2. Why is the Kelvin temperature scale needed in scientific study?
3. What causes fluids to flow?
4. What is an incompressible substance, and what are some examples of incompressible substances?
5. Explain why sailors would pay close attention to the barometer on board a ship.
6. Describe an everyday example illustrating the relationship between the volume and temperature of a gas.
7. Describe an everyday example illustrating the relationship between the volume and pressure of a gas.

9.3   Phases of Matter

Most substances can exist as a *solid*, a *liquid* or a *gas*. These three different states are usually called *phases* by physicists, although in the field of chemistry they are often called *states*. There is also a fourth phase called a *plasma*. As we will see, plasmas do occur in every day life. But we are not able to handle and touch them the way we can substances in the other three phases.

The way to understand solids, liquids, and gases is to think about what is going on at the atomic or molecular level. The key concept is the internal energy a substance has. As you know, the internal energy is the sum of the kinetic energies in the particles (atoms or molecules). At low temperatures, these energies are low, and electrical attractions between protons and electrons, or between positive and negative regions of polar molecules, hold the particles in place, making the substance a solid. This situation is illustrated in Figure 9.12. The particles vibrate and jostle against each other, but the electrical attractions are so strong the particles do not have enough kinetic energy to move away from each other. As a result, the following features hold for solids:

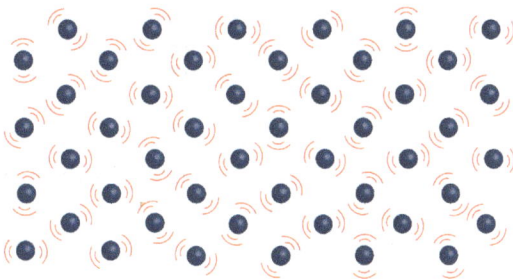

Figure 9.12. In a solid, the atoms' energies are low enough that electrical attractions hold them in place.

1. Solids are rigid and have a definite shape.
2. Solids have a definite volume.

Features of solids:

1. definite shape
2. definite volume

As a substance is heated, its temperature goes up, and so does the internal energy. With higher kinetic energies, the particles vibrate more vigorously. As the temperature increases, the particles eventually reach the point where they are energetic enough to stay apart from each other *at that same temperature* if they had a boost of energy to break them apart. This temperature is the melting point.

Figure 9.13 illustrates what happens with the energy and temperature in a substance during the *phase transition* from solid to liquid (*melting*) or vice versa (*freezing*). At the left end of the curve, the substance is a solid. As the substance is heated, its internal energy and temperature increase so the curve slopes up, indicating that increased internal energy in the substance correlates to increased temperature. When the temperature reaches the melting point (the dashed line), the particles have enough energy to stay apart at that same temperature if they are given an extra boost of energy to shake them loose from each other. This quantity of energy is called the *heat of fusion*.

At the melting point, as heat is added to the substance during melting, the energy allows one molecule after another to break free of the solid group. This is what melting is, and during melting the temperature of the substance stays the same.

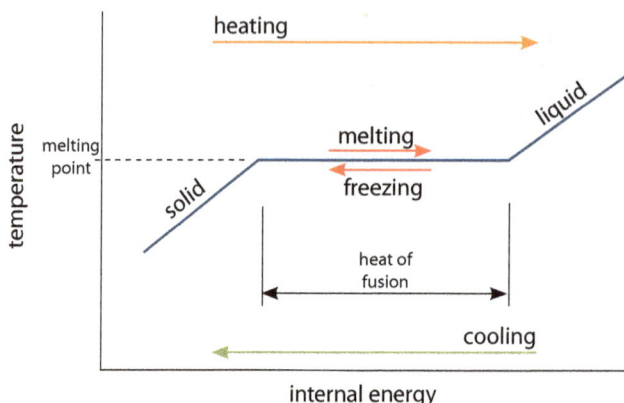

Figure 9.13. When a solid reaches the melting point, it melts completely only when an amount of energy equal to the heat of fusion for that substance is added. While this energy is being added, the temperature remains the same.

The steady temperature is indicated by the horizontal section of the curve in the figure. For water, the temperature where this happens is 0°C, and this is why ice water is always 0°C. Adding more heat allows more water molecules to break free from the crystal lattice of the ice, but it does not raise the temperature. After the full heat energy equal to the heat of fusion is added to the substance, the substance is a liquid. If more heat is added after this, the temperature goes up, as shown by the sloping part of the curve at the right end.

The diagram in Figure 9.13 is called a *phase diagram*. We can read the diagram in reverse as well. When a substance is cooled, the temperature drops until the *freezing point* is reached (the same temperature as the melting point). As more heat is removed, the liquid particles begin to cling to one another because they do not have enough kinetic energy to stay apart any more, and the electrical attractions pull them together into the solid. After the heat of fusion energy is removed, the entire substance is a solid. Further cooling after that drops the temperature below the freezing point.

Figure 9.14 illustrates the particles in a liquid. The particles are free from the structure they would have as a solid because they have high enough kinetic energies to overcome the electrical attractions between the particles. This means they are free to move around. However, notice that it appears there is still a tendency for the particles to cling together just a little bit. This is why particles in a liquid stay together in a bowl or in drops. The particles are not yet energetic enough to completely shake off the electrical attractions. As a result, the following features hold for liquids:

1. Liquids are fluid and have no definite shape; they take the shape of the container.
2. Liquids have a definite volume.

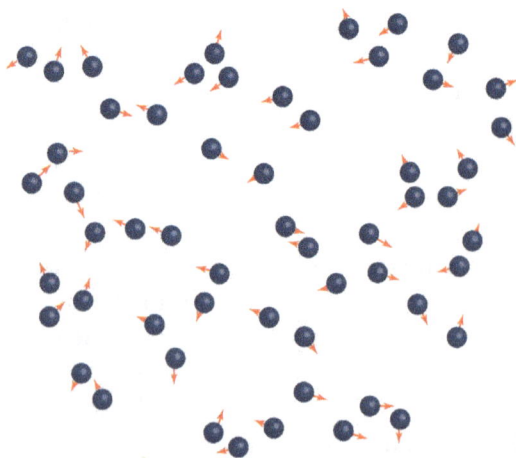

Figure 9.14. In a liquid, the particle energy is high enough that the particles break free from electrical attractions, but they are still loosely bound together.

Figure 9.15 shows another phase diagram, this time illustrating the phase transition from liquid to vapor or gas (*boiling* or *vaporization*) or vice versa (*condensation*). This diagram is a lot like the previous one, and the logic behind it is the same. For the particles to break completely free from one another so they are not clinging to one another at all, an amount of energy called the *heat of vaporization* is required. During boiling, the temperature stays the same until the entire volume of the substance has vaporized. For water, the temperature where this happens is 100°C.

Figure 9.16 illustrates the particles in a gas. They are completely free of one another because the energies the particles possess are so high that the particles just zoom around, bouncing off each other like high-speed billiard balls. Because of their high energies, they spread out more, as the figure suggests. In case you are wondering, in air at room temperature, typical molecules are moving at about 1,100 miles per hour!

We can summarize the features of gases this way:

1. Gases are fluid and have no definite shape; they take the shape of the container.

Features of liquids:

1. no definite shape

2. definite volume

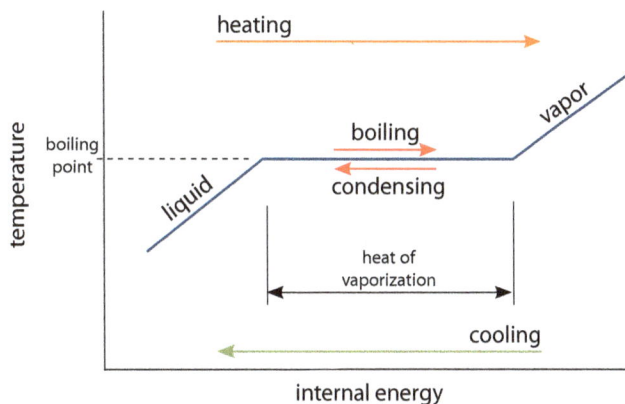

Figure 9.15. When a liquid reaches the boiling point, it vaporizes completely only when an amount of energy equal to the heat of vaporization for that substance is added. While this energy is being added, the temperature remains the same.

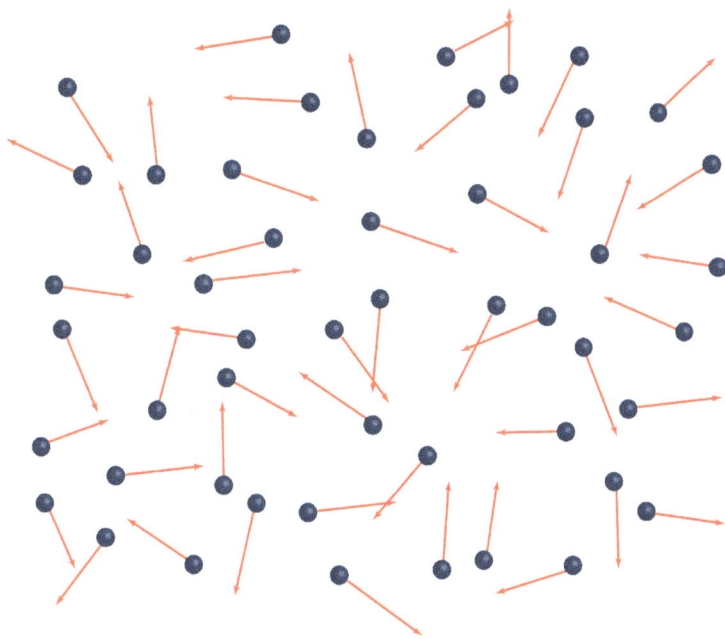

Figure 9.16. In a gas or vapor, the particles are moving extremely fast, colliding with each other and with the walls of the container (if there is one).

2. Gases have no definite volume; they fill the container.

The fourth phase of matter is called *plasma*. A plasma is an ionized gas. Hopefully, you recall from Chapter 6 that a charged atom is called an *ion*. In the presence of a very strong electric field, atoms can be stripped of some of their electrons and become ionized. If this happens in a gas, a plasma is formed.

The electric fields present in thunderstorms can produce a plasma—the bright light of a lightning bolt is the light given off by ionized air molecules, a common plasma. Neon signs, illustrated in Figure 9.17, also contain a plasma. The electricity connected to a glass tube of gas (such as neon) ionizes the gas, producing a plasma that gives off light. Plasmas are gases, and their key features are the same as those for gases. But note also that since plasmas are made of ions, they are also electrically conductive.

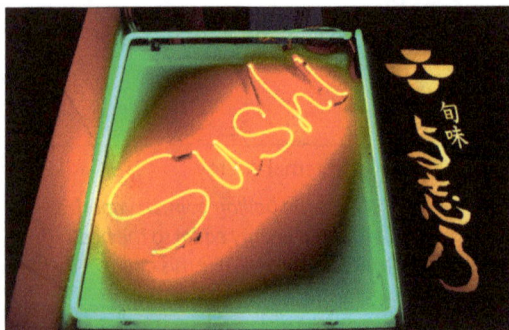

Figure 9.17. The tubes in neon lights contain a plasma that gives off light.

We close this section with a few final notes. First, the terms *gas* and *vapor* are synonymous. We use the term gas for substances like nitrogen and oxygen that are gases at normal temperatures. We use the term vapor for substances like water and mercury that are liquids at normal temperatures. When one of these is boiled, we call the vaporized substance a vapor.

Second, the quantities called the heat of fusion and the heat of vaporization depend on the electrical attractions that exist between the particles of a substance. This means these values are different for every substance. Some representative values are shown in Table 9.1. One kilojoule (1 kJ) is the amount of energy consumed by a standard 60-watt light bulb in 17 seconds. It takes a lot of heat to melt one kilogram of water—more than required for iron or copper. But when it comes to vaporization, the metals leave water in the dust for energy required.

Third, you may be wondering how evaporation occurs. Why does water evaporate at room temperature? The short answer is that in any glass of water molecules, some molecules are moving faster than the average speed, and some are moving slower. Sometimes a molecule at the surface happens to pick up enough energy from collisions with other particles to kick it up into the vapor state, and when that happens, that molecule goes bye-bye. When it

	Heat of Fusion (kJ/kg)	Heat of Vaporization (kJ/kg)
water	334	2,260
iron	247	6,090
copper	209	4,730

Table 9.1. Example values for heat of fusion and heat of vaporization.

leaves, it takes its energy with it, leaving the remaining molecules a bit cooler. This is why evaporation cools things. Your body uses this principle to help keep itself cool. When you sweat, the sweat evaporates, taking its energy with it and leaving you cooler.

Finally, some substances can change phase directly from solid to gas when heated. Solid carbon dioxide, or dry ice, is famous for this, as illustrated in Figure 9.18. This phase change is called *subli-mation*. Sublimation is different from evaporation. Sublimation is an ordinary phase change that occurs from heating dry ice up to the temperature where it changes phase; then it transitions direct-ly to the gas phase. Evaporation occurs well below the boiling point.

Figure 9.18. In water, sublimating dry ice produces a lovely fog. The fog is produced because the cold CO_2 condenses water vapor in the air above the bowl.

Learning Check 9.3

1. Describe the four phases of matter, including their key features.
2. Explain how melting and vaporization occur.
3. In your own words, explain what the heat of fusion and the heat of va-porization in a substance represent.
4. Explain the difference between evaporation and sublimation.

9.4   Calculating Density

I have mentioned density a few times already. Density is a physical property of all substances and represents the amount of matter contained in a certain volume. In this section, we will look at how to calculate density.

We will let D represent the density of a substance. We will also let m represent the mass of an object made from the substance, and V represent the object's volume. The density of a substance is calculated as

$$D = \frac{m}{V}$$ Density equation

From this equation, you can see that for a given volume, greater mass means greater density. This is like the relationship between volume and temperature we looked at in Section 9.2—a direct proportion. Also, for a given mass, greater vol-ume means lower density. This is an inverse proportion, just like the relationship between volume and pressure.

Now we will look at a few example problems. But before we do, we need to establish correct methods for solving problems. I cannot emphasize enough how important it is to use correct methods, and to develop the good habit of using the correct methods every time you work a problem. Even old dogs like me still use these methods, and I have been working physics problems for over 40 years! I have found that short cuts nearly always get you into trouble. So now is the time to learn the right method—and to use it.

We Pause Here to Talk About **Problem Solving Methods**

When solving problems in the physical sciences, it is *very* important to develop good problem solving habits. Even if a problem seems easy, it is important that you use the right methods every time. Below are the steps for the most reliable method for solving problems in physics. I will use these steps in every example problem we do from here on out.

Step 1

Write down the given information in a column down the left side of your paper. Also write down the unknown you are required to determine. Use standard variable symbols for all quantities.

Step 2

Perform any unit conversions the problem requires. Do the conversions by multiplying conversion factors to the right of your given information. For now, use this rule: look at the units the problem requires for your answer, and convert your given quantities into units that match.

Step 3

Write down the equation you need to use to solve the problem.

Step 4

Insert the values into the equation, including the units of measure.

Step 5

Use your calculator to compute the result, and write down the result with the correct units of measure.

Step 6

Double check your work to make sure you didn't make a silly mistake. Then look at your result and ask yourself if it makes sense.

In the accompanying box are the six steps to follow when solving problems in this course. Read them over and then return here. I will illustrate these six steps with our first example.

EXAMPLE 9.1

The label on a jug of antifreeze indicates that the jug contains one gallon of fluid. If the mass of the liquid in the jug is 4.215 kg, determine the density of the antifreeze. State your result in g/cm³.

We begin by writing down the given information, using symbols for the quantities. Leave enough space for the unit conversions you need to perform.

$V = 1\,\text{gal}$

$m = 4.215\,\text{kg}$

$D = ?$

The problem requires grams and cubic centimeters in our answer, so perform these unit conversions right out to the side of where you wrote down the given information.

$$V = 1\,\cancel{\text{gal}} \cdot \frac{3.785\,\cancel{\text{L}}}{1\,\cancel{\text{gal}}} \cdot \frac{1,000\,\text{cm}^3}{1\,\cancel{\text{L}}} = 3,785\,\text{cm}^3$$

$$m = 4.215\,\cancel{\text{kg}} \cdot \frac{1,000\,\text{g}}{1\,\cancel{\text{kg}}} = 4,215\,\text{g}$$

$D = ?$

The third step is to write down the equation.

$$D = \frac{m}{V}$$

Next, work to the right to insert the values into this equation. Show the units with each value.

$$D = \frac{m}{V} = \frac{4,215\,\text{g}}{3,785\,\text{cm}^3}$$

Now we compute the result and write it down with the appropriate units of measure.

$$D = \frac{m}{V} = \frac{4{,}215\ g}{3{,}785\ cm^3} = 1.114\ \frac{g}{cm^3}$$

Our final step is to check our work and to see if this makes sense. The units we have are the units required by the question. In the fraction we computed, the numerator was just a bit larger than the denominator, so we should expect to get a result a bit larger than 1.

In our next example, we will encounter some different units of measure, but the problem solving procedure is the same.

EXAMPLE 9.2

A sample of a chemical has a mass of 2.33 g and a volume of 1.67 mL. Determine the density of this chemical, and state the result in kg/L.

The given information is

$V = 1.67\ mL$

$m = 2.33\ g$

$D = ?$

Next, perform the unit conversions out to the right.

$$V = 1.67\ \cancel{mL} \cdot \frac{1\ L}{1{,}000\ \cancel{mL}} = 0.00167\ L$$

$$m = 2.33\ \cancel{g} \cdot \frac{1\ kg}{1{,}000\ \cancel{g}} = 0.00233\ kg$$

$D = ?$

Next, write the equation, insert the values with their units, and compute the result.

$$D = \frac{m}{V} = \frac{0.00233\ kg}{0.00167\ L} = 1.4\ \frac{kg}{L}$$

Finally, we check our work. The units are correct: kg/L, as required. The numerator is somewhat larger than the denominator, but not twice as large, so we expect a result between 1 and 2.

Our third example is a two-part problem. We are calculating a density, but the volume is not given. Instead, the dimensions of a box are given. We first use these dimensions to compute the volume of the box, and then use this volume in calculating the density.

EXAMPLE 9.3

A box of copy paper is 28.0 cm by 43.5 cm at the bottom and 27.0 cm deep. The mass of the paper in the box is 22.7 kg. Compute the density of the paper and state the result in kg/m³.

First, write down the information for the volume calculation.

$L = 43.5$ cm

$W = 28.0$ cm

$H = 27.0$ cm

$V = ?$

Since the result must be in units that use meters, convert each of these dimensions to meters.

$$L = 43.5 \text{ cm} \cdot \frac{1 \text{ m}}{100 \text{ cm}} = 0.435 \text{ m}$$

$$W = 28.0 \text{ cm} \cdot \frac{1 \text{ m}}{100 \text{ cm}} = 0.280 \text{ m}$$

$$H = 27.0 \text{ cm} \cdot \frac{1 \text{ m}}{100 \text{ cm}} = 0.270 \text{ m}$$

$V = ?$

Now write the equation for the volume and compute it.

$$V = L \cdot W \cdot H = 0.435 \text{ m} \cdot 0.280 \text{ m} \cdot 0.270 \text{ m} = 0.0329 \text{ m}^3$$

Now, we can proceed to compute the density. Begin by writing down the given information for this calculation.

$m = 22.7$ kg

$V = 0.0329$ m^3

$D = ?$

Next, write the density equation, insert the values with units, and compute the result.

$$D = \frac{m}{V} = \frac{22.7 \text{ kg}}{0.0329 \text{ m}^3} = 690 \, \frac{\text{kg}}{\text{m}^3}$$

Finally, we must check our work. The units are as required. It is not easy to estimate this answer, but working the calculation a second time confirms the result.

Learning Check 9.4

1. Determine the density of a brick with a volume of 0.0011 m^3, and a mass of 1,230 g. State your answer in units of g/cm^3.

2. A student uses the displacement method to determine the volume of a mineral stone. Her readings from the graduated cylinder are 66.4 mL and 90.3 mL. The stone's mass is 64.77 g. Determine the density of the stone, and state your result in g/cm^3.

3. A cylindrical wooden dowel is 44.0 cm long and 1.27 cm in diameter. Its mass is 36.0 g. Determine the density of this wood, and state your result in kg/m^3.

4. A beaker of ethanol contains 855 mL of liquid. The mass of the liquid is 0.6745 kg. Determine the density of ethanol, and state your result in grams per liter.

5. An archeologist digs up a perfect specimen of an ancient clay brick. The brick is 26.5 in long, 5.5 in wide, and 1.75 in high. The mass of the brick is measured to be 7.858 kg. Determine the density of this brick, and state your result in g/cm^3.

6. A medical researcher draws off a 475.0-mL sample of blood plasma. She finds its mass to be 486.9 g. Determine the density of the blood plasma and state your result in kg/L.

Answers   3.   646 kg/m³   6.   1.025 kg/L
1.   1.12 g/cm³   4.   789 g/L
2.   2.71 g/cm³   5.   1.88 g/cm³

9.5   Chemical Properties

So far in this chapter, we have focused on the physical properties of substances. The *chemical properties* of a substance describe the chemical behavior of the substance. By chemical behavior, we mean what kinds of chemical reactions the substance does or does not engage in.

Figure 9.19. Fire is a rapid combustion reaction, in which a fuel reacts with oxygen.

We will be discussing chemical reactions more in a later chapter. But for our present purposes, you need to know about some basic types of chemical reactions. This will make it easier to understand what we mean by chemical properties.

Many common types of reactions involve oxygen. As you may recall, oxygen gas makes up about 21% of earth's atmosphere. One of the oxygen reactions is *combustion*, illustrated by the lovely camp fire in Figure 9.19. In a combustion reaction, a substance reacts with oxygen to produce one or more other substances. In fires, the reacting substance is often a fuel containing carbon. Common carbon-based fuels are wood and the fossil fuels (coal, oil, and natural gas). In a fire, the fuel reacts with the oxygen in the air, producing carbon dioxide, water vapor, and perhaps other compounds. Another well-known fuel is hydrogen. When hydrogen burns, the reaction produces pure water vapor.

Fire is a combustion reaction, usually involving oxygen and a fuel.

When combustion of a fuel takes place all at once, an explosion occurs, depicted in Figure 9.20. Explosions are just fires in which all the fuel reacts at the same time. All the gases produced by the combustion expand extremely rapidly from the reaction site. This rapid expansion produces a violent shock wave that destroys surrounding material, vibrates the ground, and creates a massive sound wave in the air.

Both fires and explosions are examples of combustion reactions.

An explosion is an extremely rapid combustion reaction.

Figure 9.20. An explosion is an extremely rapid combustion reaction.

Figure 9.21. Rust, such as seen on this rusty chain near the Golden Gate Bridge, is a result of an oxidation reaction with iron.

Figure 9.22. Oxidation of copper produces brown or reddish brown copper oxide. The reddish oxide attracts other minerals and turns brown.

Substances that burn (combust) are said to be *flammable*. (Strangely, *inflammable* means the same thing as flammable!)

Another very common reaction involving oxygen is when metals *oxidize*. When they do, the atoms on the surface of the metal combine with the oxygen in the air to form a compound called an *oxide*. Rust is the common name for iron oxide. Iron or steel (which is mostly iron) left alone will develop the coating of iron oxide we call *rust*, illustrated by the photo of the rusty iron chain in Figure 9.21.

When copper oxidizes, it forms copper oxide, illustrated in Figure 9.22. Copper oxide is dark brown and attracts other minerals to produce the familiar brown coating on pennies.

Look at the periodic table inside the back cover of the book and find oxygen, element number 8. Right below it is sulfur, element 16. Elements in the same column (Group) in the periodic table act very similarly in chemical reactions, so sulfur acts a lot like oxygen. For this reason, sulfur forms compounds with some metals just as oxygen does.

One of these compounds is the tarnish that forms on silver—silver sulfide. Historically, silver has been a popular metal for eating utensils and ornaments. As

Figure 9.23. Silver reacts with sulfur in the air to form the black compound silver sulfide. Once the silver sulfide is removed with metal polish, the shiny silver is again visible.

Most metals will undergo oxidation reactions with the oxygen in the air.

Figure 9.23 illustrates, an object coated with silver sulfide can be cleaned with metal polish to restore the beautiful luster of silver metal.

Metal oxides are powdery substances. There are many different metal oxides, and they take on a variety of colors. Metal oxides have been used for centuries as color pigments in paint.

Corrosion is another term that is often used to describe metal oxidation. There are other forms of corrosion, but the reaction of a metal with oxygen is the most common. The term corrosion is usually used when the metal involved is gradually being destroyed by the chemical reaction.

We have looked at three common reactions involving oxygen: fire, or combustion; explosions, which are extremely rapid combustions; and the much slower oxidation of metals. I will mention two other reactions here. Biodegradable substances are not chemically stable. The compounds in the substance break down over time. The property involved here is called *chemical stability*. Finally, *toxicity* is a chemical property that describes whether a substance reacts in harmful ways with other substances in the human body.

These reactions allow us to illustrate a few chemical properties. As I mentioned, chemical properties describe chemical reactions that substances do or do not engage in. Examples of several substances and some chemical properties are listed in Table 9.2.

Substance	Chemical Property
wood	flammable/combusts
oxygen	supports combustion
plastic	chemically stable/requires long periods of time for chemical breakdown (and thus not biodegradable)
aluminum	oxidizes to form aluminum oxide, a black powder
gold	does not oxidize
gasoline	flammable/combusts
rattlesnake venom	toxic
glass	not flammable; chemically stable/not biodegradable
chlorine	toxic
hydrogen	flammable/combusts
gun powder	explosive
starch packing peanuts	chemically unstable/biodegradable
water	not flammable; chemically stable; not toxic
sodium	oxidizes to form sodium oxide
paper	flammable/combusts

Table 9.2. Examples of substances and some of their chemical properties.

Learning Check 9.5

1. Write a paragraph distinguishing between physical and chemical properties.
2. Distinguish between fires and explosions.
3. What is oxidizing?
4. Make your own list of the five different chemical reactions we have reviewed in this section. For each one, identify two or three examples on your own of a substance and one of its properties that pertains to that kind of reaction.

Scientists, Experiments, and Technology

The image to the right shows a display of paint pigments in an Indian market. The technology of paint pigments is as old as painting itself. Some of the historic pigments were later found to be toxic, and have been replaced with modern, synthetic alternatives. One example is the famous color vermilion, a beautiful, deep, orange-red made from mercury sulfide.

The second photo shows a shop in India selling vermilion pigment. In the old days natural mercury sulfide was prepared from a mineral called *cinnabar*. But since mercury sulfide is toxic, it is now almost never used—at least not in the West. The photo of the vermilion shop was taken in 2007 in India. Sadly, the continued availability of vermilion probably means that the workers preparing the pigment, and possibly the artists using it, may eventually suffer from the brain disorders associated with mercury poisoning.

Chapter 9 Exercises

Answer each of the questions below as completely as you can. Write your responses in complete sentences.

1. Describe several physical and chemical properties common to metals.
2. Write a few sentences describing the relationship between volume and temperature and between volume and pressure for gases.
3. Make a chart of the four phases of matter. In your chart, describe the phase at the molecular level. Also list the key features that characterize substances in this phase.
4. On a phase diagram, why is the curve horizontal during a phase transition?
5. Distinguish between tensile strength, compressive strength, and shear strength.
6. Identify as many physical properties as you can for these substances: rubber, oak wood, wool, brass, ceramic, water, gold, polystyrene (Styrofoam), brick, air, tin, and steel.
7. Distinguish between the properties *hardness* and *brittleness*.
8. Explain what thermal properties are, and give several specific examples of substances and thermal properties that apply to them.
9. What physical situation must be present for a fluid to flow?
10. Identify two examples of each of these: a toxic substance, a nontoxic substance, a chemically stable substance, a chemically unstable substance.
11. A one-gallon can of kerosene has a net mass of 2.990 kg. Determine the density of kerosene, stating your result with units of g/mL.
12. Determine the density of a white pine block with dimensions 2.5 in by 1.5 in by 1.5 in, assuming the mass of the block is 0.0333 kg. State your result in g/cm³.
13. A hockey puck is a cylinder of vulcanized rubber, the same material used to make car tires. A standard hockey puck is 25 mm thick and 75 mm in diameter, with a mass of approximately 165 g. Determine the density of vulcanized rubber and state your answer in kg/m³.

Answers
11. 0.79 g/mL
12. 0.361 g/cm³
13. 1,494 kg/m³

Experimental Investigation 6: Determining Density

Overview

- The goal of this investigation is to determine the densities of several different materials and compare these densities to reference density values.
- Measure the mass in grams of the five metal samples from the volume investigation (three aluminum, one brass, one carbon steel).
- Using the volume data from the volume investigation (Experimental Investigation 5), calculate the density for each sample.
- Using common reference density values for predicted values and your calculated densities for experimental values, calculate the percent difference for each sample.

Basic Materials List

- aluminum rod, 3/8 inch diameter × 3 inches long
- aluminum flat bar, 1/8 inch × 3/4 inch × 4 inches long
- aluminum angle, 1/8 in thick, 3/4 × 3/4 × 2.5 inches long
- brass rod, 1/2 inch diameter × 3.5 inches long
- carbon steel flat bar, 1/2 inch × 1/2 inch × 2 inches long
- mass balance
- calculator

    In addition to giving you an opportunity to work with the density equation, this investigation will also give you some idea of how accurate your volume measurements were from the volume investigation you performed. Our procedure in this investigation is very simple. Begin by measuring the mass, in grams, for each of the five metal samples. Record each of these, along with the units of measure, in a table in your lab journal.

    Now calculate the density for each sample, using the density equation,

$$D = \frac{m}{V}$$

In this equation, D stands for the density, m is the mass of the sample, and V is the volume. If your masses are in grams and your volumes are in cubic centimeters, then your calculated densities have units of grams per cubic centimeter, g/cm^3. Your volumes are the values you determined in Experimental Investigation 5, "Determining Volume." In that investigation, you determined each volume two different ways. Use the volume values from your previous work that you think are most accurate.

    Use your mass and volume values to calculate the density for each sample. Enter each of these density values, along with the units of measure, in your lab journal. Since three of the pieces are made of aluminum, the density values for these samples should be very close to one another.

    Now we want to compare these experimentally determined densities to the standard density values for these materials. We will use the standard values as "predicted values," and your calculated densities as "experimental values." Any time you want to compare an experimental result to a prediction, the standard way to perform the comparison is to calculate the percent difference, according to this equation:

$$\text{percent difference} = \frac{|\text{predicted value} - \text{experimental value}|}{\text{predicted value}} \times 100\%$$

This equation expresses the difference between your experimental value and your prediction as a percentage of the prediction.

You will use the percent difference equation to determine how well your density values compare with published values for the densities of these materials. Use the published values shown in the table as the predicted values to calculate the five percent difference values for your samples.

Material	Standard Density
aluminum, T-6061 alloy	2.72 g/cm³
brass, alloy 360	8.5 g/cm³
steel, CF-1018 alloy	7.85 g/cm³

The most common alloy of aluminum for making structural parts is called T-6061, so I have the density for this alloy in the table. It is likely that your aluminum samples are made of 6061 or a similar alloy. A common brass alloy is called alloy 360. The most common steel alloy for ordinary steel parts is called CF-1018, and it is likely that your steel bar is made of CF-1018.

Analysis

Make a table in your report showing the volume, mass, standard density, experimental density, and percent difference for each of the five samples. Be sure to show the correct units of measure for each entry in your table. Address the following questions in your report.

1. Do your three density values for aluminum agree? That is, are they fairly close together? If they are not, speculate on why they might be so different. What part of the process of determining the densities of the three aluminum samples did you feel was most subject to error?

2. Are your percent difference figures all below 10%? If not, then again speculate why your results might be so different from the reference values. What part of the process of determining the density did you feel was most subject to error? Do you think the problem was with your volume values, or somewhere else? For samples with a high percent difference value, go back and look at your volume data from the volume investigation. There you found volume two different ways. If your two volumes disagreed (that is, were not very close), then try using the other volume for your density calculation to see if it gives a better result. If the percent difference is lower, then your experimental result is a closer match to the prediction. This is what we mean by a "better result."

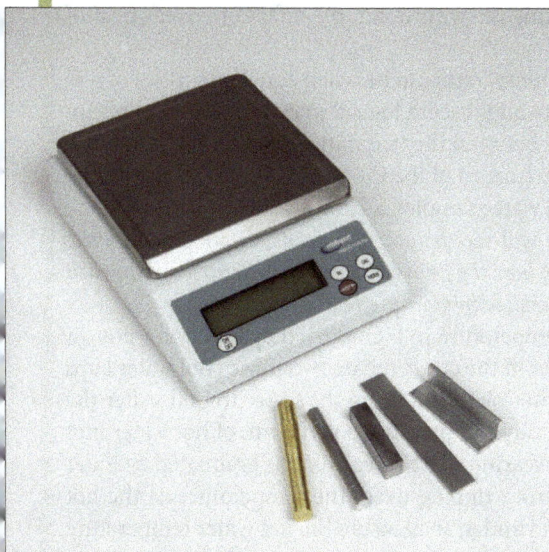

Experimental Investigation 7: Heat of Fusion

Overview

- *The goal of this experiment is to observe the effects of the heat of fusion of a substance.*
- Mix equal volumes (100 mL) of water at 20°C (room temperature) and 90°C and measure the equilibrium temperature. Repeat two more times. Repeat with water at 5°C. Then repeat with ice at 0°C.
- Collect initial and final temperature data and prepare charts showing initial temperatures and equilibrium temperature for each of the nine trials.

Basic Materials List

- digital thermometers (at least 3)
- graduated cylinder (2)
- Styrofoam cup, 16 oz (10 or so)
- refrigerator/freezer
- beaker, 600 mL (2), and 1000 mL
- hot plate and tongs
- safety glasses, hot gloves
- water, ice cubes, and ice chest

Safety Precautions

1. When pouring hot water, always use tongs to handle the hot beaker.
2. Wear safety glasses or goggles when pouring hot water, and keep your hands out of the way.
3. Use care when pouring hot water: pour slowly and carefully so hot water does not spill.

As we saw in the previous chapter, heat is needed to raise the temperature of any substance. Heat is also needed to make a substance change phases, even though the substance remains at the same temperature while doing so. The amount of heat required to melt a substance, while keeping its temperature constant, is called the *heat of fusion*.

In this experiment, you will repeatedly mix equal volumes (and thus equal masses) of water at different temperatures (six separate trials) to see how the equilibrium temperature compares to the initial temperatures. Then you will do the same thing with ice at 0°C and boiling water (three more trials). By comparing the mixing of water at different temperatures to what happens when mixing ice with water, the effect of water's heat of fusion becomes apparent.

What do you expect the equilibrium temperature to be when equal quantities of water at different temperatures are mixed? It should seem logical to expect the equilibrium temperature of the mixture to be half way between the two initial temperatures. In other words, the final temperature is the average (mean) of the initial temperatures. Now, what do you expect to happen when one of the water samples is frozen and is at 0°C? Will the final temperature still be the average? We will see. *Before you begin collecting data, write down in your lab journal your hypothesis of how the final temperature will turn out for each of the three sets of trials. Be as specific as possible in your predictions.*

Measure 100 mL of water at room temperature in a graduated cylinder and pour it into a Styrofoam cup. Insert a thermometer in the cup. Prepare two more cool water cups the same way. Fill a 600-mL beaker on a hot plate and bring the water to boil. After the water reaches 100°C, use tongs (very carefully!) to measure out 100 mL of hot water into the graduated cylinder. Have an assistant wearing a hot glove hold the graduated cylinder during pouring to prevent the cylinder from tipping over. The temperature of the hot water will be lower after pouring into the cylinder, so measure the hot water temperature in the graduated cylinder. After recording the hot water temperature, and the temperature of one of the cool water cups, pour the hot water into the cool cup and record the equilibrium temperature. Repeat this two more times. Then repeat this whole procedure using a 600-mL beaker of water that has been chilled to about 5°C.

For the trials involving ice, measure out 100 mL of water into each of three Styrofoam cups, place a digital thermometer into each cup, and place them in the freezer over night. The digital thermometers will allow you to record the temperatures inside the ice without damage to the thermometer. We want the ice to be at 0°C, but after freezing the ice temperature will be much lower than that. To warm the ice without melting it, prepare an ice water bath (1 inch deep) in an ice chest with a lid. Nestle the ice cups in the bath and let them stay there with the lid on until they have warmed to 0°C. Then remove them one by one from the bath and pour in 100 mL of hot water. For hot water in these trials, you must preheat the graduated cylinder by filling it with boiling water, emptying it, and filling again with 100 mL of water. (Without preheating, the water will not be hot enough. Try to achieve hot water temperatures in the graduated cylinder of at least 90°C.) Measure the hot water temperature in the graduated cylinder. Then pour the water into the cup with the ice. Do not record the final temperature for these three mixtures until all the ice has melted.

Cup 1 Initial Temp. (°C)	Cup 2 Initial Temp. (°C)	Equilibrium Temp. (°C)
20°C	90°C	
20°C	90°C	
20°C	90°C	
5°C	90°C	
5°C	90°C	
5°C	90°C	
0°C ice	90°C	
0°C ice	90°C	
0°C ice	90°C	

Enter all the temperature data in your lab journal in a table similar to the one shown above. In the table shown, I have entered the target temperatures. In your table, you should enter your own actual temperature data.

Analysis

Prepare a chart displaying the two initial temperatures and the equilibrium temperature for all nine trials. Your instructor will help you with setting up your graphs and labeling them properly. Use your graphs to help you compare your results to your hypothesis. In your report, address the following questions.

1. Were the results for mixtures involving only water consistent with your hypothesis? Is there a consistent mathematical relationship between the initial and final temperatures for these?
2. Is the relationship between the initial and final temperatures for the trials involving ice the same as it was for trials involving only water? Explain what you found.
3. How do the data indicate the effect of the heat of fusion of water?
4. Try to describe the mathematical relationship between the initial and final temperatures for the ice trials, and see if you can explain how it compares to the water-only trials.
5. Explain what the term *heat of fusion* means, and summarize how the heat of fusion of water affects the results of this experiment.

Chapter 10
Force and Motion

A top fuel dragster is a race car designed for one purpose: to accelerate as fast as possible in 300 meters. These cars can accelerate from rest to 100 miles per hour in as little as 0.7 seconds, reaching top speeds of over 330 mph in the length of just over three football fields. They cover the 300-meter race distance in about 3.8 seconds and use parachutes to slow themselves down after crossing the finish line. These cars are very loud.

Acceleration is a measure of how fast the speed of an object is changing. Acceleration is always caused by a force, a principle embodied in Isaac Newton's famous Second Law of Motion. In this chapter, we get into the details.

OBJECTIVES

After studying this chapter and completing the exercises, you should be able to do each of the following tasks, using supporting terms and principles as necessary.

1. Describe the contributions of Aristotle, Galileo, and Isaac Newton to our present understanding of motion.
2. Define *velocity*, *acceleration*, and *inertia*, and relate these terms to Newton's first and second laws of motion.
3. State Newton's laws of motion.
4. Give examples illustrating Newton's three laws of motion.
5. Calculate distance, velocity, and acceleration from given information.
6. Calculate acceleration and force using Newton's second law of motion.

VOCABULARY TERMS

You should be able to define or describe each of these terms in a complete sentence or paragraph.

1. acceleration
2. distance
3. inertia
4. laws of motion
5. speed
6. state of motion
7. *telos*
8. time interval
9. uniform acceleration
10. velocity

10.1   A Brief History of Motion Theory

Theories about the motion of objects began with Aristotle (Figure 10.1), a brilliant Greek philosopher who lived in the fourth century BCE. Aristotle's philosophy covered just about everything—motion, the elements, optics, ethics, and the nature of ultimate reality. Aristotle's ideas dominated discussions about physics for nearly 2,000 years.

Aristotle taught that all objects have a *telos*—a Greek word that means goal or end—toward which they are directed. This idea is rejected today by scientists, who understand the behavior of physical objects as governed by impartial physical laws apart from any inherent purpose. Aristotle's *telos* idea may have some truth to it. However, this doesn't help much in the realm of experimental and theoretical science. Today, scientists agree that if you want to know how physical objects behave, you must conduct experiments and look for the mathematical principles that fit experimental observation. So as

Aristotle: an object behaves as directed by its telos.

Figure 10.1. Greek philosopher Aristotle.

interesting as Aristotle's *telos* is for contemplating the meaning of the world, we must leave that for another course of study.

As a philosopher, Aristotle's ideas about physics were based on philosophical reasoning and not on experimentation. Aristotle taught that objects have a natural motion (part of their *telos*) that requires heavy things to go down and light things to go up. He also held that a force was required to maintain motion because without a force it was natural for objects to stop moving. And he further argued that heavy objects fall faster than lighter objects, in proportion to their weights.

The work of Galileo Galilei (Figure 10.2) brought the reign of Aristotle's ideas about motion to an end. Galileo was a mathematician, inventor, and scientist who lived in Florence from 1564 to 1642. His ideas were not the result of reasoning alone. Galileo used experiments and observations to guide his theories about the behavior of objects in the physical world. As a result, he is regarded by many as the father of modern science.

Based on his own experiments with pendulums and falling objects, Galileo determined that all falling objects accelerate at the same rate, regardless of their weights. He also determined the first mathematical relationship governing free fall: that the distance fallen is proportional to the square of the time of the fall. Galileo also first proposed that a force is not required to maintain an object's motion. Instead, objects continue moving unless a force is present to stop them. This new idea included the breakthrough concept of friction forces. We now understand the force of friction to be the force that causes objects to slow down and stop. In the absence of such a force, an object continues moving. This new idea was huge, and became the basis for Isaac Newton's first law of motion.

Like most scientists in his day, Galileo studied the motion of the heavenly bodies. In the medieval and early Renaissance periods, Church authorities were set on a view of the heavens based on Aristotle's ideas. Aristotle's thinking had been developed into an entire description of the motion of stars

Figure 10.2. Scientist Galileo Galilei, of Florence.

and planets by the Alexandrian astronomer Ptolemy, who lived in the second century CE. Some church authorities felt these ideas were biblical and regarded anyone who held other ideas as unbiblical. The system Ptolemy worked out held earth to be stationary at the center of the heavens, with the moon, the five known planets, and all the stars rotating around the earth.

Galileo: all falling objects accelerate at the same rate.

Galileo's research led him to accept a different view of the heavens. This new view was developed by the Polish astronomer Nicolaus Copernicus and made public in 1543. According to this view, only the moon orbits the earth, while the earth and other planets orbit the sun. The stars and planets only appear to orbit the earth because the earth spins around once per day on its axis.

During the Middle Ages, Christian Theologians thought Ptolemy's model was consistent with the Bible because of a few biblical passages that seem to confirm that the earth is stationary with the sun, moon, and stars going around it. Galileo was told that he should regard Copernicus' model as just that—a model—and that he must not teach that it was the truth. If you recall our discussion from Chapter 7, this is very reasonable and lines up very well with the way we view scientific knowledge today. But Galileo was such a strong supporter of the Copernican model that he kept promoting it and religious authorities eventually decided that he had violated their order not to teach that the model was true. For this, Galileo was tried in 1633. He agreed that he had violated the order and repented.

At the time of Galileo's death nine years later, scientific evidence for the Copernican sun-centered solar system was becoming undeniable. Within a few decades of Galileo's death, the controversy had passed, and virtually everyone accepted the fact that the earth orbits the sun.

Sir Isaac Newton (Figure 10.3) was born in England the same year Galileo died. He is now generally considered to be the greatest scientist who ever lived. His understanding of motion was published in 1687, in a massive and enormously famous work entitled *Principia Mathematica* (Latin for "mathematical principles"). In the *Principia*, Newton presented his three *laws of motion* and worked out a full mathematical treatment of planetary motion. He demonstrated that the planets move in elliptical orbits and presented his law of universal gravitation, as you recall from Chapter 5.

Newton's laws of motion are a theoretical model of mechanical motion in nature. This model is still used to this day to describe motion in our everyday

Newton's laws of motion remain useful models of motion to this day.

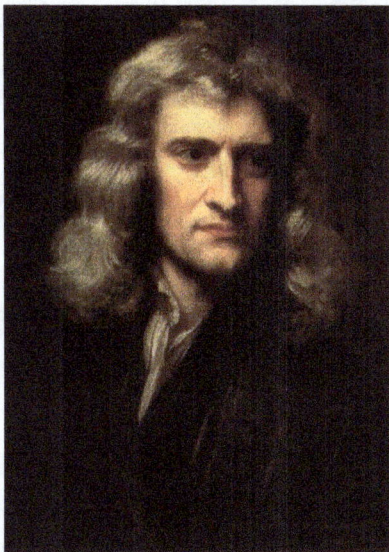

Figure 10.3. British mathematician and scientist Sir Isaac Newton.

world.[1] We will look at these laws soon. But before we do, we need to define velocity, inertia, and acceleration. So it is to this task that we proceed next.

Learning Check 10.1

1. What does the Greek term *telos* mean?
2. Describe the basic features of Aristotle's ideas about motion.
3. Describe Galileo's new theory of motion, including his breakthrough ideas.
4. Why was Galileo put on trial, and what was the outcome?
5. State two of Isaac Newton's accomplishments in theoretical physics.

10.2   Velocity

When thinking about motion, one of the first things we need to be able to talk about is how fast an object is moving. The common word for how fast an object is moving is *speed*. A similar term is the word *velocity*. For the purposes of this course, you may treat these two terms as synonyms. The difference is technical. Technically, the term velocity means not only *how fast* an object is moving, but also in what *direction*. The term speed refers only to how fast an object is moving. But in this course, we are only going to consider motion in one direction at a time. So for now, we can use the terms *speed* and *velocity* interchangeably.

For motion in one direction, the terms speed and velocity mean the same thing.

An important type of motion is motion at a constant velocity, like a car with the cruise control on (Figure 10.4). At a constant velocity, the velocity of an object may be defined as the distance the object travels in a certain period of time. This idea can be expressed very accurately using mathematics. Expressed with an equation, the velocity, *v*, of an object is calculated as

Figure 10.4. A car traveling with the cruise control on is an example of an object moving with constant velocity.

$$v = \frac{d}{t}$$ Velocity equation, for objects moving at a constant velocity

---

1   Newton's laws work fine except in two cases. When objects are moving extremely fast, Einstein's special theory of relativity must be used. And when considering events at the atomic level, the principles of quantum physics must be used.

The velocity is calculated by dividing the distance the object travels, d, by the amount of time, t, it takes to travel that distance. So, if you walk five miles in two hours, your velocity is $v = (5\text{ miles})/(2\text{ hours})$, or 2.5 miles per hour.

Notice that for a given length of time, if an object covers a greater distance, it is moving with a higher velocity. In other words, the velocity is proportional to the distance traveled in a certain length of time. When performing calculations using the SI System of units (the metric system), distances are measured in meters and times are measured in seconds. This means the units for a velocity are meters per second, or m/s.

The relationship between velocity, distance, and time for motion at a constant velocity is shown graphically in Figure 10.5. Travel time is shown on the horizontal axis and distance traveled is shown on the vertical axis. The steeper curve[2] shows distances and times for an object moving at 2 m/s. At a time of one second, the distance traveled is two meters because the object is moving at two meters per second (2 m/s). After two seconds at this speed, the object has moved four meters: $v = (4\text{ m})/(2\text{ s}) = 2\text{ m/s}$. And after three seconds, the object has moved six meters: $v = (6\text{ m})/(3\text{ s}) = 2\text{ m/s}$.

The right-hand curve in Figure 10.5 is for an object traveling at the much slower velocity of 0.5 m/s. At this speed, the graph shows that an object travels two meters in four seconds, four meters in eight seconds, and so on.

To see this algebraically, look again at the velocity equation above. If we multiply both sides of this equation by the time, t, and cancel, we have

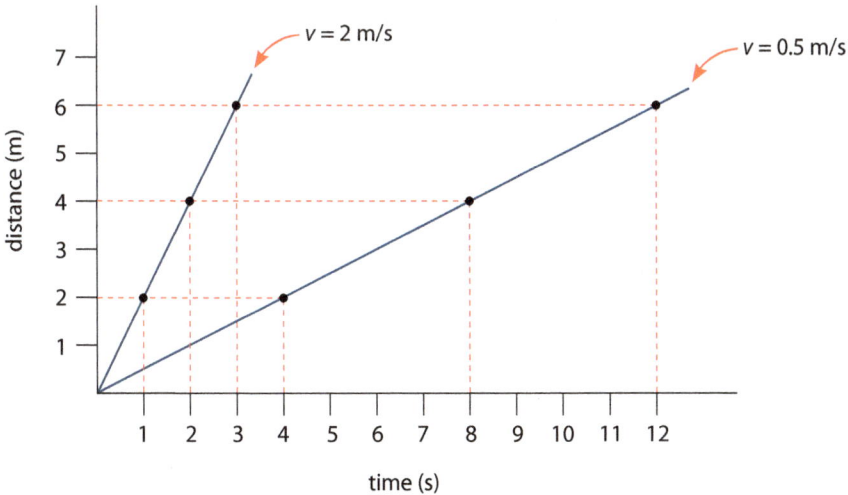

Figure 10.5. A plot of distance versus time for an object moving at constant velocity. Two different velocity cases are shown.

2   Note that when discussing graphs, the lines or curves on the graph are all referred to as *curves*, regardless of whether they are curved or straight.

$$v = \frac{d}{t}$$

$$v \cdot t = \frac{d}{\cancel{t}} \cdot \cancel{t}$$

$$v \cdot t = d$$

Or,

$$d = v \cdot t \qquad \text{Distance equation, for objects moving at a constant velocity}$$

This is the same equation, just written in a different form. It still applies to objects moving at a constant velocity. With this form of the velocity equation, we can calculate how far an object travels in a given amount of time, assuming the object is moving at a constant velocity.

Now we will work a couple of example problems. I will again be following the problem-solving method described on page 184.

EXAMPLE 10.1

An article of baggage is moving on a conveyor belt in an airport. The baggage moves 12 feet in 7.6 seconds. Calculate the velocity of the bag and state your result in meters per second.

Begin by writing down the given information and the unknown you must compute. Since the distance is given in feet but the result must be written in meters per second, leave room for the unit conversions you need to perform.

$$d = 12 \text{ ft}$$

$$t = 7.6 \text{ s}$$

$$v = ?$$

Now perform the unit conversions, using the conversion factors from Table 8.6.

$$d = 12 \ \cancel{\text{ft}} \cdot \frac{0.3048 \text{ m}}{1 \ \cancel{\text{ft}}} = 3.66 \text{ m}$$

$$t = 7.6 \text{ s}$$

$$v = ?$$

Now, write down the equation, insert the values, and compute the result.

$$v = \frac{d}{t} = \frac{3.66 \text{ m}}{7.6 \text{ s}} = 0.48 \, \frac{\text{m}}{\text{s}}$$

Finally, check your work. We have stated the result with the units required by the problem (m/s). The numerator in our calculation looks to be roughly half of the denominator and our result is very close to 0.5, so the result appears to be correct.

The next example uses a metric prefix we haven't seen in a while, micro (μ). The conversion factor for this prefix is listed in Table 8.4.

EXAMPLE 10.2

A microwave tower transmits a signal to another tower some distance away. Microwaves travel at the speed of light, which is 300,000,000 m/s. At this speed, the signal requires 72.5 µs to travel between the two towers. Determine the distance between the towers, and state your answer in kilometers.

Let's look at the unit conversions that are required in this problem. First, the answer must be in kilometers, but the given velocity is in m/s. Thus, we will convert the velocity into km/s. Next, the time given is in microseconds, so we will convert it into seconds.

Now write down the given information, leaving space for the unit conversions.

$$v = 300,000,000 \, \frac{\text{m}}{\text{s}}$$

$$t = 72.5 \, \mu\text{s}$$

$$d = ?$$

Next, perform the required unit conversions.

$$v = 300,000,000 \, \frac{\text{m}}{\text{s}} \cdot \frac{1 \text{ km}}{1,000 \text{ m}} = 300,000 \, \frac{\text{km}}{\text{s}}$$

$$t = 72.5 \, \mu\text{s} \cdot \frac{1 \text{ s}}{1,000,000 \, \mu\text{s}} = 0.0000725 \text{ s}$$

$$d = ?$$

Next, write the equation that allows you to calculate the distance, insert the values, and compute the result.

$$d = v \cdot t = 300{,}000 \ \frac{km}{s} \cdot 0.0000725 \ s = 21.75 \ km$$

Now as a check, we see we have units of kilometers, as required. Looking at the math, we have a very large number multiplied by a very small number. This should give us a result that is neither huge nor tiny, so our answer seems to make sense.

Learning Check 10.2

1. Determine the velocity of an object that moves 3.5 feet in 14 seconds at a constant velocity. State your answer in cm/s.

2. A car drives for 45 min with the cruise control set at 65 mph. Determine how far the car travels in that time. State your answer in kilometers.

3. If a certain sprinter runs at a constant speed of 10.02 m/s, how far does the sprinter run in 4.5 s? State your answer in feet.

4. Light from the sun travels at 300,000,000 m/s through space and reaches earth in 8.33 minutes. How far is the earth from the sun? State your answer in miles.

5. A bullet fired from a high-speed rifle requires 77.5 ms to travel 100 yards. Calculate the speed of this bullet and state your result in m/s.

6. A civil engineer measures the water flow in a long drainage pipe. If the velocity she measures is 8.66 ft/s, how far does the water travel in 1.5 minutes? State your answer in meters.

7. An alpha-particle fired from a decaying atomic nucleus travels 8,430 cm in 5.62 μs. Determine the velocity of the alpha-particle. State your result in m/s.

Answers		3.	148 ft	5.	1,180 m/s
1.	7.62 cm/s	4.	approx.	6.	238 m
2.	78.4 km		93,000,000 mi	7.	15,000,000 m/s

10.3   Acceleration

When an object is accelerating, its velocity is not constant—the object is either speeding up or slowing down. You may have heard the term "deceleration" for objects that are slowing down, but in physics it is more common to use the term *acceleration* for both. For our calculations, we are only going to consider cases when an object is speeding up, starting from rest.

Acceleration means changing velocity.

In this book, we also assume that the accelerations we deal with are nice and steady. This is called *uniform acceleration*. The term acceleration means that an object is not moving at a constant speed; its speed is changing. When we calculate the value of an acceleration, the value tells us *how fast* the velocity is changing.

An acceleration value is a measure of how fast an object's velocity is changing.

Figure 10.6 illustrates what happens to an object's velocity over time during the acceleration. The illustration shows a car, starting from rest at time $t = 0$. This car's velocity is increasing by 1 m/s every second. So, after one second, the velocity has increased from rest (0 m/s) to 1 m/s. After one more second, when $t = 2$ s, the velocity has increased by another 1 m/s and is up to 2 m/s. After another second, we are at $t = 3$ s and the car's velocity has increased yet another 1 m/s and is up to 3 m/s. The velocity is increasing uniformly by the same amount every second. This is why this motion is called uniform acceleration. Notice also that since the car's speed is increasing, the distance the car travels during each time interval also increases smoothly from one time interval to the next.

For an object starting from rest and accelerating uniformly, the equation for calculating the object's acceleration is

$$a = \frac{v_f}{t}$$

Acceleration equation, for objects starting from rest

In this equation, a represents the value of the object's acceleration. (We will talk about the units of measure for acceleration in just a moment.) The variable v_f stands for "final velocity," and represents the object's velocity at the end of any *time interval* starting from when the acceleration began. As

Figure 10.6. Uniform acceleration increases an object's velocity by the same amount in each unit of time.

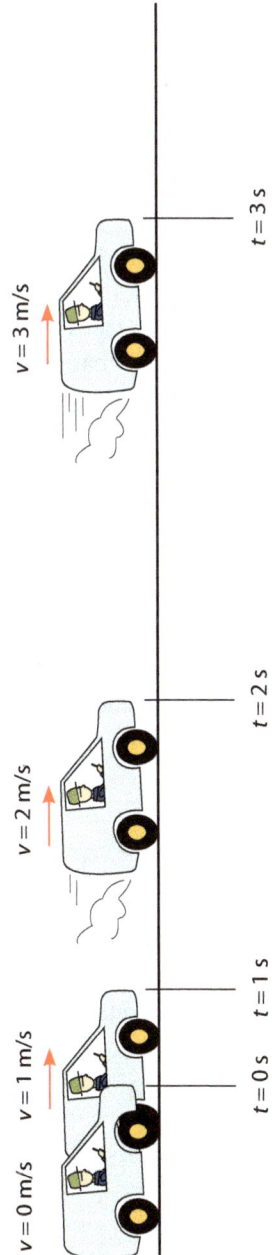

$t = 3\text{ s}$

$v = 3\text{ m/s}$

$t = 2\text{ s}$

$v = 2\text{ m/s}$

$t = 1\text{ s}$

$t = 0\text{ s}$

$v = 1\text{ m/s}$

$v = 0\text{ m/s}$

we saw in Section 10.3, the units for velocity are m/s when using the SI System base units. But the velocity can be in other units as well. The t represents the length of time for the time interval that begins with the object at rest and ends when the object is moving with a velocity v_f. The t represents the length of time for the time interval that begins with the object at rest and ends when the object is moving with a velocity v_f.

As a starter example, let's apply this equation to the motion depicted in Figure 10.6. As we do this, I will explain the units of measure that result for the acceleration. If you look at the time $t = 2$ s in the figure, you see that at that moment the car is moving with a velocity of $v = 2$ m/s. This velocity value is the final velocity for the time interval that begins at $t = 0$ and ends at $t = 2$ s, so $v_f = 2$ m/s for that time interval. Putting these values in the equation for acceleration we have

$$a = \frac{v_f}{t} = \frac{\cancel{2}\,\frac{m}{s}}{\cancel{2}\,s} = 1\,\frac{\frac{m}{s}}{s}$$

The 2 in the numerator and the 2 in the denominator cancel out to give us a value of 1. But now the units of measure look pretty weird. The seconds are *not* going to cancel out in this expression. The way to see what happens to these units is to first change the seconds in the denominator to s/1, as we know we can always do. This gives

$$a = \frac{v_f}{t} = \frac{\cancel{2}\,\frac{m}{s}}{\cancel{2}\,s} = 1\,\frac{\frac{m}{s}}{\frac{s}{1}}$$

Now the units are a fraction divided by a fraction. You have probably already learned the quick rule for dividing two fractions, which is to change the division into a multiplication by inverting the denominator fraction. When I was a kid in school, we called this rule the "invert and multiply" rule. Doing this gives

$$a = \frac{v_f}{t} = \frac{\cancel{2}\,\frac{m}{s}}{\cancel{2}\,s} = 1\,\frac{m}{s} \cdot \frac{1}{s}$$

To divide two fractions, "invert and multiply."

Now hang on—we are almost done with this. We can simplify this expression one step further by multiplying the numerator units together to get meters and multiplying the denominator units together to get seconds squared, giving us

$$a = \frac{v_f}{t} = \frac{\cancel{2}\,\frac{m}{s}}{\cancel{2}\,s} = 1\,\frac{m}{s^2}$$

Okay, now take a breath. That may seem complicated, but it really just simplifies down to a simple rule. The units for acceleration are always length over time squared. All you have to do is make sure your velocity (v_f) and your time (t) use the same time units. And in the problems we do in this book, these units are always seconds. Further, since the SI System is used so universally, and since you probably will never need to calculate an acceleration except for scientific reasons, we always use the units m/s² for acceleration. So no matter what you are given to start with in a problem, always convert the units to velocities in meters per second and times in seconds. After you do this, the acceleration units are always m/s².

Units for acceleration are always distance over time squared.

Before we work through some examples, let's look at a graphical depiction of uniform acceleration the same way we did with velocity. Figure 10.7 shows two different acceleration curves, representing two different acceleration values. The curve on the right, where $a = 1$ m/s², is the same situation as in Figure 10.6. After 1 s, the object is going 1 m/s. After 2 s, the object is going 2 m/s. After 12 s, the object is going 12 m/s. You can take the velocity that corresponds to any length of time (by finding where their lines intersect on the curve) and calculate the acceleration by dividing the velocity by the time to get $a = 1$ m/s². The other curve has a higher acceleration, 4 m/s². An acceleration of 4 m/s² means the velocity is increasing by 4 m/s every second. Accordingly, after 2 s the velocity is 8 m/s, and after 3 s, the velocity is 12 m/s. No matter what point you select on that curve, $v/t = 4$ m/s².

Just as we did with the velocity equation, we can multiply both sides of the acceleration equation by t to get

$$v_f = a \cdot t$$

Velocity equation, for objects accelerating uniformly from rest

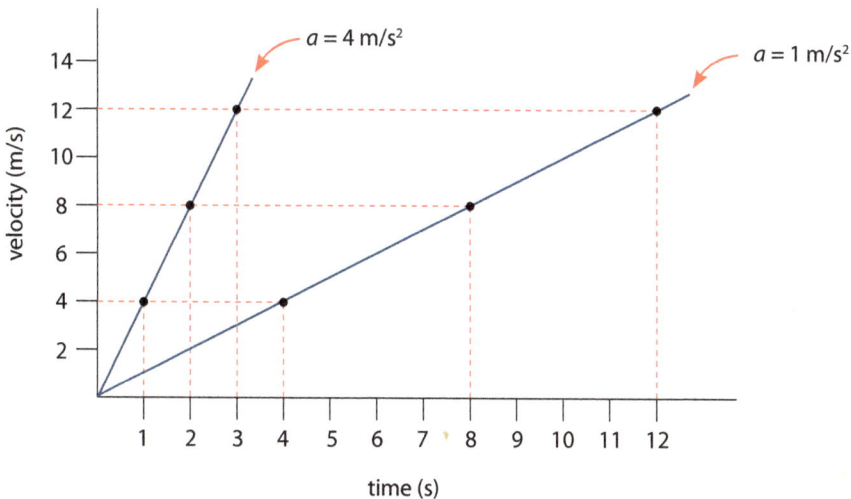

Figure 10.7. A plot of velocity versus time for an object accelerating uniformly. Two different acceleration cases are shown.

With this equation, we can calculate how fast an object with a given acceleration is going after a certain length of time. Now we will go through a few examples.

EXAMPLE 10.3

A car accelerates from rest to 35 mph in 12 seconds. Determine the car's acceleration and state your answer in m/s². Assume uniform acceleration.

Begin, as always, by writing down the given information and the unknown.

$$v_f = 35 \, \frac{mi}{hr}$$

$$t = 12 \, s$$

$$a = ?$$

Now perform the unit conversions needed on the final velocity to convert the miles to meters and the hours to seconds.

$$v_f = 35 \, \frac{\cancel{mi}}{\cancel{hr}} \cdot \frac{1{,}609 \, m}{1 \, \cancel{mi}} \cdot \frac{1 \, \cancel{hr}}{3{,}600 \, s} = \frac{56{,}315 \, m}{3{,}600 \, s} = 15.6 \, \frac{m}{s}$$

$$t = 12 \, s$$

$$a = ?$$

Now we are ready to go. Write the equation, put in the values, and compute the result.

$$v_f = 35 \, \frac{\cancel{mi}}{\cancel{hr}} \cdot \frac{1{,}609 \, m}{1 \, \cancel{mi}} \cdot \frac{1 \, \cancel{hr}}{3{,}600 \, s} = \frac{56{,}315 \, m}{3{,}600 \, s} = 15.6 \, \frac{m}{s}$$

$$t = 12 \, s$$

$$a = \frac{v_f}{t} = \frac{15.6 \, \frac{m}{s}}{12 \, s} = 1.30 \, \frac{m}{s^2}$$

Finally, we check our work. We have the time units for both quantities in seconds, as we should. The length units are in meters, as the problem requires. And on the math, the numerator is bit bigger than the denominator, so the result should be a bit larger than 1, which it is.

In our next example, we use the velocity equation. This time, look carefully at what happens with the unit cancellation when acceleration units are multiplied by time units.

EXAMPLE 10.4

To make the ride as comfortable as possible for the passengers, a high-speed train accelerates uniformly at a very gradual rate of 1.4 m/s². If this acceleration continues for 87 seconds, how fast is the train going?

Begin by writing down the given information.

$$a = 1.4 \ \frac{m}{s^2}$$

$$t = 87 \ s$$

$$v_f = ?$$

No unit conversions are required, so we proceed to write down the required equation, insert the values, and compute the result.

$$v_f = a \cdot t = 1.4 \ \frac{m}{s^2} \cdot 87 \ s = 121.8 \ \frac{m}{s}$$

As a check, the units canceled as expected to give units of m/s, which is correct for calculating a velocity. Also, we multiplied a number that is a bit more than one by another number a bit under 100. We expect the result to come out to around 100, which it does, so the answer makes sense.

As a final example, I want to illustrate further our problem-solving step of checking to make sure the answer makes sense. In this problem, I am going to include a *deliberate error*, and show how our final check at the end helps to identify the error so it can be corrected. Try to spot the error as we go, before I get to the end and point it out.

EXAMPLE 10.5

A bullet in a gun accelerates at a rate of 15,300 m/s² for 75 ms. Determine the velocity of the bullet as it exits the barrel of the gun.

I first write down the given information. Since I see that the time is not in seconds, I also perform the required unit conversion.

$$a = 15,300 \ \frac{m}{s^2}$$

$$t = 75 \ ms \cdot \frac{1,000 \ s}{1 \ ms} = 75,000 \ s$$

$$v_f = ?$$

This example contains a deliberate error— try to spot it.

Next, write the equation, insert values, and compute the result.

$$v_f = a \cdot t = 15{,}300 \ \frac{m}{s^2} \cdot 75{,}000 \ s = 1{,}147{,}500{,}000 \ \frac{m}{s}$$

Looking over this result, we see the units are all okay, and we expect to get a very large value for the result because we are multiplying two large numbers together. But wait. Take another look at that answer. This value is a lot higher than the speed of light! The speed of light is 300,000,000 m/s, a value you should work on remembering. (I would memorize it if I were you.) Nothing can go faster than the speed of light— at least so far as we know! So I must have made a silly mistake somewhere.

In fact, the error is the most common unit conversion error in the world for students: using a fractional metric prefix with the multiplier in the wrong place. There are not 1,000 seconds in 1 millisecond. A millisecond is tiny—1/1,000 of a second. The correct factor must show 1,000 milliseconds in 1 second. So going back and correcting this unit conversion, we have

$$a = 15{,}300 \ \frac{m}{s^2}$$

$$t = 75 \ ms \cdot \frac{1 \ s}{1{,}000 \ ms} = 0.075 \ s$$

$$v_f = ?$$

And now working out the rest of the problem,

$$v_f = a \cdot t = 15{,}300 \ \frac{m}{s^2} \cdot 0.075 \ s = 1{,}147.5 \ \frac{m}{s}$$

Now, that's more like it.

Always perform a "reasonableness check" on the results of your calculations. If you have velocities greater than the speed of light, or if you calculate the mass of a glass of water and find it to be greater than the mass of a Mack truck, you know you have an error, and you need to find it and correct it.

Always perform a "reasonableness check" on the results of your calculations.

Learning Check 10.3

1. When struck by a golf club, a golf ball accelerates from rest to 110 mph in about 0.5 ms. Determine the acceleration of this golf ball.

2. A football quarterback throws a pass and accelerates the ball from rest to 36 ft/s in 225 ms. Determine the acceleration of the football while it is being accelerated by the quarterback's throwing arm.

3. An M1A2 Abrams tank gun accelerates a shell at a rate of about 23,920 m/s². Assuming the shell is accelerated for 0.065 s, what is the velocity of the shell as it exits the gun?

4. In October, 2012, Felix Baumgartner set a world record for the highest free fall jump by jumping from a height of about 128,000 ft above the earth. He reached a velocity of 482 mph in 24 s. This was early in the jump and there was not much air resistance, so his acceleration was nearly uniform. Calculate Baumgartner's acceleration for the first 24 seconds of his fall.

5. Use your result from the previous problem to calculate how fast Baumgartner was falling after the first 15 seconds of the jump.

6. Imagine a toy catapult that accelerates a stone uniformly at 14.5 m/s² for 355 ms. Calculate the velocity a stone has as it leaves the catapult.

7. Consider an elevator that accelerates passengers from rest to 4.75 ft/s in 1.33 s. Determine the acceleration of the passengers in this elevator.

Answers	3.	1,555 m/s	6.	5.15 m/s
1. 98,320 m/s²	4.	8.98 m/s²	7.	1.09 m/s²
2. 48.8 m/s²	5.	134.7 m/s		

10.4   Inertia and Newton's Laws of Motion

In our everyday world, any time someone sets an object in motion, the object comes to rest. The forces of friction bring objects to rest, but this is not at all obvious. And so, for nearly 2,000 years, Aristotle's ideas, that things stop by themselves and that forces are required to keep them moving, were generally accepted.

Then along came Galileo, who put forward the opposite idea: forces are required to make objects stop moving and without a force, a moving object continues moving. This new idea reached its full expression in Newton's *Principia Mathematica*. There, for the first time, was a definition for *inertia*, the physical property that governs the way objects behave when in motion or at rest.

Inertia Inertia is a property possessed by all matter. This property causes all objects to prefer their existing *state of motion*, and resist changing it. The term *inertia* is closely related to mass. The simplest way to keep these two terms separate is this: inertia is a property of matter; mass is the way we measure inertia. Or putting it another way, inertia is a quality, mass is a quantity. When speaking about this property, people usually just say "mass."

Inertia is a property of matter that causes any object to prefer its current state of motion.

The idea of inertia can be challenging to grasp, so here is another example: Inertia is a physical property. We studied other physical properties in Chapter 9. One of the physical properties I mention there is size. Size is a property of matter, and we measure it with volume. Size is a quality, volume is a quantity with units of measure (cubic meters, etc.) In the same way, inertia is a property of matter, and we measure it with mass. Inertia is a quality; mass is a quantity with units of measure (kilograms, etc.)

Inertia is a quality; mass is a quantity.

State of Motion The phrase *state of motion* refers to an object moving with a certain velocity. This velocity could be 0 m/s, in which case the object is at rest. This is one state of motion. The velocity could also be 1 m/s, or 2 m/s, or 1,000 m/s. Motion at any specific velocity is called a *state of motion*.

The inertia of an object causes it to *prefer*, or try to remain in, its present state of motion. In the absence of a force to change an object's state of motion, it remains in the one it is presently in, neither speeding, nor slowing down, nor turning. The best example of this principle I can think of is the current state of motion of the Voyager 1 space probe, shown in Figure 10.8. Voyager 1 was launched by NASA in 1977, and has sent back to earth some amazing images of the solar system along with mountains of other valuable data. Voyager 1 is on the verge of leaving the solar system, and in the fall of 2012 some scientists thought it had. Now the thinking is that it hasn't quite left the solar system yet, but when it does, it will be the first man-made object to do so. It is now 11,300,000,000 miles from earth, and moving at 38,120 mph, heading in the general direction of the constellation Ophiuchus.

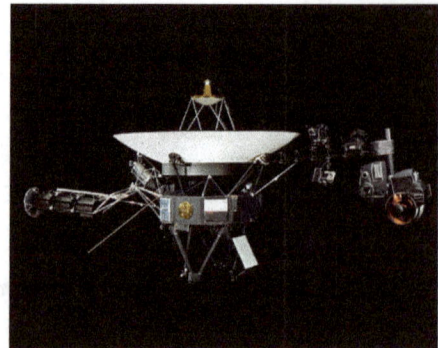

Figure 10.8. Voyager 1 space probe, now 11,300,000,000 mi from earth, and moving at 38,120 mph.

Voyager 1 has a certain state of motion—traveling at 38,120 mph and headed for Ophiuchus. It is not under any propulsion, and it is nowhere near any stars or other gravitational attractions. It is simply out in deep space, plunging through the void (and still sending data to us, until 2025 or 2030, when its sensors and transmitter run out of fuel). The space probe's inertia will keep

Voyager 1 moving in the same direction, at the same speed, forever—or at least for another 100,000 years. By then it may begin to feel the gravitational attraction of Barnard's Star in Ophiuchus, about 6 light-years from earth.

Newton's First Law of Motion In his three famous *laws of motion*, Isaac Newton encapsulated our best understanding of motion in the everyday world. Actually, we hit on the first law on the previous page; I just didn't call it that at the time. Here it is:

> *Newton's First Law of Motion*
>
> **In the absence of a force, an object remains in its present state of motion—either at rest, or moving at a constant speed in the same direction.**

This law is often referred to as the *law of inertia*. The effect of the property of inertia is that objects behave this way. Without a force acting on an object, it continues doing what it is presently doing. If it is at rest, it stays that way. If it is moving in a certain direction at a certain speed, it continues doing so. Just like the Voyager 1 space probe, without a force acting on an object, the object does not change its state of motion.

Newton's Second Law of Motion Newton's second law of motion specifies how an object behaves if there *is* a force acting on the object. The bottom line is this: when there is a force on an object, *the object changes its state of motion—it speeds up or it slows down*. That is, it accelerates. Newton's way of saying it is this:

> *Newton's Second Law of Motion*
>
> **The acceleration of an object is proportional to the force acting on it.**

Mathematically, Newton's second law of motion is expressed this way:

$$a = \frac{F}{m}$$ Newton's second law acceleration equation

Whenever I present Newton's second law of motion, I always warn people that most physics books out there write the law as $F = ma$. You can see that dividing both sides of this equation by the mass, m, gives you $a = F/m$. I prefer writing the Newton's second law equation this different way because it fits Newton's own wording of the law, which is pretty much the wording in the box.

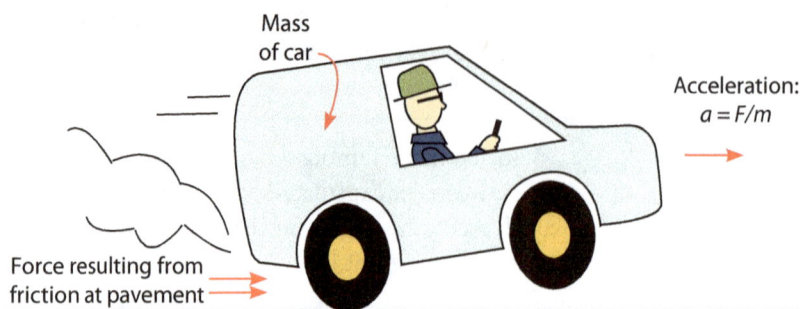

Figure 10.9. A force applied to a mass results in an acceleration equal to $a = F/m$.

Newton's second law of motion is illustrated by the car in Figure 10.9. When the driver steps on the accelerator, the power from the engine causes the tires to rotate. The friction force between the tires and the ground then acts on the car. The car has a mass, m, and the result of this force, F, acting on the mass of the car, m, is that the car begins changing its speed. The equation tells us exactly how fast the speed changes. (By the way, I know that little cars with front wheel drive don't do wheelies. I just thought the picture was more amusing this way.)

At this point, you may be thinking: "Wait a minute. Didn't we learn that acceleration was calculated as $a = v_f/t$? So why do we have another equation for acceleration, and how do we know which one to use?" Fair enough. Here is the difference.

The first equation we learned, $a = v_f/t$, is from a branch of physics called *kinematics*. In kinematics we focus on distance, velocity, acceleration, and time relationships without paying any attention to any forces that may be involved. But the Newton's second law equation is in a branch of physics called *dynamics*. The study of dynamics is all about changes in motion from forces. Both the equations calculate the acceleration, but they do it differently by using different information. The way you determine which equation to use is by looking at the information given to you in the problem. If you are given velocity and time information, use the kinematics equation. If you are given force and mass information, use the Newton's second law equation.

Use the information given in the problem to help you determine which acceleration equation to use.

One more thing. Sometimes it is handy to write Newton's second law as

$F = m \cdot a$ Newton's second law force equation

This form of the equation convenient if the problem you are working gives you the mass and the acceleration, and asks you to compute the force causing the acceleration. This equation, just like the other two equations we have seen in this chapter, can be written in different forms for different purposes. The two forms of the equation represent the same relationship between the variables involved. They are, in fact, the same equation expressed in two different ways. The two different expressions are mathematically equivalent.

And now it's time for a couple of example problems using Newton's second law of motion. When we learned the kinematics equation for acceleration, $a = v_f/t$, I indicated that you should always calculate your accelerations in units of m/s². The same thing holds now with this new equation. You may recall that the SI System unit for force is newtons (N), named after you know who. If you divide newtons by kilograms (F/m), the result is in the acceleration units we want, m/s². You just have to take my word for it on this for now. Demonstrating this mathematically is not that difficult, but it goes beyond what we need for this course. Just make sure your units for force and mass are newtons and kilograms.

EXAMPLE 10.6

In 2011, NASA's Space Shuttle program was concluded. The mass of one of the shuttles was about 1,840,000 kg. The liftoff was powered by two solid rocket boosters and three main engines. Together, these delivered a net lifting force of 12.42 MN. Calculate the acceleration of the shuttle at liftoff.

Begin by writing down the given information.

$m = 1,840,000$ kg

$F = 12.42$ MN

$a = ?$

The mass is already in the correct units, the SI base unit of kilograms. However, the force needs to be in newtons, so perform the unit conversion.

$m = 1,840,000$ kg

$$F = 12.42 \ \cancel{\text{MN}} \cdot \frac{1,000,000 \ \text{N}}{1 \ \cancel{\text{MN}}} = 12,420,000 \ \text{N}$$

$a = ?$

Since mass and force information are given, we know that the equation we need to use is the Newton's second law acceleration equation. So we proceed to write down the equation, insert the values, and compute the result.

$$a = \frac{F}{m} = \frac{12,420,000 \ \text{N}}{1,840,000 \ \text{kg}} = 6.75 \ \frac{\text{m}}{\text{s}^2}$$

Now for the final check, all the units in the computation and result are correct (N, kg, and m/s²). With a numerator of 12 million and a denominator of about 2 million, we expect the result to be about 6 or so, which it is.

Scientists, Experiments, and Technology

I will never forget being on our family summer vacation in 1969. That summer was when we listened on the radio in our car to hear the broadcast of the first landing on the moon. It still seems amazing that we accomplished that with 1960s computers and technology.

The National Aeronautics and Space Administration, or NASA, has been in the business of launching space craft since it was created by President Dwight D. Eisenhower in 1958. NASA's goals are non-military and include scientific research, space travel, and development of new technologies.

The missions to the moon were all part of the Apollo project. The moon landings were accomplished using a gargantuan launch vehicle call the Saturn V rocket (pronounced "Saturn Five"). The Saturn V is the largest rocket ever built and stands over 360 feet tall, significantly longer than a football field. A special building was constructed for vertical assembly of the rockets, and then the rockets were transported—standing up—from the Vehicle Assembly Building to the launchpad at the Kennedy Space Center in Cape

Canaveral, Florida. The image to the left shows the Apollo 10 rocket en route to the launchpad. To give you an idea of the overwhelming size of this monster, the teeny specs on the road to the left of the left tread on the transport vehicle are *people*. Each Saturn V rocket was used only once.

When the Apollo moon landing missions were concluded in the 1970s, everyone's attention turned to NASA's development of a reusable transport vehicle that could shuttle astronauts and equipment up and down to earth orbit or to space stations. The Space Shuttle program kicked off with its first operational launch in 1982. After that, launches occurred several times per year (except for two two-year gaps due to fatal accidents) until the program's conclusion in 2011. Images on the next page show the shuttle during launch in Florida and landing in California. After landing, the vehicles were transported back to Florida by flying them *on the back of a Boeing 747 airplane*, if you can believe that!

The Space Shuttle was used to launch the Hubble Space Telescope (HST) into space in 1993. This telescope has proved to be one of the most valuable research instruments of all time, furnishing researchers with thousands of stunning images of objects from deep in space. The HST has revolutionized the sciences of astronomy and astrophysics. The HST is still operational, but is scheduled to be succeeded by another space telescope around 2018. A lovely image of the HST, taken from Space Shuttle Discovery, is shown below.

NASA has had to face several fatal disasters, and these can never be forgotten by those of us who were around when they happened. But NASA's accomplishments are impressive and have had an impact on technology that we cannot really even begin to estimate.

Pause now and ask yourself what this result *means*. It means that the velocity of the shuttle at liftoff is increasing by 6.75 m/s every second. After one second, the shuttle is moving at 6.75 m/s. After two seconds, its velocity is 13.5 m/s, and so on. That's zero to 60 mph in under 4 seconds. Not bad for a vehicle that weighs over 2,000 tons. This is actually a bit faster than the 4.2 second time for a Porsche 911. Think about that.

And now for our final example.

EXAMPLE 10.7

A toy car with a mass of 35 g accelerates down a hill at a rate of 6.13 m/s². Determine the force causing this acceleration.

Write down the given information.

$$m = 35 \text{ g}$$

$$a = 6.13 \ \frac{m}{s^2}$$

$$F = ?$$

The mass units must be in kilograms, not grams. So perform the unit conversion.

$$m = 35 \ \cancel{g} \cdot \frac{1 \text{ kg}}{1,000 \ \cancel{g}} = 0.035 \text{ kg}$$

$$a = 6.13 \ \frac{m}{s^2}$$

$$F = ?$$

Now use the Newton's second law force equation. Write down the equation first, then insert the values, and then compute the result.

$$F = m \cdot a = 0.035 \text{ kg} \cdot 6.13 \ \frac{m}{s^2} = 0.215 \text{ N}$$

As a final check, the units are all as they are supposed to be, and the result looks reasonable.

Newton's Third Law of Motion   Newton's third law of motion seems simple, and it is. But the implications of this law are very profound. The law may be stated as follows:

> ### Newton's Third Law of Motion
>
> When one object pushes on another with a certain force, the second object pushes back on the first one with equal force.

Figure 10.10. At contact, the bat exerts a force on the ball (lower arrow), and the ball exerts an equal and opposite force on the bat (upper arrow).

As with the first two laws of motion, the third law applies to everything. As shown in Figure 10.10, when a baseball bat hits a baseball, the baseball hits the bat just as hard. This is why the bat can hurt the batter's hands, or even break. As another example, when a helicopter rotor pushes the air forcefully down, the air pushes back just as forcefully on the rotor, lifting the helicopter, as illustrated in Figure 10.11. In this photo, the fact that the helicopter blades are pushing the air down is demonstrated by the violent disturbance of the surface of the water. But when the rotor pushes the air down, the air pushes the rotor up, and this upward force is what holds up the helicopter.

For another example, let's go back to the accelerating car of Figure 10.9. For the car to accelerate, something must be pushing on it—this is what Newton's second law says. What is it, exactly, that is pushing on the car? A close-up of the car's tires on the road is shown in Figure 10.12. The car's engine forces the tires to rotate. Because of the friction between the tires and the road surface, the tires exert a horizontal force on the road surface, toward the left in the figure. Newton's third law says that the road surface pushes back, just as hard, and in the exact opposite direction, toward the right in the figure. This force pushing to the right accelerates the car to the right. So what is pushing on the car to make it accelerate? The *road* is.

Newton's third law also applies to objects that are not in motion. A car sitting on the pavement exerts a force downward on the pavement equal to the car's

Figure 10.11. The helicopter blades force the air down, disturbing the water. The air pushes back up on the helicopter blades with equal force, lifting the helicopter.

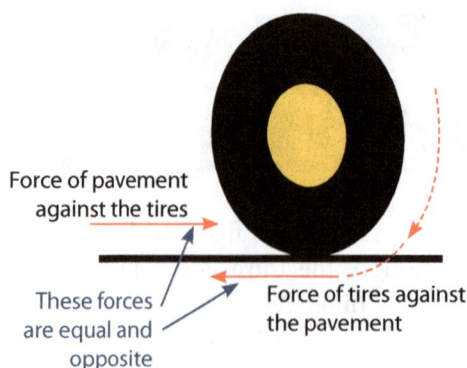

Force of pavement against the tires

These forces are equal and opposite

Force of tires against the pavement

Figure 10.12. Because of friction, the tires exert a horizontal force on the surface of the road. The surface of the road pushes back on the tires equally and oppositely, accelerating the car.

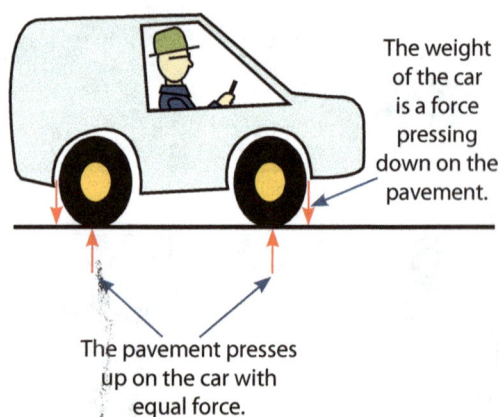

The weight of the car is a force pressing down on the pavement.

The pavement presses up on the car with equal force.

Figure 10.13. The third law applies also to objects not in motion.

weight, as illustrated in Figure 10.13. The pavement exerts an equal an opposite upward force on the car, which holds the car up.

Learning Check 10.4

1. What is a state of motion?
2. Write a paragraph distinguishing between mass and inertia.
3. If an object starts from rest and its velocity increases uniformly at a rate of 5 m/s every second, how fast is it going after 4 seconds?
4. For each of the three laws of motion, write down five examples illustrating the law. For each of your 15 illustrations, write a sentence explaining how your example illustrates the law in question.
5. Calculate the acceleration of a 125-g mass being acted upon by a force of 13.5 kN.
6. Hummingbirds can accelerate at over 27 m/s² (fastest creature on earth). What force is needed to accelerate a 3.0-g hummingbird?
7. Fighter jets on aircraft carriers are launched by a catapult. The aircraft reaches 160 mph in two seconds, an acceleration of 36 m/s². Calculate the total force needed to do this on an F14 Tomcat, which has a mass of 27,700 kg.
8. An electric rail gun can accelerate a 3.4-kg projectile with a force of 1.7 MN. Calculate the resulting acceleration.

Answers		5.	108,000 m/s²	7.	997,200 N
3.	20 m/s	6.	0.081 N	8.	500,000 m/s²

Chapter 10 Exercises

Answer each of the questions below as completely as you can. Write your responses in complete sentences.

1. Write a couple of paragraphs describing the contributions of Aristotle, Galileo, and Newton to motion theory.
2. Assume a certain object is accelerating at a rate of 10 m/s². Describe what is happening to the object's velocity.
3. Explain why Newton's first law of motion is called the law of inertia.
4. Use Newton's third law of motion to explain how a ski boat propels itself. Assume the boat has a propeller (like the blades of a fan).
5. You have probably seen times when a car was in mud, on snow, or on ice. In such cases, the wheels of the car can spin without moving the car at all. Explain why the car does not accelerate in a situation like this, even though the wheels are spinning.
6. Any time an object is at rest on planet earth, there must be at least two forces acting on the object. Explain why this is so.
7. Give three different examples of states of motion.
8. Galileo and Newton both realized that in the absence of a force to change its motion, an object keeps moving. If this is so, why do we need to have the engine on in a car cruising down the highway? Why can't we just get up to speed, turn off the engine, and maintain our state of motion all the way to our destination?
9. The top speed of the F14 Tomcat fighter jet is 1,544 mph. At this speed, how far does the jet fly in 10 minutes? (Assume the speed is constant.)
10. How much force is required to accelerate a large truck with a mass of 4,500 kg at a rate of 2.5 m/s²?
11. At the hummingbird's acceleration of 27 m/s², how fast is the bird going after 250 ms? State your answer in m/s.
12. A force of 356 µN is applied to particle with a mass of 94 mg. Determine the acceleration of the particle.
13. A proton in an experiment travels 13.6 cm in 0.293 µs (constant speed). Determine the velocity of this proton, and state your answer in m/s.
14. A car accelerates uniformly from rest to 65 mph in 0.32 min. Determine the acceleration of the car.

Answers	11. 6.75 m/s	14. 1.51 m/s²
9. 414,049 m (257 mi)	12. 3.79 m/s²	
10. 11,250 N	13. 464,164 m/s	

Experimental Investigation 8: Inertia and Force

Overview

- *The goal of this experiment is to observe how equal forces affect objects with different masses.*
- Assemble a jig that allows a small spring to be compressed by four different fixed amounts. Use the jig to launch balls of different masses straight up from the jig.
- Have an assistant use a pole or board held vertically to mark the maximum height achieved by the different balls.
- Prepare a graph of maximum height vs. mass for each of the four spring compressions.

Basic Materials List

- golf ball, table tennis ball, and 1-inch diameter wooden ball
- launching jig
- spring
- mass balance
- small screwdriver
- measuring pole and measuring tape
- safety goggles

Safety Precautions

The person launching the balls needs to wear eye protection. Protect your eyes by wearing safety goggles.

The force exerted by a compressed spring depends only on how stiff the spring is and how far the spring is compressed. If a given spring is repeatedly compressed by the same amount, then the force exerted by the spring in each case is the same. In this experiment, we use this principle to apply equal forces to balls with different masses. This allows us to observe how the inertia of each ball affects the way the ball responds to a given amount of force.

You have learned that Newton's second law of motion can be expressed in the equation

$$a = \frac{F}{m}$$

According to this equation, if we use the same amount of force to accelerate different masses, then the larger the mass is the lower the acceleration is—an inverse proportion. We are going to use a spring in a launch jig to launch three balls straight up. The three balls are all about the same size, so the air affects them each about the same. But they have different masses, so their accelerations are different. The faster a ball accelerates while the spring is releasing, the faster the ball is moving as it leaves the ground. And the faster it is moving when it leaves the ground, the higher it goes.

In an experiment like this, velocity and acceleration are difficult to measure. So we will use the height each ball achieves to represent the acceleration the ball experienced when it was launched. The height is a bit tricky to measure too, but I think you will have fun doing it.

The launching jig is shown in the photo. There are four spring holes in the jig, with four different depths. To launch one of the balls, press the ball down on the spring with the sides of your thumbs or two fingers, pressing the ball all the way to the wooden block. Then snap your thumbs down suddenly to release the ball. This is tricky to do, and sometimes the ball does not go very straight, or even up, so plan on launching each ball 10 or 15 times from each spring position. When a launch does not work, just retrieve the ball and do it again.

To prevent the spring from flying out of the hole along with the ball, there is a small hole in the side of the jig near the bottom of each spring hole where you can insert a small screwdriver. After the spring is in the hole, insert the screwdriver and the spring is prevented from flying out.

While you or one of your teammates is launching the balls, another team member can mark how high the balls go. Use a long pole or 1 × 4 held next to the launch jig. For the lighter balls, and the greater spring compressions, the person monitoring the height may need to stand on a step stool or chair. Use your fingers to indicate on the pole the maximum height achieved for a given ball and a given spring compression. After you have seen the ball make it to that same maximum height three times, then measure the height above the launching jig with a measuring tape and record the value (in inches) in your lab journal.

Also measure and record the mass of each ball (in grams), and the height the spring extends above the jig for each hole (in mm). The spring height above the jig is the amount the spring is compressed for launch at each hole.

Analysis

Prepare a graph with four separate curves on it, one for each amount of spring compression. On the graph, the horizontal axis represents the mass of the ball. The vertical axis represents the maximum height achieved. Your instructor will help you in scaling the axes and plotting the points on each of your four curves. In your report for this experiment, address the following questions. Use the mass of the table tennis ball as a reference for these questions.

1. Compared to the table tennis ball, approximately how massive are the other two balls (twice the mass, five times the mass, etc.)?
2. According to the Newton's second law equation, if a mass is twice as much, what should happen to the acceleration? If a mass is five times as much, what should happen to the acceleration? In your case, for the masses you measured for the wooden ball and golf ball, how should their accelerations compare to that of the table tennis ball?
3. Using the maximum height achieved as an estimate for each ball's acceleration, how do the maximum heights compare to that of the table tennis ball (50% as high, 20% as high, etc.)? Do the height comparisons match with what you expected from the mass comparisons? Explain.

Chapter 11
Compounds and Chemical Reactions

The four compounds shown above are all examples of metallic salts. The green one contains nickel, the yellow one sodium, the red one potassium, and the blue one copper.

Salts are one of the most common types of compounds around. A salt forms any time a metal reacts with an element in Group 17 in the Periodic Table of the Elements. Salts also form when metals react with acids and when acids react with bases.

So much terminology! In this chapter, we will take a careful walk through the terminology—and the associated concepts—with a goal of understanding some of the basics of chemistry and chemical reactions.

OBJECTIVES

After studying this chapter and completing the exercises, you should be able to do each of the following tasks, using supporting terms and principles as necessary.

1. Identify common compounds and elements in chemical equations.
2. Explain what acids, bases, and salts are.
3. Identify the strength of acids and bases from their pH.
4. Describe the two main goals atoms seek to fulfill in chemical reactions.
5. Distinguish between ionic and covalent bonds, and identify the kinds of elements (metals or nonmetals) that participate in each.
6. State two pieces of evidence to support John Dalton's theory that "atoms combine in whole number ratios to form compounds."
7. Given appropriate supporting information, identify combustion, metal oxidation, and acid-base reactions.
8. Define the terms *oxidation* and *reduction*, and explain how they are involved in redox reactions.

VOCABULARY TERMS

You should be able to define or describe each of these terms in a complete sentence or paragraph.

1. acid	9. halide	17. pH indicator
2. alkali	10. halogen	18. polyatomic ion
3. aqueous solution	11. hydrogen ion	19. precipitate
4. base	12. hydroxide	20. product
5. caustic	13. ionic bond	21. reactant
6. chemical equation	14. neutralization	22. redox reaction
7. coefficient	15. oxidation	23. reduction
8. covalent bond	16. pH	24. salt

11.1 Tools for Chemistry

Any introductory study of chemistry—even a very basic one—takes the student into a bewildering world of names and terminology. If you are at all like me, you will appreciate an *organized and methodical* approach to learning about chemistry. That is what we are going to have in this chapter. As we begin this study, the first step is to acquire some tools to help us talk about compounds and chemical reactions. Let's get started.

Chemical Equations The ingredients that go into a chemical reaction are called *reactants*. Reactants are pure substances, either

223

elements or compounds. The new substances formed by a chemical reaction are called *products*. As with the reactants, the products of a chemical reaction are pure substances, either elements or compounds.

In chemical reactions, the reactants go in, and the products come out.

A *chemical equation* is a way of showing the reactants and products in chemical reactions. Chemical equations are similar to algebraic equations, and are very helpful for following the action in a chemical reaction. Figure 11.1 shows an example chemical equation for the reaction of hydrogen and oxygen to form water. In this figure, the chemical equation itself is shown in red and blue. The reactants are on the left side of the equation, and the products are shown on the right. An arrow, usually pointing to the right, shows which direction the reaction is going.

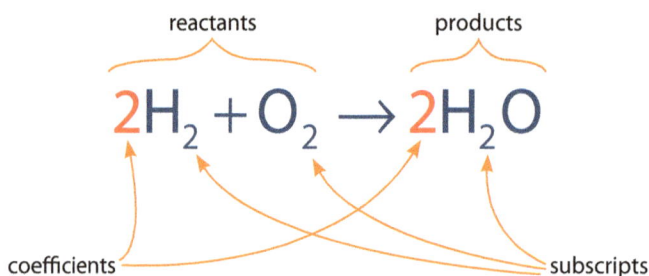

There are two kinds of symbols shown in this equation. First, the symbols for the reactants and products involved in the reaction are shown in blue. From the left, we have diatomic hydrogen (a gas), a molecule composed of two hydrogen atoms; diatomic oxygen (also a gas), a molecule

reactants — products

$$2H_2 + O_2 \rightarrow 2H_2O$$

coefficients — subscripts

Figure 11.1. The parts of the chemical equation.

composed of two oxygen atoms; and water, a molecular compound. Each of these substances is represented in the equation by its chemical formula. The chemical formula for hydrogen is H_2. The formulas for oxygen and water are O_2 and H_2O, respectively. The subscripts shown in the equation are part of the formulas of the elements and compounds in the reaction.

The second kind of symbol is the *coefficients*. The coefficients are in the equation to indicate the proportions of the different reactants and products involved in the reaction. The example reaction shown is read this way: *Two molecules of hydrogen react with one molecule of oxygen to form two molecules of water.*

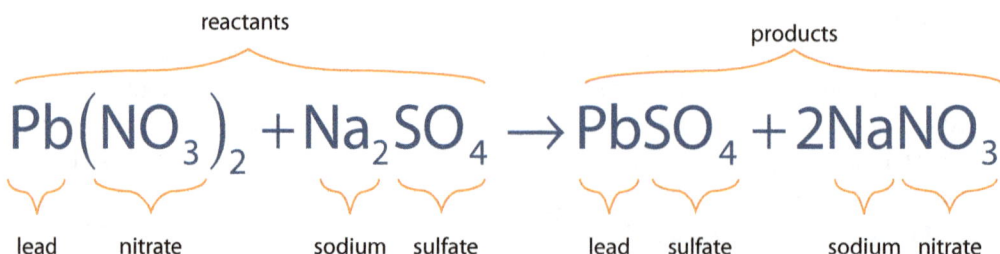

reactants — products

$$Pb(NO_3)_2 + Na_2SO_4 \rightarrow PbSO_4 + 2NaNO_3$$

lead   nitrate   sodium   sulfate   lead   sulfate   sodium   nitrate

Figure 11.2. Another example chemical equation.

The coefficients in a chemical equation indicate the proportions of reactants and products in the reaction.

Figure 11.2 shows another example of a chemical equation. Again, reactants are on the left and products are on the right. In this reaction, the two reactant compounds come apart and exchange their parts with each other to form two new compounds. The first reactant compound is lead nitrate, with the formula $Pb(NO_3)_2$. This substance reacts with sodium sulfate, Na_2SO_4. The reaction products are lead sulfate, $PbSO_4$, and sodium nitrate, $NaNO_3$. If you turn to a Periodic Table of the Elements (check the inside back cover of this book), you will see that lead and sodium are metal elements (atomic numbers 82 and 11, respectively). Nitrate and sulfate are *not* elements. What they *are* is what we are going to hit next.

Ions and Polyatomic Ions

Many chemical reactions involve ions. You have seen several times before in this text that an *ion* is an atom that has either gained or lost one or more electrons. As a result of gaining or losing electrons, the atom no longer has an equal number of positive charges (protons) and negative charges (electrons), and so the atom is a charged particle. Just about any kind of atom can become an ion, but the most common of all—and thus involved in many kinds of chemical reactions—is the hydrogen ion, illustrated in Figure 11.3. Look at the periodic table again to remind yourself that hydrogen is atomic number 1. A hydrogen atom only has one electron and one proton. So if a hydrogen atom loses its one electron, all that is left is the lone proton.

Other ions appear often in solutions. Recall from Chapter 6 that when crystals dissolve in solutions, the atoms dissociate. That is, the crystal comes apart and the individual ions from the crystal mix in with the molecules of the solvent. When sodium chloride dissolves, the sodium atoms are in the solution as ions with a +1 charge, and the chlorine atoms are in the solution with a –1 charge.

Figure 11.3. The world's most common ion—a hydrogen ion, also known as a proton.

The fact that so many compounds dissociate and dissolve in water makes water a terrific stage for chemical reactions. Solutions in water are called *aqueous solutions.* Lots of reactions occur by mixing aqueous solutions together and letting the ions combine with each other.

An aqueous solution is a solution with water as the solvent.

There is another very common form of ion that plays a major role in chemical reactions—the *polyatomic ion.* "Poly" means *many,* and "polyatomic" means *including many atoms.* A polyatomic ion is a group of nonmetal atoms joined together in a molecule with a net charge.

To explain this, first recall from Chapter 6 that nonmetals often combine together to form molecules. Examples that we looked at include water, H_2O, made of

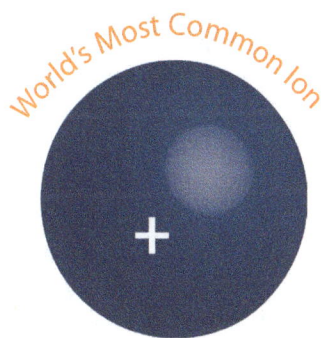

hydrogen and oxygen; methane, CH_4, made of carbon and hydrogen; and sulfur dioxide, SO_2, made of sulfur and oxygen. These are all molecules with several nonmetal atoms in the molecule.

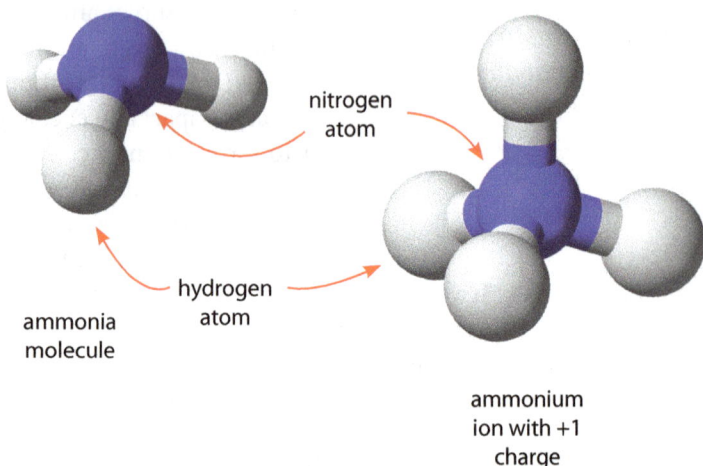

Figure 11.4. An ammonia molecule with an extra hydrogen ion attached becomes an ammonium ion.

Polyatomic ions are very similar objects, except for one thing: polyatomic ions have a net charge. They are not electrically neutral. Here's an example of one polyatomic ion to help you see how this works. Figure 11.4 shows an ammonia molecule (NH_3) side by side with an ammonium ion (NH_4)$^+$. These two molecules are exactly the same, except the ammonium ion has an extra hydrogen ion attached to it. Now, in the ammonia molecule, the number of protons and electrons in all the atoms is balanced, so the molecule has no net electrical charge. But when an extra hydrogen ion is attached, the molecule now has an electrical charge of +1. This molecule is an ion made of many atoms—a polyatomic ion.

A polyatomic ion is a group of nonmetal atoms joined together in a molecule with a net charge.

In chemical reactions, polyatomic ions act just like ions made of single atoms. There are dozens of these different polyatomic ions. A few of them are listed along with their formulas and charges in Table 11.1. Some of these ions appear regularly in compounds called *acids* and *bases*, which we discuss next. Note that the important thing here is not for you to memorize this list, although it is important to begin familiarizing yourself with these names. Our goal here is simply for you to know that these "many-atomed" ions exist, there are a lot of them, and they are very common. They play a big role in common chemical reactions.

Name	Formula	Charge
ammonium	NH_4	+1
bicarbonate	HCO_3	−1
carbonate	CO_3	−2
hydroxide	OH	−1
nitrate	NO_3	−1
phosphate	PO_4	−3
sulfate	SO_4	−2

Table 11.1. A few common polyatomic ions. Those shown in yellow can form acids. The one shown in orange is typically found in bases.

Now look back at Figure 11.2. You can now see that two of the polyatomic ions are in the reactant compounds, nitrate (NO_3) and sulfate (SO_4). These same two polyatomic ions are in the reaction products, but they have switched places. The lead nitrate has become lead sulfate, and the sodium sulfate has become sodium nitrate.

Acids and Bases One more tool that will help you understand chemical reactions is to know what *acids* and *bases* are. There are several different ways to define them, but here is a definition that will work for the moment: acids are compounds that produce hydrogen ions (+1 charge) in aqueous solutions, and bases are compounds that produce hydroxide ions (−1 charge) in aqueous solutions.[1] Figure 11.5 illustrates two acids and two bases dissociating to free up hydrogen ions or hydroxide ions. This figure also illustrates the meaning of the subscripts in chemical formulas—the proportions of ions in the compound are indicated by the subscripts.

Acids donate hydrogen ions (+1); bases often donate hydroxide ions (−1).

Several common acids are formed by hydrogen combining with one of the polyatomic ions (such

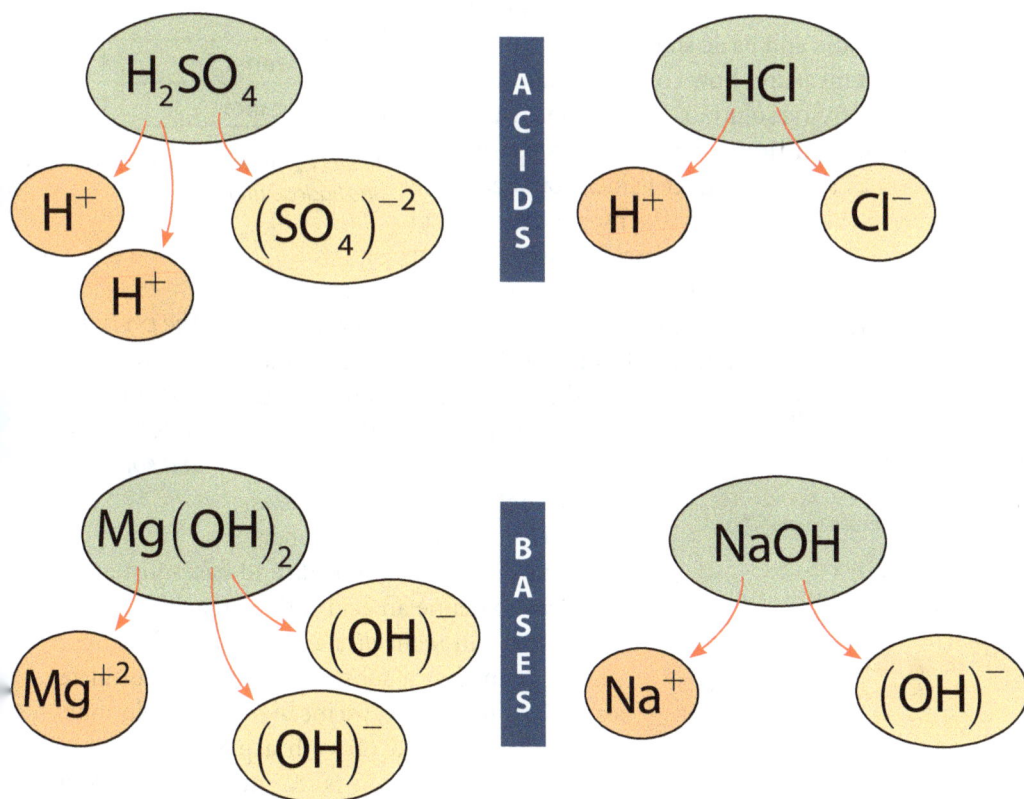

Figure 11.5. In aqueous solution, acids dissociate to provide hydrogen ions into the solution (top). Bases dissociate to provide hydroxide ions into the solution (bottom).

---

1   There are bases that do not contain hydroxide, but many common ones do. We'll look at some other bases shortly.

as those marked in yellow in Table 11.1). As Figure 11.5 illustrates, sulfuric acid (H_2SO_4) is two hydrogen ions joined to a sulfate ion. Other acids are composed of hydrogen joined to one of the elements in Group 17 (the 17th column) of the periodic table. Examples are hydrochloric acid (HCl) and hydrofluoric acid (HF).

Acids and bases have different strengths, measured by a parameter called the *pH*. The pH of a solution is a value ranging from 0 to 14 that indicates how acidic or how basic a solution is. On this scale, the pH of pure water is 7. This value is said to be "neutral," neither acidic nor basic. Solutions with a pH lower than 7 are acidic, and the lower the pH, the stronger the acid. Solutions with a pH above 7 are basic, and the higher the pH, the stronger the base. Table 11.2 lists the pH values for a number of common acidic and basic solutions.

The pH measures how concentrated the hydrogen ions are in the solution. Each value lower down the scale has ten times as many hydrogen ions in a solution as the next higher value. Both acids and bases can be very corrosive. The more acidic an acid is (lower pH), and the more basic a base is (higher pH), the more corrosive the solution is. So the pH also indicates how dangerous the solution is. Years ago when I was working on my car, I got a little battery acid on my shirt tail. The next day my shirt had holes in it where the acid destroyed the cloth.

Substance	pH
battery acid	0
stomach acid (HCl)	1
lemon juice, vinegar	2–3
orange juice	3–4
acid rain	4
black coffee	5
urine, saliva	6
pure water	7
sea water	8
baking soda	9
ammonia solution	10–11
soapy water	12
bleach	13
oven cleaner	13–14
drain cleaner	14

Table 11.2. pH values for common acidic and basic solutions.

The pH is a measure of the hydrogen ion concentration in a solution.

Now that you know about the strength of acids and bases and the pH scale, we can generalize our definitions for acids and bases just a bit. One way to define an acid is that an acid is a substance that can *neutralize* a base. Similarly, a base can be defined as a substance that can neutralize an acid. By neutralize, we mean that two solutions, mixed together, end up with a pH of 7—that is what a neutral solution is. Ammonia (NH_3) and calcium carbonate ($CaCO_3$) are both bases because they can neutralize acids, even though neither of them contains the hydroxide ion. We discuss more about neutralization soon when we get to chemical reactions.

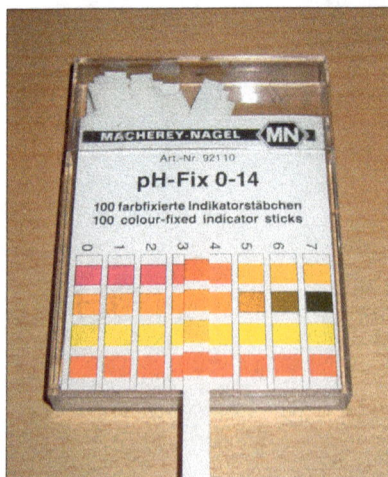

Figure 11.6. pH indicator paper. The paper strip is compared to the color chart to determine pH. A pH of 4 is indicated here.

When bases contain the hydroxide (OH) ion, they are often said to be *alkaline*. So an *alkali* is a base that contains hydroxide. Since many common bases do contain the hydroxide ion, the term *alkaline* is frequently used in place of the term *basic*. Another common term that means the same thing as alkaline is *caustic*.

Acids and bases neutralize each other.

Chemists use several different *pH indicators* to measure the pH of a solution. A pH indicator is a compound that changes color when exposed to substances of different pH values. This allows the pH to be determined visually. A common indicator is pH indicator paper, such as shown in Figure 11.6. One of the paper strips is moistened with the solution, and the colors are compared to the chart to determine the pH of the solution. In the figure, the colors indicate a pH of 4.

A fun and interesting fact is that some plants serve as natural pH indicators. The flowers of the hydrangea are blue in acidic soil and pink in alkaline soil, illustrated by the photos in Figure 11.7. Purple cabbage is another natural pH indicator.

Figure 11.7. The hydrangea is a natural pH indicator, displaying blue flowers in acidic soil, and pink flowers in alkaline (basic) soil.

Learning Check 11.1

1. What is a polyatomic ion?
2. What are acids and bases? Give two separate definitions.
3. What does the pH scale measure?
4. What do the coefficients in chemical equations indicate?
5. What are reactants?
6. What is an aqueous solution?
7. Which is the stronger acid, lemon juice or black coffee?
8. Which is the stronger base, baking soda or bleach?
9. Should we apply the term caustic to battery acid or to oven cleaner? Explain your answer.

11.2 How Compounds Form

When compounds are formed or broken apart, a chemical reaction is happening.

As you recall from Chapter 6, compounds are formed by chemical reactions. Chemical reactions can also break down compounds by separating the elements in the compound from one another. So in the most general sense, when elements combine together into a compound, or when elements in a compound are separated, a chemical reaction is happening. Substances are motivated to react by two main goals.

Main Goal #1 First, most elements are happier with a different number of electrons than they have when they are electrically neutral. In other words, elements are a lot like humans: no one is ever satisfied with what they have; people usually want something different! The electrons in atoms reside in orbitals, as you may recall from Section 1.3, and the atoms like to have their orbitals filled. If they have partially filled orbitals, they don't like it. So they try either to gain electrons or to lose electrons to get to an arrangement of electrons that suits them better.

Now, there are two ways open to an atom to fulfill this goal of having the right number of electrons—to get rid of extra electrons, or to grab some if they need just one or two more. Reactions involving *electron transfer* like this result in *crystals*.

Figure 11.8 illustrates this using egg cartons as an analogy for atoms. The two egg cartons represent two atoms, and the eggs represent the atoms' electrons. Assume that egg cartons want to be perfectly full if they can, with no eggs left over. Assume further that if the egg carton can't find a way to be full, then it will be happy for now if it can have only full rows of eggs, with no partially filled rows. If transferring one or more eggs from one carton to another helps each carton towards its goal, they do it.

When electrons are transferred from one atom to another, both atoms become ions because neither atom has a balance of protons and electrons any more. Their new electrical charges attract them together in a crystal lattice full of identical positive and negative ions. The bonds between the atoms in this case are called

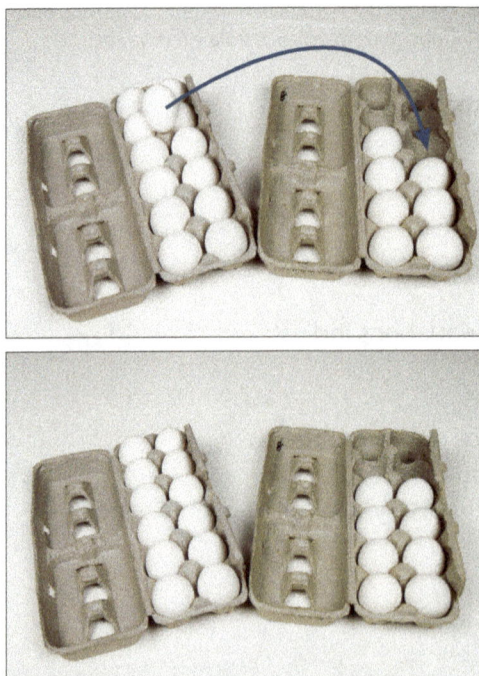

Figure 11.8. Relocating eggs in cartons as an analogy for electron transfer between atoms.

ionic bonds. Bonds between metal and nonmetal atoms are typically like this—ionic bonds in a crystal lattice.

Bonds between metal atoms and nonmetal atoms are typically ionic bonds.

Figure 11.9 illustrates the bonding sequence in more detail. In this figure, the black dots represent empty places the atoms want to fill with electrons. Atoms wanting more electrons are usually nonmetals, represented by green circles in the figure. The circles with negative signs represent electrons that aren't fitting into the filled orbitals in the atoms. The atoms would be happy to get rid of these if they could. The atoms in this situation are the metals, represented by the yellow circles. The sequence in the figure shows the electron transfer, the resulting ions, and the crystal formation that occurs as all the electrical forces find their equilibrium points.

The second way for an atom to have a more desirable number of electrons is for the atom to *share* electrons with one or more other atoms. This is the kind of reaction that makes *molecules*, and the bonds between the atoms in this case are called *covalent bonds.* All the molecule images from back in Chapter 6 represent atoms bonded together by the sharing of electrons.

Bonds between nonmetal atoms are typically covalent bonds.

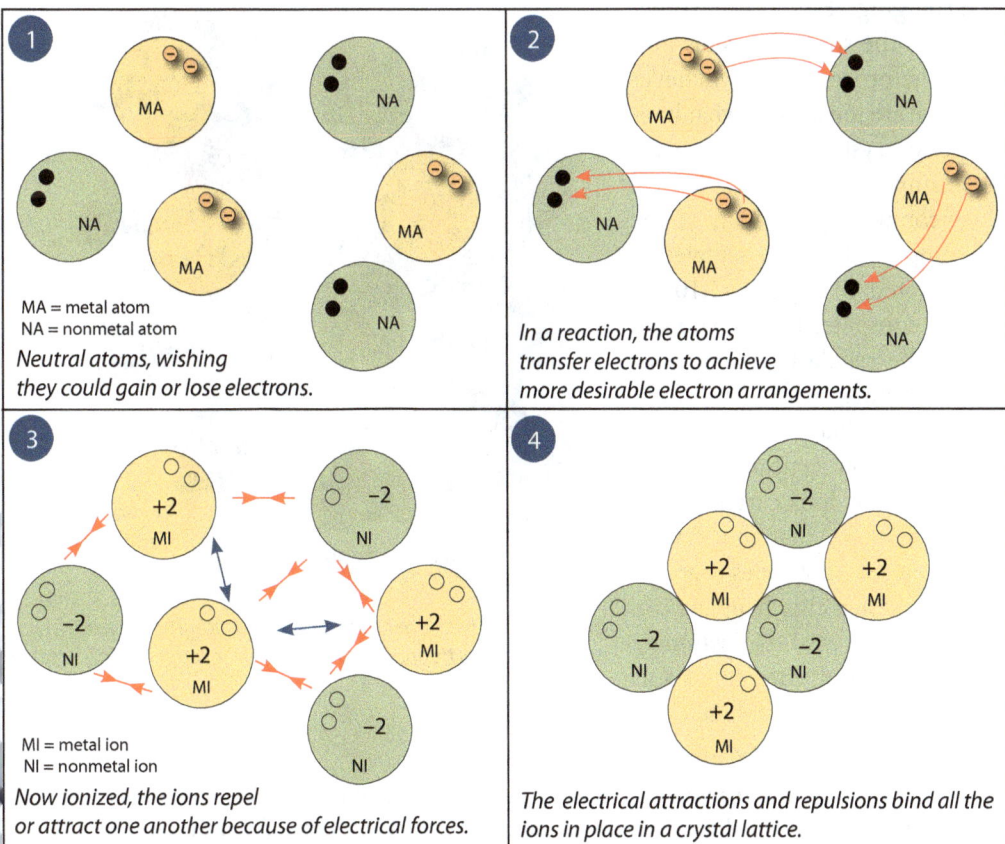

1 MA = metal atom
NA = nonmetal atom
Neutral atoms, wishing they could gain or lose electrons.

2 *In a reaction, the atoms transfer electrons to achieve more desirable electron arrangements.*

3 MI = metal ion
NI = nonmetal ion
Now ionized, the ions repel or attract one another because of electrical forces.

4 *The electrical attractions and repulsions bind all the ions in place in a crystal lattice.*

Figure 11.9. In ionic bonding electron transfer creates ions, leading to crystal formation.

In forming covalent bonds, some of the electron orbitals in the atoms sort of meld together, and the shared electrons reside in these combination orbitals. The net effect is to create a strong, covalently bonded, molecular package. Bonds between nonmetal atoms are typically like this—covalently bonded atoms in molecules.

> **Main Goal #2**   The second main goal applies to atoms as well as to polyatomic ions, which is: to live in a world of electrical neutrality. If there are any extra electrical charges around, the charged objects seek oppositely charged objects to attach to.

We saw this process at work in the ionic bonding shown in Figure 11.9. After the electron transfer, the atoms are happy with their electron arrangements, but now they are *ions*. So what do these ions do next? They all pile in together in a big, geometrical crystal lattice that evens out the electrical charge.

Another way electrical charge drives chemical action involves polyatomic ions. A polyatomic ion is a molecule, held together by covalent bonds between the atoms. In other words, the atoms are sharing electrons to satisfy their desire for electron arranging. But again, sometimes the sharing creates ions in the process. And whenever there are ions, guess what happens: crystal lattice! A lovely example of this, which we will encounter again later, is the compound copper sulfate, illustrated in Figure 11.10. In this computer model, you can see the regularly spaced copper atoms, which are ions, and the sulfate (SO_4) ions with their yellow, central sulfur atoms. If you look carefully, you can also see that this crystal has water molecules trapped in it. When this happens, we say the crystal is *hydrated*,[2] and the result for copper sulfate is the beautiful blue crys-

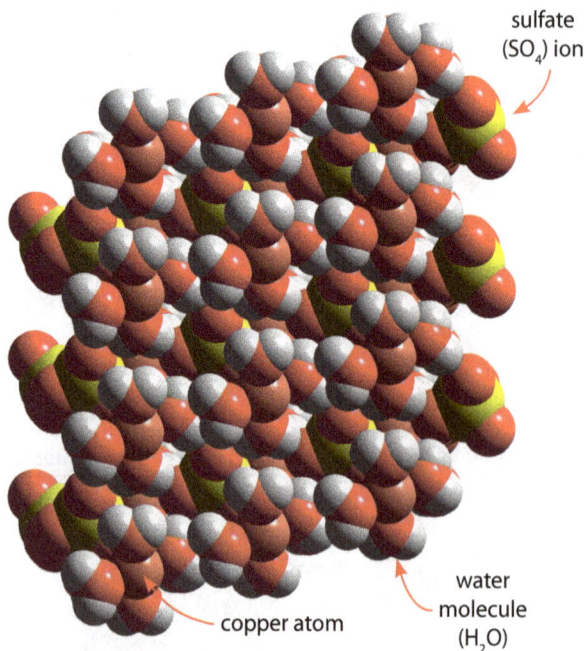

Figure 11.10. The copper sulfate crystal lattice. The presence of water molecules trapped in the lattice gives copper sulfate crystals their stunning blue color.

sulfate (SO_4) ion

water molecule (H_2O)

copper atom

---

2   In this case with the water molecules, the full name of the crystal compound becomes copper sulfate pentahydrate, with the formula $CuSO_4 \cdot 5H_2O$. Throughout the lattice, there are four water molecules attached to every copper atom, and one more water molecule attached to each sulfate ion, giving five H_2O molecules in the lattice for each unit of $CuSO_4$.

tals shown on page 226. Without the water molecules, copper sulfate turns to a light blue powder.

To wrap up this section, I'll just remind you of an important fact about compounds we first saw back in Section 6.4. Regardless of whether a compound is formed as a crystal or as molecules, there are always definite ratios of the numbers of atoms of different elements involved. John Dalton first proposed this in 1803 as part of his atomic model (Section 1.4). As you can see from the geometry in crystals and the clustering in molecules, he was correct.

Dalton: Atoms combine in whole number ratios to form compounds.

Learning Check 11.2

1. Describe the two main goals atoms are pursuing when they form compounds.
2. Describe two alternatives atoms have for acquiring a more desirable arrangement of electrons.
3. Distinguish between ionic and covalent bonds, including the types of elements involved in each (metal or nonmetal).
4. What evidence have we seen to support John Dalton's theory that atoms combine together in whole number ratios to form compounds?
5. What kinds of bonds are holding together the atoms in a polyatomic ion?

11.3   Chemical Reactions

The number of different chemical reactions is beyond counting. Even if we leave out the highly complex reactions that take place in living organisms (which themselves are beyond number), we still have so many different ways elements can combine that it is staggering. This richness in nature makes our world fascinating and full of endless variety, but it can be daunting to the young student of chemistry!

But beginning your study of chemistry is not about memorizing thousands of different chemical reactions. Initially, all you really need to be concerned with is recognizing some general types of reactions and representative examples of these types. Just knowing this will help you to understand many of the common reactions taking place in the world around you. In this section, we look at several different basic kinds of reactions. In each category, we consider some representative examples. In the examples, we look at the chemical equation for the reaction, as well as some images of the substances involved.

Salts   Lots of chemical reactions result in the formation of salts. This term is so important that we should begin with a short explanation.

Turn please to the Periodic Table of the Elements found on the inside back cover of this book. The columns in the periodic table are called *groups*. The elements in Group 17 are important for our discussion here. The elements in Group

17—fluorine, chlorine, iodine, bromine and so on—are called *halogens*. Compounds formed with halogens are sometimes called *halides*.

The term *salt* is used to describe a class of compounds that can be formed in a couple of different ways. One way salts are formed is by a metal combining with a halogen. The salts formed from the reaction of a metal with a halogen are called *halide salts*. The most obvious example of a halide salt is sodium chloride (NaCl), also known as table salt. Other examples are potassium chloride (KCl) and silver iodide (AgI). Silver iodide is very sensitive to light and is one of the main compounds involved in film photography.

Salts are also formed by metals combining with some polyatomic ions. From Table 11.1, the ions that produce salts are nitrate, phosphate, and sulfate. Examples of salts formed from these ions include sodium phosphate (Na_3PO_4), silver nitrate ($AgNO_3$), and magnesium sulfate ($MgSO_4$). Magnesium sulfate is also known as Epsom salt.

We encounter several salts in the coming paragraphs, so I placed this short section here so you will know about salts before we plow in and can watch for them as we go. We define salts more carefully when we get to the section on acid-base reactions.

Combustion Reactions

As you recall from Section 9.5, combustion is a chemical reaction between a fuel and oxygen. One of the most common combustion reactions is the burning of methane. The chemical equation for this reaction is

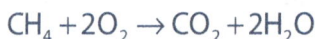

$$CH_4 + 2O_2 \rightarrow CO_2 + 2H_2O$$

This equation shows methane (CH_4) reacting with oxygen (O_2) to produce carbon dioxide (CO_2) and water (H_2O). Graphically, the reaction can be depicted as shown in Figure 11.11. In this image, the carbon, oxygen, and hydrogen atoms are represented by

Figure 11.11. Combustion reaction of methane with oxygen to produce CO_2 and water.

black, red, and white spheres, respectively. Notice how the coefficients in the equation represent the proportions of molecules shown in the figure. Both oxygen and water have coefficients of 2, and in the figure each of these substances is represented by two molecules.

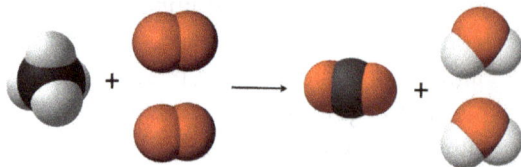

The natural gas we use for cooking and heating is composed primarily of methane. Methane combustion produces a pure blue flame, illustrated in Figure 11.12.

A combustion reaction we have seen before is the burning of hydrogen to make water. The chemical equation for this reaction is

$$2H_2 + O_2 \rightarrow 2H_2O$$

This equation shows two molecules of hydrogen (H_2) reacting with one molecule of oxygen (O_2) to produce two molecules of water (H_2O). The burning of hydrogen is illustrated in Figure 11.13. In this image, I am touching a balloon full of hydrogen with a burning wooden stick, causing the hydrogen to explode. This is not a fire; it's an explosion. (You can see a fragment of blue balloon flying off to the right.) The only product of

Figure 11.12. Methane flame.

this reaction is *water*—that's right, water. Students often have a hard time believing that a fiery explosion can produce water, but it does. The water is in the vapor phase because of the heat of the explosion, but it is water nonetheless.

Unlike the burning of methane, burning hydrogen does not produce any CO_2, a greenhouse gas. Most combustions, especially those involving the fossil fuels coal, oil, and gasoline, produce CO_2. These combustions also typically produce carbon monoxide (CO), and various nitrogen compounds such as nitric oxide (NO) and nitrogen dioxide (NO_2). These are all toxic pollut-

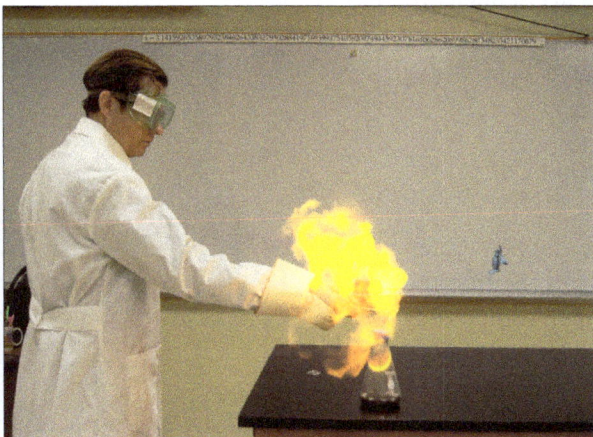

Figure 11.13. Hydrogen flame.

The reaction of hydrogen with oxygen produces nothing but pure water (and heat!).

ants, harmful to the environment and to us. But burning hydrogen produces nothing but good old H_2O (and a lot of heat!).

Oxidation Reactions

In Section 9.5, we also looked at some common reactions involving the oxidation of metals. The most well-known metal oxidation is probably the reaction of iron with oxygen to produce iron oxide, or rust. The equation for iron rusting is

$$4Fe + 3O_2 \rightarrow 2Fe_2O_3$$

The compound Fe_2O_3 is called iron(III) oxide. This is what rust is, and although I know you already know what rust looks like, I can't resist showing you the tremendous photo of a rusting train in Figure 11.14.

Figure 11.14. An excellent display of iron(III) oxide.

Some interesting oxidation reactions involving copper result in the formation of copper carbonate ($CuCO_3$) and copper hydroxide ($Cu(OH)_2$). On page 168, I noted that copper carbonate was called *verdigris*. Verdigris is a general term for the green coating that forms on copper when the copper is exposed to the air. In addition to the copper carbonate, verdigris also generally involves a few other compounds, such as copper hydroxide and copper chloride. These compounds are all part of the natural green coating that forms on copper, famously on the Statue of Liberty (Figure 11.15).

The natural chemical reaction is

$$2Cu + H_2O + CO_2 + O_2 \rightarrow Cu(OH)_2 + CuCO_3$$

In this reaction, the reactants are copper, water, carbon dioxide, and oxygen. You can see that the reaction products are the two copper compounds I mentioned above—copper hydroxide and copper carbonate. When this reaction takes place near sea water (as with the Statue of Liberty), copper chloride (a halide salt) is also produced because of the sodium chloride salt in the air.

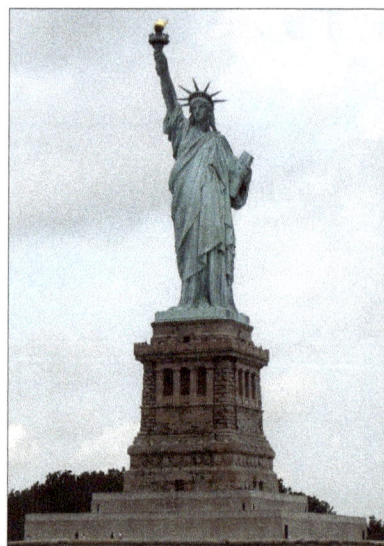

Figure 11.15. World's most famous display of verdigris.

Redox Reactions

Before we leave off talking about oxidations, I will define oxidation a bit more carefully. This will be easier now that you know something about oxidation. Strictly speaking, the term oxidation does not mean "reacting with oxygen," although oxidation reactions like rusting often do involve oxygen. Instead, it means "losing electrons." Recall from Section 11.1 that metals form ionic bonds. They do this by becoming ions, and when metals ionize, they typically do it by losing electrons. This is what oxidation is—losing electrons. The reason this is called oxidation dates back to the 18th century when chemists were thinking that this always involved a reaction with oxygen. But some oxidation reactions do not.

Whenever an element loses electrons, another element gains them. The word for gaining electrons is *reduction*. When compounds are formed, nonmetals are usually the elements gaining the electrons. Recall again that ionic bonds form by one element transferring electrons to another, and that ionic bonds involve a metal and a nonmetal. Now you see that what is happening is that the metal is oxidized because it loses electrons. The nonmetal is reduced because it gains electrons. There is a well-known mnemonic

LEO the lion says GER!
LEO: Lose electrons—oxidize
GER: Gain electrons—reduce

Figure 11.16. Everyone's favorite mnemonic for remembering oxidation and reduction.

to help you remember which is which, shown in Figure 11.16.

Oxidation and reduction always occur together.

Reduction and oxidation always happen together. Whenever one element is oxidized, another is reduced. Since oxidation and reduction always happen at the same time, scientists call these reactions *redox* reactions. Rusting iron, oxidizing copper, and even combustions and explosions are commonly called oxidations. But these reactions are really all part of the class of reactions called redox reactions.

Deeper study of redox reactions goes beyond what we are going to do here. But you might appreciate knowing that redox chemistry is all over the place. Figure 11.17 shows a uranium mine in Utah. The rock formed when groundwater conditions were oxidizing, producing red rock. When groundwater chemistry changed to reducing conditions, certain layers were whitened as the reducing water slowly seeped through the rock. Figure 11.18 shows an example of a situation called *acid mine drainage*. Oxidation of minerals like iron sulfide produces drain water that is both acidic and toxic. The water then dissolves minerals in surrounding rocks.

Redox chemistry is also prevalent in biology. As just one example, photosynthesis is a redox reaction. Photosynthesis is the process plants use to produce their green leaves. Photosynthesis reduces the carbon dioxide in the air to

Figure 11.17. Redox chemistry in groundwater produces color variations in rock formations.

Figure 11.18. Oxidation of iron sulfide produces acidic water, which then dissolves minerals in rocks. This process is colorful, but environmentally devastating.

237

provide energy for the plant and oxidizes water into oxygen, which the plants release into the atmosphere.

Finally, redox reactions are everywhere in industry; the box on the next page provides an example.

Precipitation Reactions

Let's go back to verdigris and copper carbonate. The next Experimental Investigation involves producing both copper carbonate and copper hydroxide, but by different chemical reactions than the ones nature uses to cover the Statue of Liberty with its verdigris coating. (These are more convenient than the natural reaction because they don't take 100 years.) We begin the experiment by combining an aqueous solution (that is, a solution in water) of copper sulfate and an aqueous solution of sodium hydroxide to produce sodium sulfate and copper hydroxide. The reaction is

$$CuSO_4 + 2NaOH \rightarrow Na_2SO_4 + Cu(OH)_2$$

You can recognize two of these compounds as salts, the two containing the sulfate ion. All that happened in this reaction was that everyone changed partners—the copper and sodium metals dissociated from the polyatomic ions they were attached to (sulfate and hydroxide) and bonded with the other polyatomic ion.

The beautiful blue copper hydroxide is shown in Figure 11.19. Note here that we make this reaction happen by combining two *solutions*—one of copper sulfate and one of sodium hydroxide. But when the reaction occurs, the copper hydroxide does not remain in solution. It *precipitates* out of the solution. This makes it easy to separate the copper hydroxide from the solution. All we have to do is pour the liquid through a filter. The water and the sodium sulfate in solution pass through the filter. But the copper hydroxide *precipitate* does not.[3]

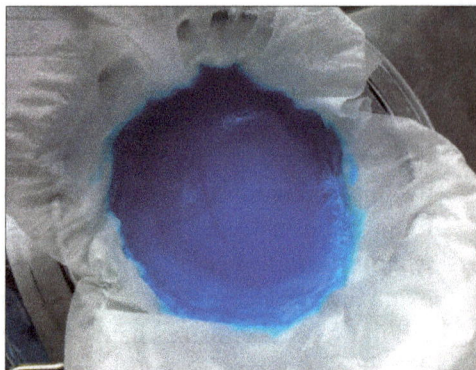

Figure 11.19. Copper hydroxide.

When a compound comes out of solution it precipitates. After doing so, it is called a precipitate.

We use another precipitation reaction to produce copper carbonate. This reaction involves the same two salts as the previous one. One of the reactants is sodium bicarbonate ($NaHCO_3$), a compound that contains the bicarbonate ion. Sodium bicarbonate is baking soda, a compound found in every cook's kitchen. The equation for this reaction is as follows:

---

3   Note that the word *precipitate* is used in chemistry both as a verb and a noun.

Scientists, Experiments, and Technology

Redox reactions play a major role in industry. Industrial processes all begin with raw materials such as aluminum and copper. In nature, these metals are usually found in ores such as bauxite (aluminum) and malachite (copper). Bauxite contains aluminum hydroxide ($Al(OH)_3$), and malachite contains copper carbonate ($CuCO_3$). To use the metals, they have to be extracted and purified.

The ores contain a lot of foreign material other than the metal compounds, so the metal compounds must first be separated from other compounds in the ore. Various refining and smelting processes then come into play to convert the hydroxide or carbonate compound into an oxide. Once the metal is an oxide, it can undergo reduction to become a pure metal. The facilities that perform these operations are huge, dirty, hot, loud, and dangerous—and very, very interesting!

For example, copper carbonate is literally roasted, converting the $CuCO_3$ into copper oxide, CuO. The copper oxide is then reduced by using carbon monoxide (CO) to pull away the oxygen atoms, converting the CO into CO_2. This process reduces the CuO to produce pure copper, and it oxidizes the CO to make CO_2.

In the case of aluminum, the $Al(OH)_3$ is converted to aluminum oxide, Al_2O_3 (also called alumina), by heating. But the reduction reaction for aluminum is a lot more difficult than for copper because aluminum is more reactive. The aluminum oxide is melted and mixed with a molten aluminum-fluoride salt. Then a huge electric current of about 100,000 amps is passed through this extremely hot, molten, metal brew, and the aluminum metal is reduced. The process not only generates CO_2, but also the acid gas hydrogen fluoride (HF). This acid is neutralized with a base to make the salt sodium fluoride (NaF).

You will learn more about acid-base reactions and the acid hydrogen fluoride on the next two pages.

$$CuSO_4 + 2NaHCO_3 \rightarrow CuCO_3 + Na_2SO_4 + CO_2 + H_2O$$

Once again, the metals traded places. But this time some of the hydrogen, carbon, and oxygen atoms were cut loose to form two other familiar compounds: carbon dioxide and water. You will definitely see the carbon dioxide produced by this reaction—it fizzes like a shaken up soft drink! The copper carbonate produced by the reaction precipitates out of solution. As we did with the copper hydroxide, we just filter the solution to capture the lovely blue-green copper carbonate, shown in Figure 11.20. There is not much use for our copper hydroxide or copper carbonate compounds, but they sure are nice to look at.

Figure 11.20. Copper carbonate.

Acid-Base Reactions

At the beginning of this section, I gave you some examples of salts. I delayed defining the term until later, but now is the time for a definition. When an acid and base react, the acid and base neutralize each other, and the products of the reaction are a salt, water, and sometimes CO_2. So a better definition for salt is that a salt is what you get when an acid is neutralized by a base. As with all the other kinds of reactions, there are many different acid-base reactions. Here, we take a look at a few of them.

In Section 11.1, I mention that acids produce hydrogen ions in water and bases produce hydroxide ions in water. Most of the time, you can identify an acid by the hydrogen atom(s) at the beginning of the chemical formula. Likewise, you can usually identify a base by the hydroxide ion(s) at the end of the chemical formula. There are exceptions, such as the base ammonia (NH_3), but this is a pretty good rule of thumb.

One well-known base is sodium hydroxide (NaOH), historically called *lye*. Lye is a strong base and the main ingredient in drain cleaner. Long ago, lye was used to make a crude, harsh soap. Sulfuric acid (H_2SO_4) is a very strong acid. When sodium hydroxide and sulfuric acid react, the products are the salt sodium sulfate (Na_2SO_4) and water. The reaction is

$$2NaOH + H_2SO_4 \rightarrow Na_2SO_4 + 2H_2O$$

Sodium sulfate used to be prized for its medicinal properties, and for this reason the man who discovered it back in 1625 named it *sal mirabilis*, or miracle salt. Sodium sulfate is a white, crystalline salt, just like sodium chloride.

A common everyday example of an acid-base reaction is the use of a base to neutralize stomach acid. Two common products used for this purpose are Tums and Rolaids. Now, as a teenager you may not yet have had opportunity to make use

Scientists, Experiments, and Technology

The art of glass etching with acid was invented in the 19th century. Hydrofluoric acid (HF) was used for the process because it is a very corrosive acid that bonds with the silicon compounds in glass. Artisans would apply a mask of wax to protect parts of the glass that were not to be etched and apply the acid to the exposed glass.

Hydrofluoric acid is classed as a weak acid, but that is only because it doesn't ionize very much in water. The acid is extremely dangerous, and even a small amount on the skin can be fatal.

A contemporary and very similar use for this acid is etching wafers of silicon in the microfabrication of computer chips. As shown in the graphic, the orange layer is a supporting base layer, or substrate, made of silicon. On top of this is a layer of silicon dioxide (shown in green), which is glass. In the upper image, a protective mask (shown in yellow) is applied using photographic techniques. Then the acid is used to etch away the portions of the silicon dioxide layer not protected by the mask. In this way, the transistors and diodes in the computer chip are fashioned into the surface of the chip. Contemporary technologies allow for feature sizes (such as the width of the peninsula on the right side of the image) in the range of 30 nm, which is less than 10% of the wavelength of light. With features this small, engineers are able to fit billions of devices in the area of a postage stamp. This is why you can fit billions of bytes of data on a small jump drive on a key chain.

of these products, but as you get older you probably will. Your stomach produces hydrochloric acid (HCl) to digest your food. Sometimes, depending on what a person has been eating and drinking, one can feel a burning sensation in the stomach caused by an oversupply of stomach acid. When that happens, you are ready for an acid-base neutralization reaction.

Tums is made of calcium carbonate ($CaCO_3$), with some other ingredients added to make it taste good. Calcium carbonate is a mild base that is found everywhere in nature—sea shells, egg shells, limestone, chalk—all these are made of calcium carbonate. When you munch a tablet or two of Tums, here is what happens:

$$CaCO_3 + 2HCl \rightarrow CaCl_2 + CO_2 + H_2O$$

Those suffering from excess stomach acid experience almost instantaneous relief when some of that hydrochloric acid is converted into harmless calcium chloride salt. Rolaids is also made primarily of calcium carbonate, but Rolaids also contains a bit of the mild base magnesium hydroxide ($Mg(OH)_2$). This compound also reacts with stomach acid to produce a salt, as shown in this chemical equation:

$$Mg(OH)_2 + 2HCl \rightarrow MgCl_2 + H_2O$$

The magnesium hydroxide neutralizes the hydrochloric acid to produce magnesium chloride—a salt—and water.

When mixed in water, magnesium hydroxide forms a thick, milky mixture. This mixture goes by the name Milk of Magnesia, and is often used for its medicinal value.

Learning Check 11.3

1. Write a paragraph explaining the difference between acids, bases, and salts.
2. What is precipitation, and when does it occur?
3. Explain what oxidation and reduction are.
4. Why would someone claim that hydrogen is the ultimate clean fuel?
5. Identify the reactions below as combustion, metal oxidation, or acid-base.

   a. $C_3H_8 + 5O_2 \rightarrow 3CO_2 + 4H_2O$

   b. $NaOH + HF \rightarrow NaF + H_2O$

   c. $2KOH + H_2SO_4 \rightarrow K_2SO_4 + 2H_2O$

   d. $4Al + 3O_2 \rightarrow 2Al_2O_3$

6. For item 5a above, draw a picture representing that reaction similar to the picture shown in Figure 11.11. The propane molecule (C_3H_8) looks like this:

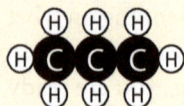

Chapter 11 Exercises

Answer each of the questions below as completely as you can. Write your responses in complete sentences.

1. Explain what a salt is.
2. How do we know that atoms always combine in whole number ratios?
3. When iron oxide (Fe_2O_3) is heated to 1,250 °C and surrounded with carbon monoxide (CO), the iron oxide becomes pure iron metal (Fe). Is the iron being reduced or oxidized in this process? Explain how you know.
4. Describe the two main goals atoms seek to fulfill when engaging in chemical reactions.
5. From the pH table below, identify each substance as a strong acid, weak acid, strong base, or weak base.

blood, 7.4	egg yolk, 6	stomach acid, 2
hydrochloric acid, 0	rainwater, 5	baking soda, 9
sea water, 8	Milk of Magnesia, 10	saliva, 6
sodium hydroxide, 14	egg white, 8	Drano, 13

6. From the table below, indicate which compounds are acids, which are bases, and which are salts. Write a sentence for each group of compounds explaining how you were able to identify the compounds in that group.

KCl	$Mg(OH)_2$	$Ca(OH)_2$
NaF	H_3PO_4	LiBr
KOH	$CaCl_2$	$Fe(NO_3)_3$
H_2SO_4	HCl	NH_3
$CuSO_4$	HF	NaCl
HNO_3	$MgCl_2$	NaOH

7. Distinguish between ionic and covalent bonds.
8. From the table below, look up each pair of elements in the periodic table and write down their names. Then indicate what kind of bond forms if the given elements combine to form a compound.

O and N	Hg and S	Ni and O
Fe and O	C and O	Cu and Cl
Ca and Cl	H and O	Na and F
K and P	S and O	Li and Br

Experimental Investigation 9: Observing Chemical Reactions

Overview

- Synthesize (make) three of the compounds contained in the verdigris finish on the Statue of Liberty.
- Watch six chemical reactions (conducted by an adult instructor) and document your observations.
- *The goals of this investigation are to gain experience identifying types of compounds in equations, and to associate chemical formulas and chemical equations with visual observations of what the compounds and reactions look like.*

Basic Materials List

- copper sulfate, sodium hydroxide, sodium bicarbonate, hydrochloric acid
- distilled water
- graduated cylinder, 100 mL
- beaker, 250 mL (2)
- flask, 250 mL (2)
- plastic spoons, large funnel (2), filter paper, paper towel
- plastic gloves, hot gloves, protective eyewear
- frying pan (small), spatula, scraper
- burner and burner tripod.

In this investigation you are going to observe six separate chemical reactions, making careful notes of what you observe in your lab journal. The reactants and products of these reactions are compounds we discussed in Chapter 11. Also, the reactions are all inter-related and fun to watch. It should be interesting to see the chain of events that leads from one set of compounds to the next.

Safety Precautions

1. Sodium hydroxide is a very corrosive base, and hydrochloric acid is a very corrosive acid. Do not let either substance contact your skin. Wear gloves and protective eyewear when handling NaOH and HCl. Rinse all glassware thoroughly in water.

2. Wear hot gloves and protective eyewear when working with the burner and hot pan.

Let's begin with some notes on symbols for chemical equations. To help keep track of the compounds in reactions, it is customary to place parentheses after each compound. Inside the parentheses is a symbol indicating what form the compound is in. Here are the symbols used in this Experimental Investigation:

- (aq) means the compound is in an aqueous solution. This could refer to a compound like a salt that is dissolved in water. Also, acids and bases are solutions in water, so they are denoted with this symbol.
- (s) means the compound is a solid.
- (l) means the compound is a liquid.
- (g) means the compound is a gas.

Also, sometimes we need to make it clear in a chemical equation that the reaction doesn't just happen by itself. For example, the reactants may need to be heated to a certain temperature in order for the reaction to take place. This is indicated by placing symbols above and below the reaction arrow like this:

$$\xrightarrow[290°C]{\Delta}$$

The triangle above the reaction arrow means the reactants are being heated, and the temperature underneath indicates the temperature the reactants must reach for the reaction to take place.

As mentioned in the chapter, the verdigris coating on the copper of the Statue of Liberty contains copper hydroxide, copper carbonate, and copper chloride. We *synthesize* (make) each of these in the reactions you observe.

REACTION 1

We begin with this reaction:

$$CuSO_4(aq) + 2NaOH(aq) \rightarrow Na_2SO_4(aq) + Cu(OH)_2(s)$$

This equation says that aqueous solutions of copper sulfate and sodium hydroxide react to produce an aqueous solution of sodium sulfate and solid copper hydroxide. Copper sulfate is a salt: the blue crystals pictured on page 226. Sodium hydroxide, also known as lye, is a strong base. The reaction produces another salt, sodium sulfate, as well as *solid* copper hydroxide. Any time mixing solutions together produces a solid, you have a precipitation reaction. The precipitate you get should resemble the compound pictured in Figure 11.19.

Your instructor will conduct the reactions as follows. For each solution, use the graduated cylinder to measure out 75 mL of distilled water. Distilled water is used because tap water contains dissolved salts, which produce all kinds of deposits we do not want. Weigh out 10 g of copper sulfate crystals and add them to the water in a 250-mL beaker. Stir this with a plastic spoon until the crystals are dissolved. (Students can help with this task. It takes about 15 minutes for the crystals to dissolve, so be patient!) Weigh out 10 g of sodium hydroxide pellets, being careful not to let them come in contact with your skin. Add the sodium hydroxide pellets to the water in a 250-mL flask. Swirl the flask until the pellets have completely dissolved. The solid $CuSO_4$ and NaOH are shown in the two pictures below.

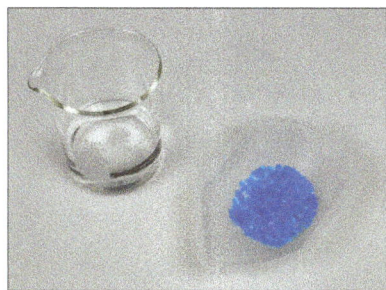

The two solutions appear as in the image to the right. Carefully pour the NaOH solution into the $CuSO_4$ solution. Students should observe carefully what happens and document their observations in their lab journals. Note the appearance and color of the reaction products, along with other details. Pour the contents of the beaker into a large funnel lined with filter paper or paper towel. Leave the $Cu(OH)_2$ precipitate to dry for a day or two. As it dries, examine it further.

REACTION 2

The chemical equation for our second reaction is as follows:

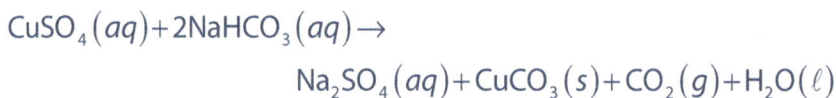

$$CuSO_4(aq) + 2NaHCO_3(aq) \rightarrow$$
$$Na_2SO_4(aq) + CuCO_3(s) + CO_2(g) + H_2O(\ell)$$

Again, a $CuSO_4$ solution is one of the reactants. The other reactant is a solution of sodium bicarbonate, also known as baking soda. The reaction again produces the aqueous solution of Na_2SO_4 and a precipitate. The precipitate this time is copper carbonate, pictured in Figure 11.20. The reaction also produces a lot of gas, and it should be obvious to you from what you see that a gas is being produced. The gas is carbon dioxide. The reaction also produces some water.

Make the $CuSO_4$ solution in the 250-mL beaker as before. Weigh out 10 g of sodium bicarbonate. Measure out 75 mL of distilled water in the graduated cylinder, and transfer it to the flask. Add the $NaHCO_3$ and swirl for a moment until the $NaHCO_3$ is more or less dissolved. (Sodium bicarbonate doesn't dissolve very well in water.) Now pour the $NaHCO_3$ solution—a small amount at a time—into the $CuSO_4$ solution. If you pour too much or too fast, this reaction will "boil over" the beaker and make a mess. Pour in small increments until the $NaHCO_3$ solution is gone.

Students document their observations as before. Pour the contents of the beaker into a large funnel with filter paper, and let it drain for a few minutes.

REACTION 3

The chemical equation for our third reaction is as follows:

$$CuCO_3(s) \xrightarrow[290°C]{\Delta} CuO(s) + CO_2(g)$$

This equation says we are going to cook the copper carbonate we made in the previous reaction. When the copper carbonate reaches 290°C, it converts to copper oxide, releasing carbon dioxide gas as well.

To heat the copper carbonate, just heat it in a frying pan over a propane or methane gas flame. (A hot plate does not get hot enough.) After draining off the excess water, pour the blue $CuCO_3$ precipi-

tate from Reaction 2 into the frying pan over a high flame, as shown at the bottom of the previous page.

As this material is heated, first observe how the water separates from the copper carbonate. Then make your observations as the copper carbonate converts to copper oxide. The complete reaction takes 10 or 15 minutes of heating over a high flame. The final product of this reaction, copper oxide, is a dark brown powder, shown to the right.

REACTION 4

The chemical equation for our fourth reaction is as follows:

$$CuO(s) + 2HCl(aq) \rightarrow CuCl_2(aq) + H_2O(\ell)$$

One of the reactants this time is the copper oxide powder produced by the previous reaction. CuO does not dissolve in water, but combined with hydrochloric acid (HCl), an aqueous solution of copper chloride (a salt) is produced, along with some more water.

Place the CuO powder in a 250 mL beaker, as shown to the right. Add small amount of HCl solution, just enough to get all the CuO to react completely. Swirl the beaker. Students write down their observations.

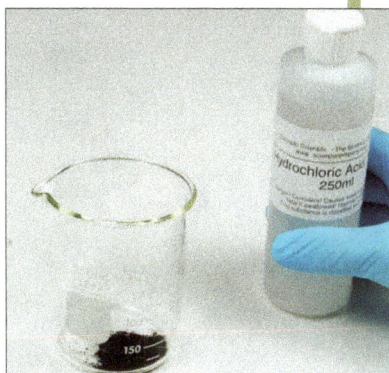

REACTION 5

The previous reaction has left us with an acidic solution in the beaker because we put more HCl in than was necessary to convert the CuO into $CuCl_2$. To neutralize the acid prior to disposal, we add some sodium bicarbonate, a rather mild base. Here is the reaction:

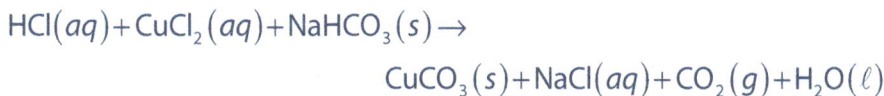

$$HCl(aq) + CuCl_2(aq) + NaHCO_3(s) \rightarrow$$
$$CuCO_3(s) + NaCl(aq) + CO_2(g) + H_2O(\ell)$$

All the reactants and products in this equation are familiar. But look what happens— one of the products is table salt, and another one is copper carbonate again!

REACTION 6

We left the copper hydroxide from Reaction 1 in the funnel to drain and dry out. After a day or so, it should be clear that something has happened to the copper hydroxide. It is no longer a beautiful blue. See if you can speculate what it might be—your instructor will know the answer. Once you have a hypothesis formed, use the reactions you have seen here to suggest a reaction that will enable you to put your hypothesis to the test. You will be surprised and pleased with what you find.

Chapter 12
Waves, Sound, and Light

Ocean waves carry massive amounts of energy. In fact, history is full of examples of ships that have run aground and then have been smashed to pieces by the pounding of the ocean waves. Strangely though, in the absence of a current, mild waves passing beneath a floating object do not take the object anywhere. Floating objects are only carried by a water wave if the waves get big enough, or if the object's shape allows the wave to propel the object, as with a surfboard.

OBJECTIVES

After studying this chapter and completing the exercises, you should be able to do each of the following tasks, using supporting terms and principles as necessary.

1. Explain what a wave is.
2. Define *frequency, period, wavelength,* and *amplitude.*
3. Explain the three main types of waves and give examples of each.
4. Describe the difference between mechanical waves and light waves.
5. Define and describe *reflection, refraction,* and *diffraction,* giving at least two examples of each.
6. State the frequency range of human hearing.
7. Describe the Doppler effect and explain its cause.
8. Explain why radio waves cannot be heard by human ears.
9. Explain why X-rays can be used in diffraction techniques like crystallography, but waves of visible light cannot.

VOCABULARY TERMS

You should be able to define or describe each of these terms in a complete sentence or paragraph.

1. amplitude	8. incident	15. reflection
2. circular wave	9. longitudinal wave	16. refraction
3. crystallography	10. mechanical wave	17. sonar
4. diffraction	11. medium	18. transverse wave
5. Doppler effect	12. normal line	19. ultrasonic
6. echolocation	13. oscillation	20. wave
7. frequency	14. period	21. wavelength

12.1   What is a wave?

Nature is full of waves; they are *literally* everywhere we look. I can write "literally" because light itself is composed of waves, and when we look at something we see it because the light waves enter our eyes.

But there are many other examples of waves in nature. Sound is waves of changing air pressure. Stringed instruments are played by making waves on the strings. Earthquakes are waves of ground motion. And oceans and lakes are covered with water waves formed by the wind.

With the exception of light waves, all other waves are formed when a *medium* of some kind is disturbed by a force. Light waves are different. Light waves are formed when atoms emit packets—quanta—of energy. As we saw in Chapter 2, light waves can propagate (that is, travel) without a medium, right through the

All waves except light are formed when a medium is disturbed by a force.

vacuum of empty space. They can also propagate through a medium, as when light passes through air, water, or glass.

So except for light, all other waves travel in a medium of some kind. Sound waves travel in air, water, and other media. Waves on strings travel on the string. Earthquake waves propagate through the ground. For each of these different kinds of waves, the wave is formed when the medium is disturbed by some kind of force. Waves such as these that require a medium are called *mechanical waves*. All waves except light are mechanical waves because light is the only wave phenomenon that can propagate without any medium.[1]

Waves involve oscillation—something going up and down or back and forth. When we think of how to describe an oscillation, we think of things like how fast the oscillation is and how large the oscillation is.

To get an idea of what we mean by the speed and size of an oscillation, take a look at Figure 12.1. In this figure, a mass is attached to a spring. At rest, the mass is at the rest position. If the mass is then pulled down to the lowest position and released, the mass begins oscillating up and down between the highest position and the lowest position.

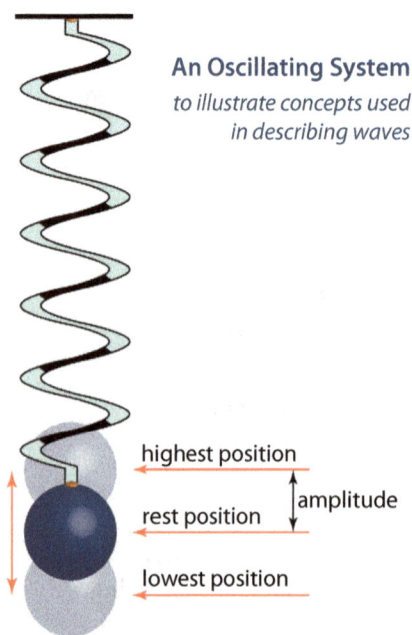

An Oscillating System *to illustrate concepts used in describing waves*

We measure the speed of an oscillation like this by how many full cycles of the oscillation the mass completes each second (cycles per second). The term for this variable is the *frequency*. So the frequency, f, of an oscillation is defined as

$$f = \frac{\text{number of cycles completed}}{1\,\text{second}}$$

The SI System unit for frequency is *hertz* (Hz), named after Heinrich Hertz, a German physicist who did a lot of early work with Maxwell's Equations to develop the science of radio waves. If an object is oscillating with a frequency of, say, 82 Hz, this means the object is completing 82 cycles of its motion every second. As it turns out, this frequency corresponds to the lowest note on a guitar. If you pluck the lowest string on a guitar, the string oscillates at 82 Hz, and the sound wave produced has a frequency of 82 Hz.

highest position

rest position

lowest position

amplitude

Figure 12.1. When pulled down and released, a mass on the end of a spring oscillates up and down between the highest position and the lowest position.

---

1   There may also be *gravitational waves*, ripples in the curvature of spacetime that travel through the vacuum of outer space. Einstein predicted the existence of gravitational waves in 1915. So far they have not been directly detected, but experiments attempting to detect them are now underway.

The frequency of a wave indicates the number of oscillation cycles the wave completes each second.

We measure how large an oscillation is with a variable called the *amplitude*. The amplitude of the oscillating mass is indicated in Figure 12.1. The amplitude is the distance between the rest position and the highest (or lowest) point of the motion. In the oscillating mass example, the amplitude is a length, so its units of measure are meters. But more generally, the units of the amplitude depend on what is oscillating, and this could be many different things—air pressure, electric field strength, the height of the ground (in an earthquake), electric current, and so on. The amplitude of the wave relates to how much energy the wave is carrying. Stronger waves carrying more energy have greater amplitudes. Weaker waves carrying less energy have smaller amplitudes.

When we represent oscillation graphically, we use graphs like the one you saw back in Chapter 2, in Figure 2.7. That diagram is shown again in Figures 12.2 and 12.3. As you can see from these two diagrams, the way we identify the separation between the peaks of a wave depends on whether the horizontal axis in the graph represents space or time. Let's examine this more closely. Refer to these two diagrams as you read the next two paragraphs.

Waves propagate through space. This goes for all waves, whether we are talking about light, sound, or some other kind of wave. So one way of speaking about waves is in terms of the length of the wave in space, called the *wavelength*. Just as we did with light in Chapter 2, we often refer to the wavelength of a wave, which is the *spatial* separation between the peaks of the wave. The wavelength of green light in a vacuum is about 540 nanometers (nm), and in glass it is about 360 nm. In air, the wavelength of the lowest note on a guitar (82 Hz) is about 4.2 m.

Another way of speaking about waves relates to the fact that we measure the "speed" of an oscillation with the frequency, which relates to time. As mentioned above, the frequency is the number of cycles the oscillation completes in one second. This means the separation between the peaks of an oscillation is one second divided by the frequency. For example, if the frequency of a wave is 5 Hz, this means the wave completes five cycles every second, and the length of time taken to complete each one

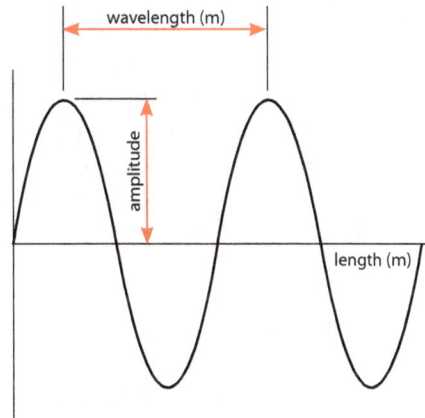

Figure 12.2. A graphical representation of a wave in space.

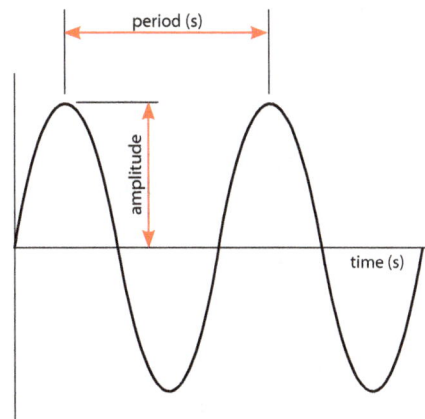

Figure 12.3. A graphical representation of a wave in time.

is 1/5 of a second, or 0.2 s. This amount of time—the time required for the wave to complete exactly one cycle—is called the *period* of the wave. The period is an amount of time, so it is measured in seconds.

With the background about waves we have covered so far in this chapter, we can now define what a wave is. A *wave* is a disturbance in space and time that carries energy from one place to another. Electromagnetic waves carry the energy from the sun to earth. The shock wave of air pressure propagating out from an explosion can carry enough energy to blow out the windows of buildings miles away. The energy in the wave of an earthquake or tsunami can level a city. But happily, the sound energy produced by the strings of a cello can bring tears of joy to your eyes as you listen to your daughter play.

A wave is a disturbance in space and time that carries energy from one place to another.

Learning Check 12.1

1. What are some examples of waves in nature? List at least five.
2. What is oscillation?
3. Explain what the frequency and period of a wave are.
4. Why are there two different ways to describe the separation between the peaks of a wave?
5. What does the amplitude of a wave tell us?
6. What is a wave?

12.2   Types of waves

In the previous section, I made the point that waves are formed by something causing a disturbance—a force of some kind in a medium gives the particles in the medium a shove. This shove then propagates away from the source of the wave. Waves radiating outward from where a stone hits the water are a perfect and very visible example.

There are three different basic ways that an oscillating motion can produce a wave, so there are three names for the waves produced by these different ways of forming wave motion.

Transverse waves are formed when the source of the wave is oscillating in a direction perpendicular to the direction the wave propagates. This concept is illustrated in Figure 12.4. In this figure, a man is whipping the end of a cord up and down, causing waves to travel down the cord away from his hand. Since the man's hand is oscillating up and down but the wave is traveling horizontally to the right, this is a transverse wave. The situation of whipping a cord applies generally to waves on strings. All waves on strings travel down the string. But to create a wave on a string, the string is disturbed by strik-

In a transverse wave, the source of the wave oscillates perpendicularly to the direction of wave propagation.

direction of
disturbance
causing the wave

direction the wave is propagating

These two directions
are perpendicular

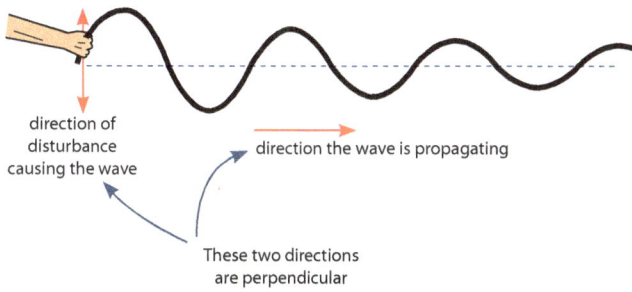

Figure 12.4. Whipping the end of a cord up and down forms a transverse wave—the direction the wave source is oscillating is perpendicular to the direction of wave propagation.

ing or plucking the string in a direction perpendicular to the string. Examples are guitar strings struck sidewise with a pick or with the fingers, or piano strings struck from the side by the hammers triggered by the piano keys. Concert stringed instruments work the same way. The musician applies a tacky substance called rosin to the hairs on the bow. This creates a lot of friction between the bow hairs and the strings of the instrument. As the bow is drawn across the strings, it tends to pull the strings with it. The action of this friction moves the strings from side to side.

To show you a bit more clearly that waves on strings travel down the string, Figure 12.5 shows two images of waves on a string. The waves are being formed by the black gadget in the foreground of the pictures. This device uses an electromagnet to whip the end of the string up and down at a rate of 60 Hz. The up and down motion is just like the motion of the man's hand in Figure 12.4. The upper photo in

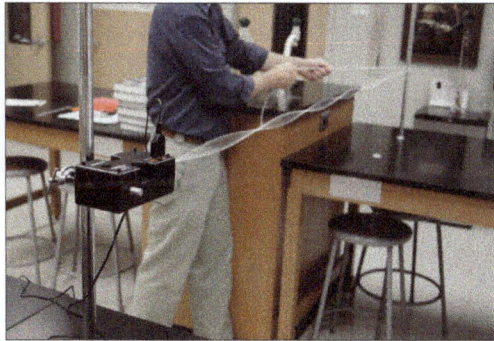

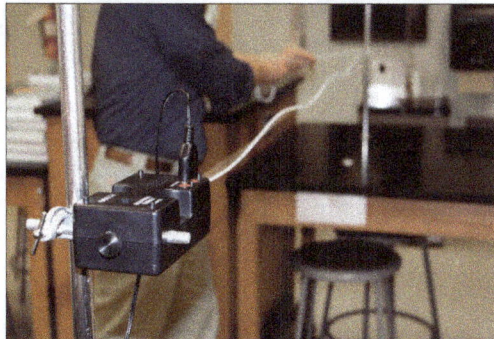

Figure 12.5 shows what the string looks like to the naked eye. The lower photo was taken with a fast shutter speed. Since the electromagnet is whipping the end of the string up and down 60 times per second, your eyes can't see that the waves are actually propagating down the string away from the electromagnet. But in the high-speed image in the lower photo, it is clear that the waves are traveling down the string.

Another important example of transverse waves is electromagnetic waves—visible light and all other forms of electromagnetic radiation. Light waves consist of oscillating electric and magnetic fields, and the direction of the field lines is perpendicular to the direction the light is propagating. This is difficult to imagine and even harder to draw, but I have given it my best shot in Figure 12.6.

The second type of wave formation results in *longitudinal waves* (pro-

Figure 12.5. Transverse waves being formed on a string by the up and down motion of a string vibration device.

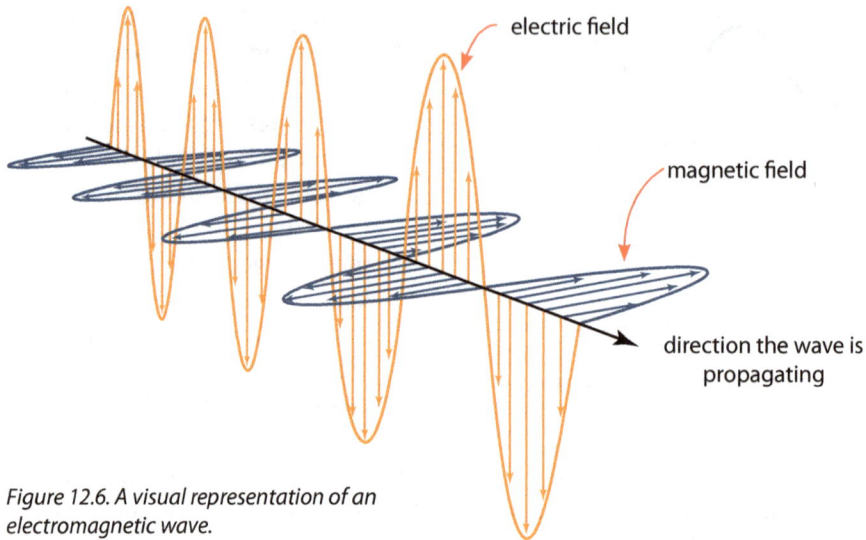

electric field

magnetic field

direction the wave is
propagating

*Figure 12.6. A visual representation of an
electromagnetic wave.*

nounced with a soft g). To produce longitudinal waves, the source of the wave is oscillating in a direction parallel to the direction of wave propagation.

The number one example of longitudinal waves, and the best way to illustrate the concept, is sound waves. And the easiest way to think about how sound waves are formed is to look at how a loudspeaker produces sound. Two loudspeakers are shown in the photograph of Figure 12.7. Two of the main parts of a speaker are the cone, usually made of paper or plastic, and a magnet mounted on the back. Inside the magnet is the third main part, a coil of wire attached to the back of the cone. Electricity passing through the coil of wire makes the coil of wire produce its own magnetic field, just as we saw in Section 5.2. Since the coil is inside another magnet, the two magnetic fields push and pull against each other, forcing the loudspeaker cone to vibrate back and forth.

*Figure 12.7. Front (right) and rear (left) of an
audio loudspeaker.*

The diagram in Figure 12.8 shows what happens in the air as a result of the motion of the loudspeaker. As the loudspeaker moves back and forth, it pushes out waves of air molecules. The waves of air molecules bump against layers of air molecules farther out, and then even farther out, propagating the wave away from the speaker. As indicated in the diagram, the back and forth motion of the loudspeaker cone is parallel to the direction the wave is propagating. Thus, a longitudinal wave is produced.

In a longitudinal wave, the source of the wave oscillates in a direction that is parallel to the direction of wave propagation.

Finally, the third way waves are formed results in what we call *circular waves*. Circular waves are the waves formed on the surface

air molecules

direction
the wave is
propagating

oscillating
motion of the
loudspeaker

These two
directions are
parallel.

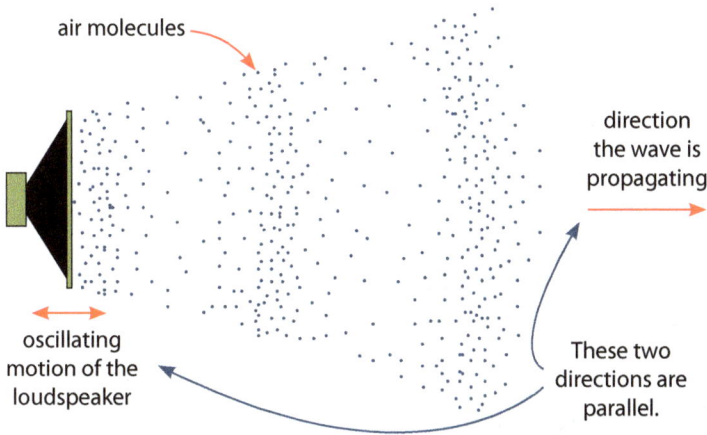

Figure 12.8. A visualization
of the air molecules in a
sound wave emerging from
a loudspeaker.

of water by the action of the wind. To see why water waves
are called circular, look at the diagram in Figure 12.9. The
diagram illustrates the very interesting motion of an object
floating on the water as the waves pass by underneath it. As
the waves pass by, the object goes nowhere, but simply moves
about in a vertical circle on top of the water.

The circular motion of a floating object gets changed up
a bit for waves with larger amplitudes. If the wave amplitude
is great enough, the floating object does not quite return to
its starting place at the top of the circle. Instead, as it returns
to the top it arrives at a point slightly to the right of where it

*Waves on water
are called circular
waves, because of
the circular motion
of a floating object
as the waves pass
by underneath it.*

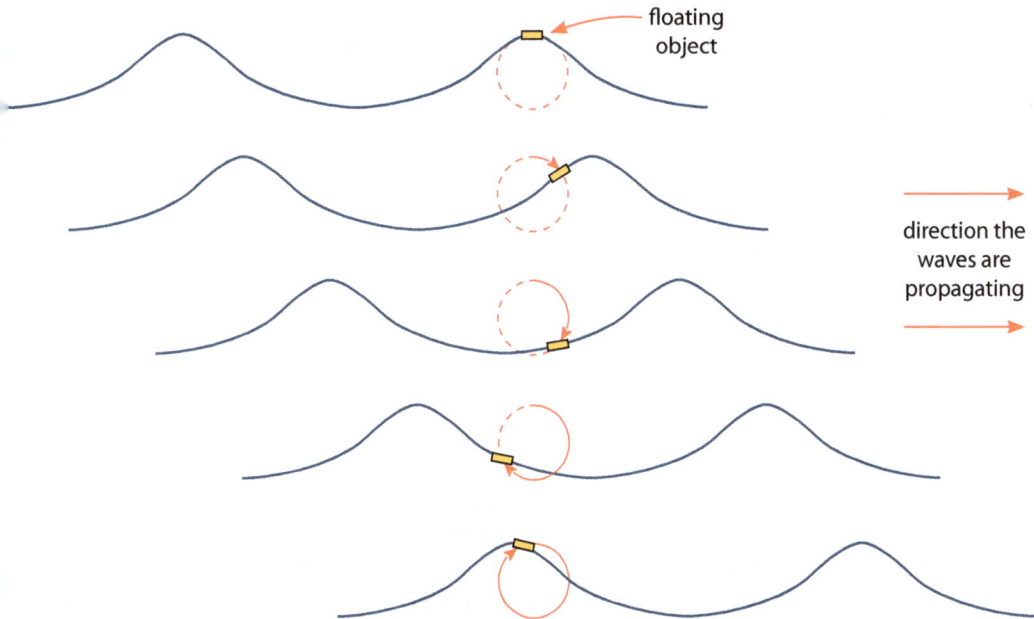

floating
object

direction the
waves are
propagating

Figure 12.9. Motion of a floating object as calm water waves pass by underneath it.

started. This drift to the right means the object moves in a slight spiral, drifting slightly in the direction of the wave propagation with each loop.

Learning Check 12.2

1. Explain the difference between the ways transverse waves and longitudinal waves are formed.
2. Why are waves on strings transverse waves?
3. What is a circular wave, and where did this kind of wave get its name?
4. Explain why sound waves are longitudinal waves.

12.3 Common Wave Phenomena

As waves propagate, hit surfaces, and pass obstructions in their paths, they exhibit common, predictable phenomena. Some of these phenomena are most visible with light (pardon the pun), but all waves act the same ways. Here, we look at three common wave phenomena—reflection, refraction, and diffraction.

Reflection *Reflection* is the most well known of the things waves do. The reason you can see yourself in a mirror or in a window glass is because ambient light from the environment reflects off your body toward the glass. When it arrives at the glass it reflects again, this time hitting you in the eye.

Reflection is simple to understand, and the basic rule is illustrated in Figure 12.10. In this diagram, I am using a ray of light to illustrate the path the light takes as it reflects. You can think of the ray as a narrow beam, like the beam from a laser pointer. The ray approaching the mirror is called the *incident ray*. As the ray leaves the mirror after reflection it is called the *reflected ray*.

The angles these rays make with the reflecting surface are measured with respect to the dashed line. This line is perpendicular to the reflecting surface, and is called the *normal line*. (In math and physics, the term *normal* means *perpendicular*.) The angle the incident ray makes with the normal line is called the *angle of incidence*; the angle the reflected ray makes is called the *angle of reflection*.

The law of reflection is that *the angle of incidence equals the angle of reflection*. All waves obey this law,

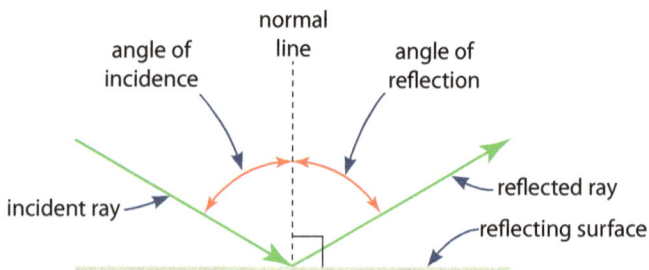

Figure 12.10. When a wave reflects, the angle of incidence equals the angle of reflection.

The law of reflection: the angle of incidence equals the angle of reflection.

not just light waves. Another example is sound waves at an outdoor musical performance. The sound waves obey this same law as they reflect off the side of a nearby building.

Refraction

As waves propagate, they often pass from one medium into another. As an example, imagine that a friend is shouting to you while you are under water. The sound waves originating from your friend's voice originate in the air, but when they strike the surface of the water, they pass from the air medium and enter the water medium. Light waves traveling from you to your friend, or vice versa, must also pass from one medium to another, from water into air or from air to water. Whenever a wave passes from one medium into another, if the wave's velocity changes, the wave's direction changes. This change of direction due to the wave speeding up or slowing down is called *refraction*.

For example, in glass, the velocity of light is about 66% of its velocity in air. Since the velocity of light is different in these two media, the path of a ray of light bends as the light passes from air into glass or vice versa.

Figure 12.11 shows a laser beam refracting as it passes into a block of glass. The incident ray is the bright red ray coming down from the top of the photo. As it strikes the glass, some of the light reflects off to the right, obeying the law of reflection. But some of the light enters the glass. You can clearly see that as it does so, it change direction. Its path, the faint red line angling down at the bottom of the photo, bends toward the normal line. (I added the dashed, white normal line and the labels to the photo. The rest of the image is an actual photograph.)

The most common demonstration of refraction is the straw in a drinking glass, shown in Figure 12.12. Light coming to the camera from the upper part of the straw does not pass through the water. But the light coming to the camera from the lower part of the straw has first passed through the water, then through air, and that has changed the direction the light waves are traveling. The way we perceive the locations of objects with our eyes is according to the direction the light waves are coming from. Wherever that is, that is where we see the object located. The refraction causes the light waves on the lower part of the straw to appear to come to the camera from a location to the left of where the straw really is.

When passing from one medium to another, waves refract—change direction—if the wave velocity is different in the two media.

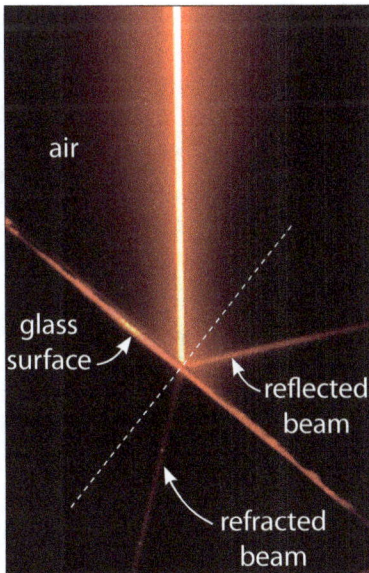

air

glass surface

reflected beam

refracted beam

Figure 12.11. When the bright red laser strikes the surface of a glass block, the beam passing into the block refracts toward the normal line. Some of the beam reflects off the surface.

To illustrate this further, here's another interesting example. For both light and sound, the velocity of the waves in air is quite different from the wave velocity in water. In air, the speed of light is 300,000,000 m/s. But light travels more slowly in water, about 225,000,000 m/s. So when light passes from air into water, it slows down. The speed of sound in air is 342 m/s, but in water, the speed of sound is 1,480 m/s. So when sound waves pass into water, they speed up dramatically.

Figure 12.13 illustrates what happens to light waves and sound waves as they strike the surface of a body of water. On the left are incident waves of light. Just as with the image of the laser on the previous page, some of the light reflects, obeying the law of reflection. Some of the light enters the water, and the light waves refract as they enter the water. Just as with the laser and the prism, the light's path refracts toward the normal line.

Figure 12.12. A very common demonstration of refraction.

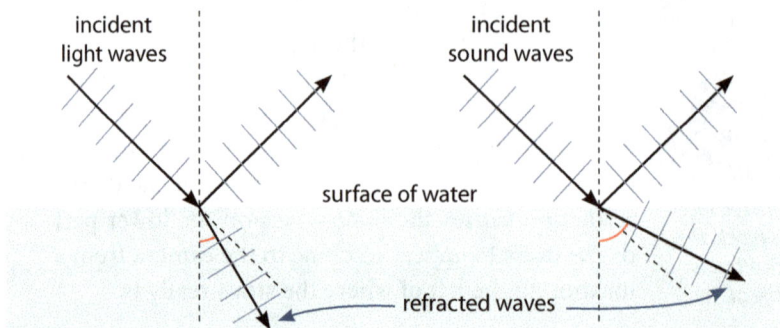

On the right side, sound waves strike the water's surface. Some of the sound reflects, obeying the law of reflection. Some of the sound enters the water, and refracts as it does so. But there is an interesting difference in the refraction of the light and the sound. As you can see, the light refracts toward the normal line, but the sound refracts away from the normal line. The difference between the two angles inside the water is indicated by the red arcs in the graphic. Light refracts toward the normal line because it slows down as it enters the water. But since sound travels faster in water, it refracts away from the normal line. This general principal always holds for refraction. Waves refract toward the normal line if they travel slower in the new medium. They refract away from the normal line if they travel faster in the new medium.

As another illustration of reflection and refraction, consider a reflection in an ordinary glass mirror. You might have noticed that sometimes if you look very carefully at the reflected image, you can actually see two images. The main image is accompanied by a faint second image—sometimes called a ghost image—a fraction of an inch away from the main image. Figure 12.14 is a photo of a paper cup and its reflection in an ordinary bathroom mirror. The arrow indicates the ghost image on the left edge of the reflection. If you want to try this yourself, you won't see the

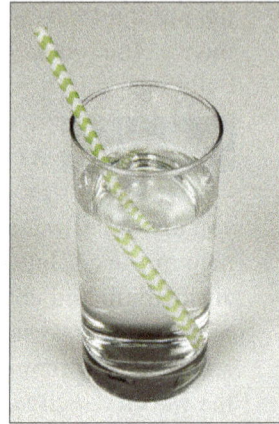

incident
light waves

incident
sound waves

surface of water

refracted waves

Figure 12.13. Upon entering water, light slows down and refracts toward the normal line. Sound speeds up, and refracts away from the normal line.

ghost by looking straight on at the mirror. You need to look at the glass with a large angle of reflection.

Waves refract toward the normal line when they slow down and away from the normal line when they speed up.

A bathroom mirror is a plate of glass with highly reflective metal coating applied to the back of the glass. The main image you see in such a mirror is from the reflective coating. The ghost image is coming from the reflection from the front surface of the glass. The ghost is faint because most of the light passes into the glass and reflects off the coating on the back. Only a small amount of light reflects off the front of the glass.

The diagram in Figure 12.15 illustrates the cause of the ghost image. Here you can see the double light pathways, one from reflection off the front glass of the mirror, and one from the silver coating on the back of the mirror. Notice that the light from the back has refracted twice before reaching your eyes. As it enters the glass, the light slows down and refracts toward the normal line. As it exits the glass, the light speeds up, so it refracts away from the normal line. The red arcs in the graphic indicate where refraction occurs. As a result of the double refraction, the refracted ray travels on a path that is parallel to the path of the ray reflecting from the front of the glass. So the main image you see in a mirror is the doubly refracted one. The faint ghost is from the small amount of light that reflects off the front of the glass.

Figure 12.14. A ghost image seen on the reflection of a paper cup in a bathroom mirror.

Our next illustration of refraction is perhaps my favorite, for the simple reason that it is so astonishing. Figure 12.16 shows two photographs of the same bowl. To take these photographs, I positioned the camera on a tripod so it would not move. In front of the camera I placed the bowl *with the quarter in the bottom* and adjusted the position of the bowl and the camera so the quarter could not be seen.

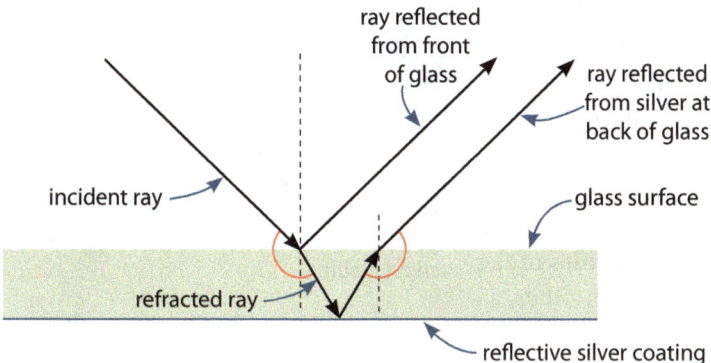

Figure 12.15. Reflection and refraction on the surface of a glass mirror. Red arcs indicate where refraction is occurring.

Then I took the upper photo. Next, without moving anything, I gently filled the bowl with water, being very careful not to disturb the quarter at all. Then I took the second shot. The only difference between the two photos is that in the upper photo the bowl is empty, and in the lower one it is full of water. The light reflecting off the quarter refracts so much that the quarter suddenly becomes visible without moving anything at all.

Here is one more illustration. The atmosphere of the earth refracts the light coming from the sun. At sunrise or sunset, the sun's rays enter the atmosphere and refract downward. The result is that the sun you see at sunset has, in fact, already set. As Figure 12.17 illustrates, the sun is actually below the viewer's horizon, indicated by the dashed line. But the atmospheric refraction of the sun's light enables the viewer to see it—just like the quarter in the bowl. The refraction of light in the atmosphere also causes the flattened appearance of the "harvest moon," illustrated in Figure 12.18.

Figure 12.16. The quarter, the bowl, and the camera are in exactly the same place in both photos, but with water in the bowl, the light reflecting off it refracts so much that we can see the quarter.

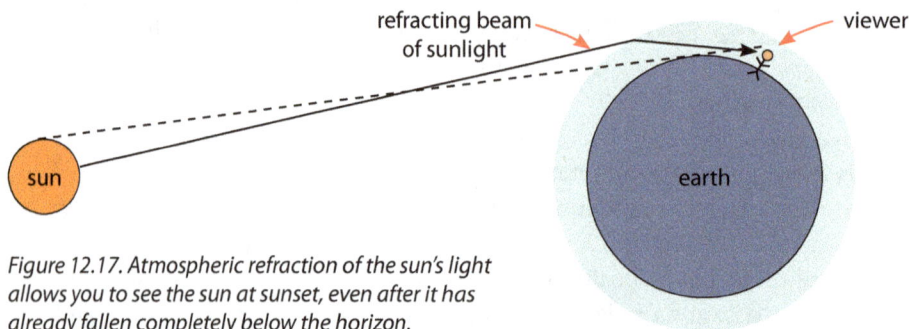

Figure 12.17. Atmospheric refraction of the sun's light allows you to see the sun at sunset, even after it has already fallen completely below the horizon.

Diffraction A third behavior waves exhibit is the phenomenon called *diffraction*. Diffraction simply means that waves bend and spread out as they pass by corners of objects in their paths. An easy way to demonstrate diffraction is to have people stand near the corner of a wall, on opposite sides of the corner and out of site of each other, as shown in Figure 12.19. Person A speaks softly through a tube, such as the cardboard tube from a roll of paper towels. The sound waves diffract as they emerge from the

Figure 12.18. Atmospheric refraction flattens the harvest moon.

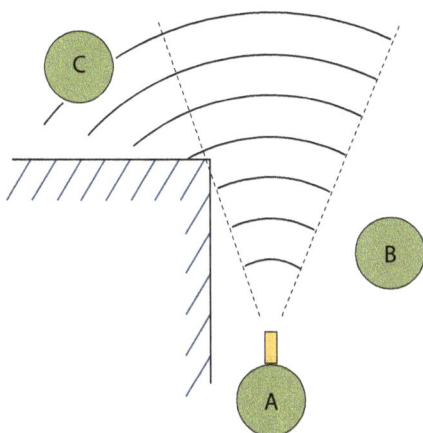

Figure 12.19. Diffraction of sound waves at the corner of a wall, causing waves to spread around the corner as they pass.

tube, which is why they begin spreading out. They are not spreading out fast enough for person B to hear what is being said. But when the sound waves reach the corner of the wall, they diffract around the corner and person C hears them. Without the diffraction at the corner of the wall, person C would not be able to hear any better than person B can.

The diffraction of sound also occurs at your mouth when you speak. If the sound waves coming out of your mouth did not begin spreading out, the sound would propagate like a beam from a flashlight, and no one would be able to hear you unless you aimed your sound beam at them. (I know—strange to think about.) Of course, the light in the flashlight beam is spreading out too, but the wavelengths of sound waves are vastly longer than the wavelength of light, and this makes sound diffraction much more pronounced.

Another example of diffraction, one that has been of great benefit to scientists, is described in the box on the next page.

> *Diffraction is the bending and spreading out of waves as they pass corners or obstructions.*

Learning Check 12.3

1. What is the law of reflection?
2. What is refraction and what causes it?
3. Why do light and sound refract in opposite directions when crossing the boundary between air and water?
4. Define and give an example of wave diffraction.

12.4   Sound and Human Hearing

In Section 2.3, we took a good look at electromagnetic waves in general and visible light in particular. We examined the different regions in the electromagnetic spectrum and discussed the wavelengths and colors of light in the visible spectrum. Light is obviously of enormous importance to us, so it was important to get into some detail. Clearly, sound waves are also of great importance to us, so we need to take a closer look at human hearing and sound waves in general.

When we discussed visible light in Chapter 2, we spoke in terms of wavelengths. But for discussions of sound, it is conventional to speak in terms of frequencies. As you recall, the frequency of a wave is the number of cycles the wave completes each second. We measure frequencies in hertz (Hz), which means cycles per second. For

Scientists, Experiments, and Technology

Have you been wondering how scientists are able to figure out the structures of the different crystals? Look again at the three different calcium crystals in Figure 6.2 or the copper sulfate crystal in Figure 11.10. These atomic structures are far too small to see with the eye or any microscope using light. So how did scientists figure those structures out? Answer: By the use of X-Ray crystallography.

In fact, copper sulfate was the first crystal ever used for X-Ray crystallography. German physicist Max von Laue, shown to the right, first had the idea for this research technology in 1912 and applied the idea to a copper sulfate crystal. In 1914, he was awarded the Nobel Prize in Physics for his discovery.

For significant diffraction to occur, the wavelength of the diffracting waves must be roughly the same size as the obstruction causing the diffraction. This explains why we notice sound waves diffracting around the corner of a wall, but not light waves. The wavelength of sound waves is about the same as the size of the wall—in the range of a few meters. Light waves are about ten million times smaller, so light makes a sharp shadow at the corner of a wall, instead of diffracting. But the spacing between atoms in crystals is in the range of 0.1 nm to 10 nm, 100 times smaller than the wavelength of visible light. The wavelength of X-Rays is also about 100 times smaller than the wavelength of visible light, so X-Rays diffract beautifully when they hit a crystal.

The technique for X-Ray crystallography is shown in the image below. A beam of X-Rays hits a crystal, causing the beam to diffract as it passes between the layers of atoms in the crystal. The diffracted Rays emerge from the crystal and form patterns of spots on film or on the CCD sensor in a digital camera. Mathematical analysis is then used to infer the crystal structure from the pattern of dots.

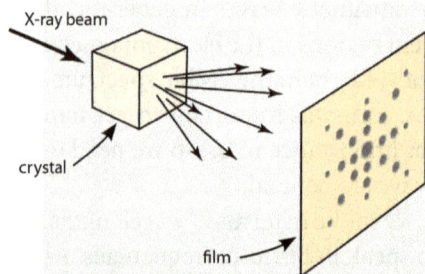

X-Ray crystallography is not just useful for crystals. The same technique is used to determine the atomic structure of large biological molecules. Most famously, Francis Crick and James Watson worked out the double-helix structure of the DNA molecule in 1953, based on an X-Ray image taken in 1952 by Rosalind Franklin and Raymond Gosling.

sound waves, the main metric prefix we use is *kilo–*, meaning 1,000. So a sound wave with a frequency of 2 kHz is a wave that completes 2,000 cycles per second.

The frequency range of human hearing is 20 Hz to 20 kHz.

The range of human hearing extends from about 20 Hz to 20,000 Hz, or 20 kHz. Actually, only young people can hear sounds up in the 15 kHz to 20 kHz range. By the time we reach middle age, most people find that they cannot hear much above 15 kHz. Table 12.1 lists some frequencies for common sounds.

27.5 Hz	The lowest note (A) on a piano keyboard.
41.2 Hz	The lowest note (E) on a bass guitar.
82.4 Hz	The lowest note (E) on a guitar.
440 Hz	The A above middle C on a keyboard, often called "A 440," and used as a reference for tuning pianos and other concert instruments.
100 – 3,000 Hz	Frequency range of the human voice.
1,000 Hz	A low whistle.
4,186 Hz	The highest note (C) on a piano keyboard.
6 kHz – 8 kHz	Frequency range for the "s" sounds we make between our teeth when we say a word like "Susan." (This sound is called *sibilance*.)

Table 12.1. Frequencies and frequency ranges for some common sounds.

Next, we look at several examples of how sound waves are used in the world around us. You are probably aware that bats are essentially sightless and that they navigate and hunt for food by the use of reflected sound waves. (Dolphins do this, too.) Bats emit the sound waves from their mouths, and they are able to interpret the locations of things around them by listening to the echoes—reflections—of the sound waves. This process is called *echolocation*. The sound waves bats emit are *ultrasonic*—above the frequency range of human hearing—so we can't hear them. Bat sounds range from about 14 kHz up to well above 100 kHz! This is astonishing, because to form a sound wave with a frequency of 100 kHz, something in the bat's anatomy must be oscillating at a rate of 100,000 cycles per second. (It is mind boggling to think of any part of a bat's sound-producing apparatus moving that fast. The wings of a hummingbird are incredibly slow by comparison, moving at up to 80 beats per second.) The sounds bats make are rapid clicks and pops, ranging from 10–20 clicks per second while hunting around for insects to eat, up to a buzzing 200 clicks per second as a bat closes in on a snack.

Humans also use reflected sound waves for depth detection under water. This technology is called *sonar*. Sonar was first developed after the *Titanic* disaster, just before World War I. As you may know, the *Titanic* sank because it collided with an unseen underwater iceberg. Within a month of that collision, the first sonar systems had been developed. The technology took off in the mid-20th century. Ships and submarines now routinely use sonar to locate underwater obstructions or oth-

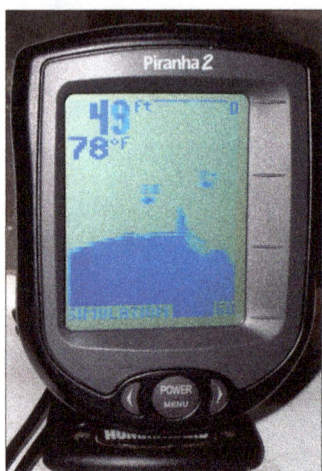

Figure 12.20. A sonar fish finder used in sport fishing.

er ships. Sonar devices emit "pings" with frequencies in the middle of the human hearing range. Sensors in the water listen for the echo and use the echo information to construct a map of what's down there. Fishermen even use sonar now too for finding fish. Figure 12.20 shows the display of one of these devices.

These days, every expectant mother knows about the visual images of unborn babies that can be formed using *ultrasound* technology. The images can be used to diagnose problems with the baby and can detect the sex of the baby by 11 to 13 weeks into the pregnancy. Medical ultrasound works basically the same way as sonar, but uses much higher frequency sound waves. Sonar pings can be heard, but medical ultrasound uses frequencies in the range of 1 to 10 megahertz.

The list of technologies using sound waves goes on and on, but here are two more examples that use sound waves to smash things. Ultrasonic cleaning systems are now widely used for cleaning surgical instruments, jewelry, small parts, and other items. Using sound waves in the 40 kHz to 200 kHz range, these systems introduce pressure waves in water that cause the formation of microscopic vacuum bubbles. When these tiny vacuums collapse, they do so with such force that they break down dirt, grease, or other deposits clinging to the parts to be cleaned.

And kidney stones can be crushed and flushed from a person's body—without surgery—by the use

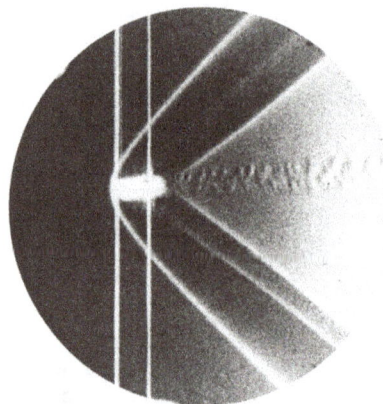

Figure 12.21. A shock wave at the front of a bullet traveling at supersonic speed.

of focused, acoustic *shock waves*. Briefly, shock waves are sound waves produced by any object moving faster than the speed of sound. In the photo of Figure 12.21, a bullet is traveling faster than the speed of sound, and you can see the shock wave in the air streaming away from it. (This famous photograph was taken in 1888 by physicist Ernst Mach. Mach's name is now used to designate speeds faster than the speed of sound. "Mach 2" means a velocity of twice the speed of sound, approximately 684 m/s, or 1,530 mph.)

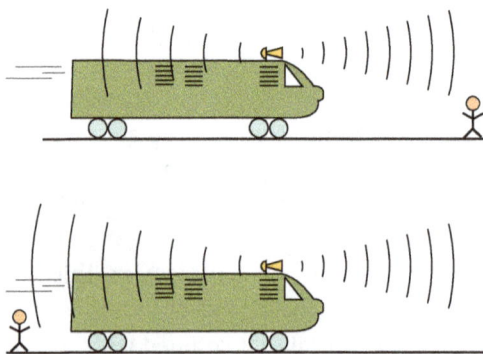

Figure 12.22. The wavelength of the sound waves is stretched or compressed due to the motion of the train, creating the Doppler effect.

We close this chapter with a brief look at the familiar *Doppler effect*. You have heard this effect yourself, I'm sure, when the horn from an emergency vehicle or train has been sounding while the vehicle passed you by. You hear the pitch of the horn drop as the vehicle passes. The diagram in Figure 12.22 illustrates what is happening.

As the train approaches you, the motion of the train compresses the sound waves together. This causes the wavelength in air to be shorter, which our ears perceive as a higher frequency—a higher pitch. Once the train passes and is moving away from you, the sound waves from the horn are being stretched out by the train's motion, increasing the wavelength. Our ears hear this as a lower frequency, and a lower pitch.

Learning Check 12.4

1. Are the signals broadcast by radio stations sound waves? If so, why can't we hear them in the air?
2. Write a paragraph explaining the Doppler effect in your own words.
3. Describe three examples of the use of sound waves in technology.
4. Dogs can hear up to about 45 kHz. Using this fact, develop an explanation for why dogs can hear a dog whistle, but humans can't.

Chapter 12 Exercises

Answer each of the questions below as completely as you can. Write your responses in complete sentences.

1. Write out descriptions of reflection, refraction, and diffraction. Clearly distinguish them from each other and give examples of each.
2. Define the terms oscillation and frequency and explain how they are related.
3. Is the wave from an earthquake a transverse wave or a longitudinal wave? Explain your answer.
4. Define what a wave is and explain how the amplitude of a wave relates to this definition.
5. Referring to Figures 12.13, 12.16, and 12.17, draw a diagram showing how light from the quarter in the bowl is refracted by the water so the camera can see it. There will be a key difference between your diagram and the one in Figure 12.17. Explain what that is.
6. When you hear an emergency vehicle go past with its horn blaring, does the pitch go up or down as it passes? Explain your answer.
7. Explain the difference between mechanical waves and light waves.
8. Explain why X-Rays can be used with crystallography but visible light cannot.

Experimental Investigation 10: Refraction

Overview

- Measure the angles of incident and refracted laser beams as the laser refracts through acrylic glass.
- Use Snell's law to calculate the index of refraction of the acrylic material.
- Repeat these two steps with a transparent tray of water as the refracting medium.
- Compare the experimentally determined indices to standard reference values.
- *The goal of this experiment is to measure the indices of refraction for acrylic glass and water, and to compare the experimental values to published reference values.*

Basic Materials List

- laser pointer
- acrylic glass prism or block
- ring stand and clamp holder
- Hot Wheels car bubble packaging (to use as a water tray)
- graph paper, protractor, ruler
- scientific calculator
- thick, flat books (1 or 2) (to position the prism)

Safety Precaution

Never look into a laser, and never shine the beam or a beam reflection into anyone's eye.

I realize you probably haven't studied trigonometry yet, but in this experiment we will make use of one trigonometric calculation, the sine of an angle. I will first explain what this means. Hopefully you have learned about triangles and measuring angles in degrees. A right triangle is a triangle with one 90° angle in it, like the one shown below. For the angle marked θ in this triangle, the sine of that angle, written as $\sin\theta$, is defined as $\sin\theta = a/c$. In any right triangle, the sine of an angle in the triangle is simply the ratio of the lengths of the side opposite the angle and the longest side in the triangle, called the *hypotenuse*.

One of the important laws in the study of optics and light is *Snell's law*. When a ray of light (like a laser beam) strikes a surface and refracts, Snell's law relates together the angles of the incident and refracted beams (θ_1 and θ_2) and the indices of refraction (n_1 and n_2) for the two media in which the ray is propagating. Snell's law is:

$$n_1 \sin\theta_1 = n_2 \sin\theta_2$$

The index of refraction of air is very close to one, so when the incident ray is in air, we can write Snell's law as

$$n_2 = \frac{\sin\theta_1}{\sin\theta_2}$$

In this equation, n_2 is the index of refraction for the refracting medium, and θ_1 and θ_2 are the angles the incident and refracted rays make with the normal line. With this equation, we can determine the index of refraction of a material. All we have to do is measure the two angles and put them in the equation to calculate n_2.

As shown in the photo, set up a laser pointer on a ring stand with a clamp, and adjust the beam so it hits a prism on a sheet of graph paper at an angle. Turn off the lights so you can see the laser beam inside the prism. Now make four marks on the paper. Put a

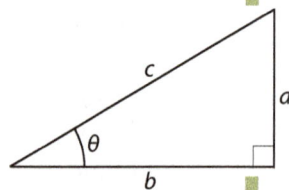

mark directly under the beam near the edge of the paper closest to the laser. Put another mark directly under the spot where the beam strikes the prism. Put a third mark directly under the spot where the beam exits the prism. Finally, holding the prism down with your finger, draw a line on the paper along the front edge of the prism—that is, at the bottom of the side where the laser beam refracts. I have indicated these four marks with blue arrows in the sketch. In the photo below, you can see the laser beam as it passes through the prism while a person uses a pencil to make the third mark.

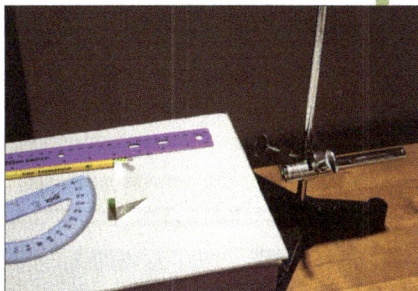

Once the marks are made, turn on the lights and turn off the laser. Use a straight edge to connect your marks to show the ray paths and the front edge of the prism. Use a protractor to draw in the normal line, right at the place where the incident beam hits the front surface of the prism (mark 2). Now measure and record the two angles, and use them to calculate the index of refraction for the acrylic.

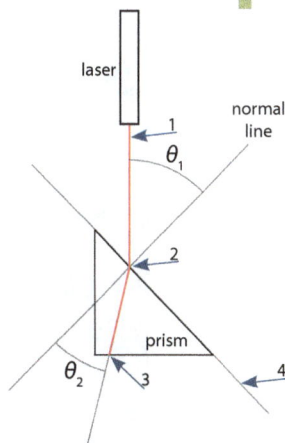

The next part of the experiment is to repeat the whole procedure using water in a transparent tray in place of the prism. The packaging for a Hot Wheels car makes a nice water tray, if a piece of tape is placed across the top to hold the sides vertical. The next photo shows the water tray in place. The bottom edge of the tray curves under a bit, so you can't mark the front edge by drawing a line across the bottom of the tray. Instead, place a ruler along the tray edge, then move the tray and use the ruler to mark the front edge.

The standard reference values for the indices of refraction are 1.49 for acrylic glass and 1.33 for water. Use these reference values as the predictions and your measurements as the experimental values, and calculate the percent difference for both your indices.

Analysis

In your report, address the following questions.

1. For an incident angle of 17°, an error of only 1° in measuring that angle can cause your index of refraction to be off by about 5%. In light of this, do your percent difference figures seem reasonable, given the degree of accuracy with which you were able to measure the angles? Explain.
2. What were the most difficult aspects of the experiment?

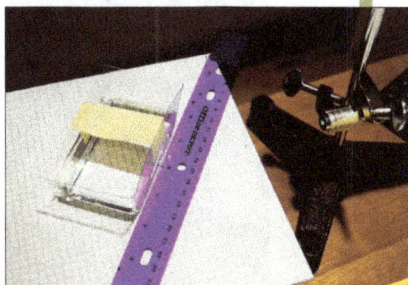

Lightning is a massive discharge of static electricity. Ice crystals in clouds cause a layer of negative charge to accumulate on the bottom of the cloud. Channels of negatively ionized air (a plasma) called leaders *begin branching toward the ground. When the leaders get close to the ground, channels of positively ionized air called* streamers *begin shooting up. When a leader attaches to a streamer, a huge current of positive charges flows up the channel from the ground to the cloud. This is the main strike. In the image above, an unattached streamer may be seen at left.*

OBJECTIVES

After studying this chapter and completing the exercises, you should be able to do each of the following tasks, using supporting terms and principles as necessary.

1. Explain why electrical effects under ordinary conditions are not usually noticed.
2. Describe the three ways static electricity can form.
3. Distinguish between static electricity and electric current.
4. Describe what happens during a static discharge.
5. Use Ohm's law to compute voltages and currents in single-resistor DC circuits.
6. Distinguish between series and parallel resistances.

VOCABULARY TERMS

You should be able to define or describe each of these terms in a complete sentence or paragraph.

1. alternating current
2. ampere
3. arc
4. conduction
5. conduction electrons
6. conductor
7. direct current
8. electric current
9. electricity
10. friction
11. ground
12. induction
13. insulator
14. multi-resistor network
15. ohm
16. Ohm's law
17. parallel resistance
18. resistance
19. resistor
20. schematic diagram
21. series resistance
22. static discharge
23. static electricity
24. volt
25. voltage

13.1   The Nature of Electricity

In this text, we have already discussed electrical concepts quite a bit. We have repeatedly encountered protons and electrons and their charges (positive and negative). We have also come across ions several times, and we have looked at electric fields in some depth. All this focus on things electrical is for good reason—electrical charges and fields are everywhere and dominate everything.

As you know from Chapter 5, electrical charges produce electric fields, and electric fields exert forces on the electrical charges close enough to be affected by them. Ordinary matter, as you also know, is composed of enormous multitudes of protons, electrons, and neutrons. Normally, the number of charged particles in matter is so huge and the particles are so evenly distributed that we do not notice the electrical effects of these charges at all. Just consider that in the single grain of salt pictured in Figure 13.1, a tiny crystal less than half a millimeter wide, there are about 69,000,000,000,000,000,000 protons and 69,000,000,000,000,000,000 electrons (not to mention 74,000,000,000,000,000,000 or so neutrons). *This* is what I mean by large numbers of evenly distributed charge.

These unfathomable numbers of protons and electrons (or polyatomic ions, as the case may be) are all pushing or pulling on one another. Ordinarily, all these forces result in the charges locking themselves together into the grid of a crystal lattice or into tightly packaged molecules, each still exerting minute electrical attractions on the others.

Figure 13.1. The number of positive and negative charges in a single grain of salt is unimaginably large.

But there are two scenarios in which the effects of these charged particles make themselves known to us in our human-scale world. When we talk about *electricity*, it is these two scenarios we are talking about. The first is when a large number of like charges accumulate in one place, giving an overall positive or negative charge to a human-scale object. This effect is called *static electricity*. The second is when large numbers of like charges are flowing together in some direction—inside a copper wire, for example. This effect is called *electric current*. Our task for the next two sections is to look into the details of these two forms of electricity.

Large numbers and even distribution of charged particles in matter mean that matter normally does not exhibit electrical characteristics.

Learning Check 13.1

1. With such large numbers of charged particles around in nature, why aren't we constantly experiencing electrical attractions or repulsions?
2. Besides protons and electrons, what other type of "charged particle" have we looked at that can play a role in electrical phenomena?
3. Name and describe the two main forms of electricity.

13.2   Static Electricity

The term *static* is from Greek and means "standing still" or "at rest." So the phrase "static electricity" denotes electrical charges that are at rest. As mentioned in the previous section, static electricity is an accumulation of a large number of like charges in one place. Obviously, to accumulate they have to move to get there. And you know from our studies in previous chapters that all particles are constantly vibrating or translating. So when we say static, we don't mean *completely* free of motion; static is a relative term. What the term gets at is that of the two forms of electricity, one of them is caused by charges in motion (electric current), and the other is caused by a crowd of particles accumulated in one place. This crowd of accumulated charges is static electricity.

Static electricity is an accumulation of charge.

In Experimental Investigation 3, you observed that there are three ways the accumulation of charges can happen. The first is by *friction*, or rubbing. When you rubbed the Styrofoam cup on your head, you were using the friction between the cup and your hair to accumulate electrons on the cup. It just so happens that the electrons in the molecules human hair is made of have some freedom of movement, so it is easy to collect a bunch of them together on the cup.

Figure 13.2. The plastic slide readily allows electrons to accumulate on the backside—and then all over—the sliding child. The child's hairs standing up and pushing each other apart are evidence of an accumulation of static charge.

There are lots of materials that will supply electrons from rubbing. Many of these are plastics of various sorts. The nylon in carpet and car upholstery supplies electrons to humans as they walk across the carpet or slide into a car seat. You may have experienced a spark of static electricity when you touched a door knob in a house after walking on

carpet. Sometimes you can get a little shock from the door handle in a car. As shown in Figure 13.2, plastic slides also supply electrons by rubbing.

A second way charges can accumulate is by *conduction*. In this context, the term conduction is a reference to electrons flowing in a material that is electrically conductive. As you recall, one of the physical properties of metals is their high electrical conductivity compared to other materials. Because of this high electrical conductivity, metals are called *conductors*. Examples of good electrical conductors are gold, platinum, silver, copper, and aluminum. Materials that do not conduct the flow of the electrons are called electrical *insulators*. Dry air is a pretty good insulator, as are glass and most plastics.

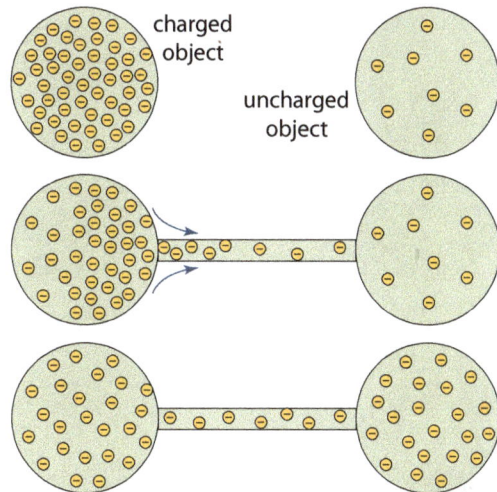

Figure 13.3. Initially, the concentration of electrons is much greater on the left object than on the right. When they are joined by a conductive material such as a wire, electrons immediately spread out by flowing—conducting— into the object with fewer of them.

Two of the ways static electricity forms are friction and conduction.

Figure 13.3 illustrates how the process of conduction can result in a charged object. At the top are two metal spheres. The one on the left with the high concentration of electrons represents a charged object. The sphere on the right with few electrons shown represents an uncharged object. If the objects are connected by a conducting bar or wire, electrons immediately conduct from the left to the right until every electron is as far away from the others as it can get. Afterwards, since the sphere on the right has a high concentration of charges, it is charged—it has an accumulation of electric charge on it. As you recall, this is exactly what happened in Experimental Investigation 3 with the aluminum leaves of the electroscope. When the excess electrons were placed on the top of the copper wire loop, they conducted down into the aluminum leaves. Our evidence for what had happened was the leaves pushing apart from each other, as shown in Figure 13.4.

Figures 13.5 and 13.6 illustrate two more instances of conduction. Both figures use this symbol:

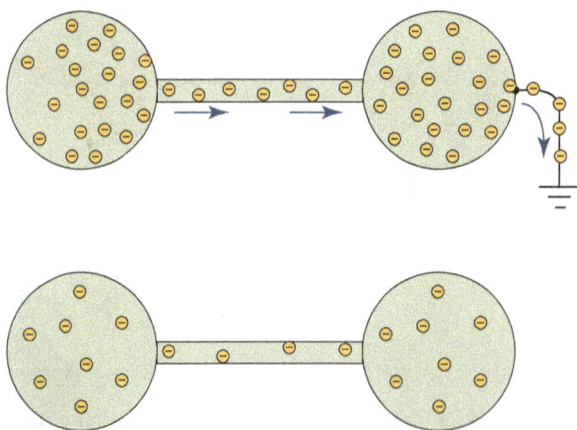

Figure 13.4. Electrons conducting onto the two electroscope leaves cause the leaves to push apart.

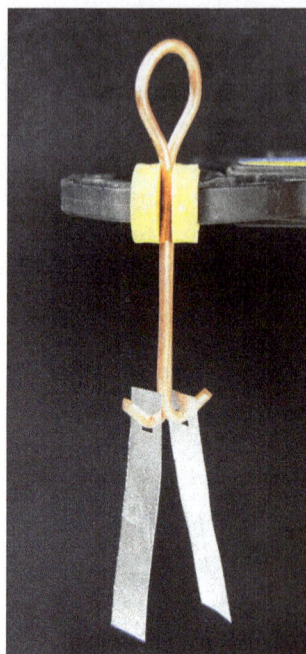

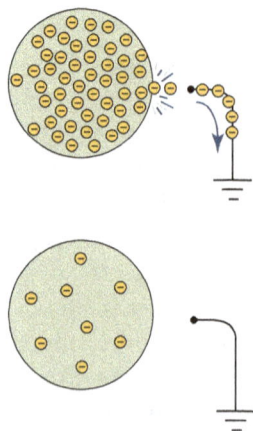

Figure 13.6. A grounded conductor brought near enough to a charged object results in a static discharge.

Figure 13.5. When connected to "ground," excess charges drain off, leaving the objects uncharged.

This is the symbol for *electrical ground*. The earth can be considered to be an electrically neutral object that is so massive it can never be charged. So when any charged object is connected to the earth—the ground—excess charges flow off the charged object and into the earth. The ground symbol is used in electrical diagrams to show any place where a connection to ground has been made.

Figure 13.5 illustrates what happens if the charged spheres from the bottom of Figure 13.3 are grounded. When the ground connection is made to the charged spheres, the excess electrons flow off the spheres and into the ground, leaving the spheres and the connecting wire between them uncharged.

Figures 13.6 and 13.7 illustrate what happens if a highly charged object is brought near enough to a grounded metal object. When a highly charged object is brought near—but not touching— a grounded conductor, a strong electric field develops between the two. If this electric field is strong enough, then an exciting sequence of events called a *static discharge* occurs. These events all oc-

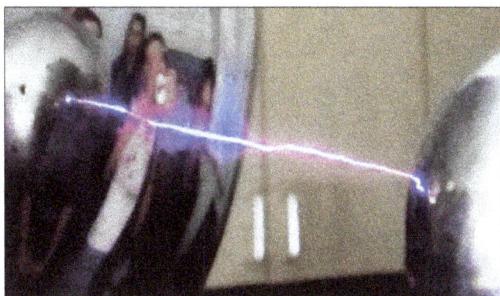

Figure 13.7. The arc of a static discharge.

cur in the merest fraction of a second, but here's a detailed, blow-by-blow sequence of what happens:

1. The air molecules in the region ionize, because of the strong electric field, forming a plasma.
2. The energy that tore the electrons from the air molecules (ionization) also excites the atoms in these molecules, causing them to emit a burst of blue-violet visible light (see the nitrogen plasma on page 107).
3. The energy released by the ionization creates a shock wave in the air, producing a sharp snap or zapping sound.
4. The ions are positively-charged particles, so the ionized air becomes an attractive pathway for electrons to flow in.
5. The static accumulation of electrons rushes through the conductive pathway in the air to ground, discharging the static accumulation.
6. As the ions put themselves back together after the discharge, some of the oxygen atoms combine in triples instead of the usual pairs for diatomic oxygen. An O_3 molecule is the gas *ozone*, and if you are around something that is arcing or sparking a lot, you can often smell the clean, refreshing smell of ozone in the air.

This crazy sequence of events is the back story behind a spark or arc, like the one in Figure 13.7. Welding arcs are more or less the same thing. Of course, the granddaddy of all arcs is lightning.

A large enough electric field ionizes nearby air, forming a conducting channel that results in a spark, or static discharge.

Lightning has been studied extensively, but much about it is still unknown. (It's hard to study things when you don't know when they will happen and they last less than a half second!) The sequence of events above applies roughly for lightning but with a few additional details. Refer to the image on page 268 and to Figure 13.8 as you read these items.

Figure 13.8. Static charge distributions prior to a lightning strike.

1. The motion of ice crystals in a cloud causes a static build-up of electrons on the bottom of the cloud (and positive ions on the top of the cloud, but we'll just stick to the bottom).

2. The presence of a huge static accumulation on the bottom of a low cloud attracts positive ions in the ground, causing them to accumulate at the surface. We now have a major electric field between the ground and the cloud (field lines pointing from positive to negative, as always).

3. Channels of negative charges called "leaders" begin branching down through the air from the cloud to the ground. These leaders are composed of ionized air, a plasma that emits purple light and a zapping or crackling sound as described before.

4. When a branch of leaders gets close to the ground, a channel of positive ions from the ground call a "streamer" begins branching up toward the leader.

5. Leader and streamer connect, which is called "attachment." All this build-up has taken maybe a quarter of a second.

6. There is now a conducting channel all the way from the cloud to the ground, and the main "return strike" occurs—the mighty crack of lightning. One would think that the electrons in the cloud rush to the ground, since that is what lightning looks like. But the downward motion we see is the descending leaders over the course of a few hundred milliseconds. The main strike is a rush of positive ions flowing up from the ground (or building or tree) to the cloud that lasts only a few microseconds. The electric current in a lightning strike is typically in the league of 30,000 amperes, but can be over 100,000 amps in a big strike.

In the photo on page 268, you can see several descending leaders. Only one attachment occurs, right at the top of the poor tree, appointed to go up that night in a blaze of glory. (Bad night to be a squirrel nesting in that tree.) To the left of

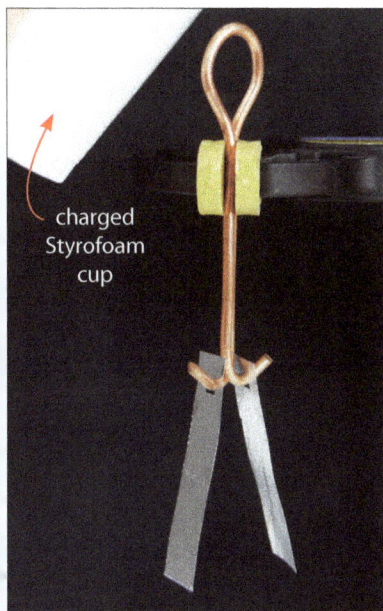

charged
Styrofoam
cup

Figure 13.9. The charge on the cup
induces a static accumulation of charge
in the leaves of the electroscope.

the houses on the ground, you can see a streamer going up from the top of a power pole. There also appears to be another unattached streamer coming up from the tree.

The third way static accumulations can occur is through *induction*. The word *induce* means "exert an influence that will make it happen." You saw induction occur in Experimental Investigation 3, when the charged cup was brought near to the electroscope but was not allowed to touch it. This is illustrated in Figure 13.9, where the repulsion of the electrons on the cup forces the conduction electrons in the copper wire to crowd down into the lower part of the copper, and right on down into the aluminum foil leaves.

Since the cup is not touching the electroscope wire, the influence of the electrons on the cup is temporary. No additional electrons are conducting onto the copper wire, so when the cup is withdrawn, the electrons in the wire and foil leaves relax and return to occupy all parts of the copper wire and leaves evenly.

Static electricity is formed by induction when a charged object comes near another object, forcing the charges in the second object to crowd together.

Finally, the distribution of positive ions on the ground during a lightning storm (Figure 13.8) is another example of an induced static charge accumulation. The static electricity in the clouds got there by a complicated process of friction between ice crystals in the cloud. But the positive ions accumulated at the top of the ground are being drawn there—induced—by the negative charges in the cloud.

Learning Check 13.2

1. What is static electricity?
2. Describe the three ways static electricity can form.
3. For each of the three potential causes of static electricity, describe an example that illustrates static electricity being formed this way.
4. As briefly as you can, explain what an arc (or spark) is.
5. What is an electrical insulator?

13.3   How Electric Current Works

In the most general sense, electric current is the flow or movement of electric charge. The rush of positive ions up from the ground to a cloud during a lighting strike is a massive electric current, even though it only lasts for a few microseconds. However, even though positive charges may get involved when a plasma is present, usually it is the negatively charged electrons that are free to move. This is particularly the case with the freely moving *conduction electrons* in metals. So whenever we talk about electric circuits, we are talking about flowing electrons.

Electric current is flowing charges. In wires, the flowing charges are negative electrons.

Electrons flowing in wires form an *electric current*. Electricity is a mystery to many people because we can't see it. But happily, electric current flowing in wires is very similar to the way water molecules flow in a water pipe. This allows us to use the water flow as an analogy for electrical current flow. Figure 13.10 illustrates the arrangement of parts in the water and electrical systems for comparison.

In the diagram on the left, a water pump is connected to a loop of water piping. The water pump pushes water in the same way an ordinary fan pushes air—with rotating blades. The water pump forces the water to flow in the piping around the loop. I have placed a water filter in the system to give the system something to do, rather than just pumping water in a loop. You may have seen water filters like this in aquariums—they are full of sand and bits of charcoal. So as you can easily imagine, the water filter in the pipe makes it harder for the pump to push the water through the pipes. But that's okay; that's just the price we have to pay for getting the water cleaned.

On the right in the figure is a diagram of a simple electric circuit. In this case, a battery is connected to a loop of copper wiring. (The wiring is covered in blue insulation.) In the electric circuit, the battery forces electric current to flow through the wires. In the loop there is a light bulb. The current has to flow through the tiny curl of wire in the light bulb, which is called a *filament*. (The filament is made of very thin wire, and as the electrons flow through it the filament becomes very hot—

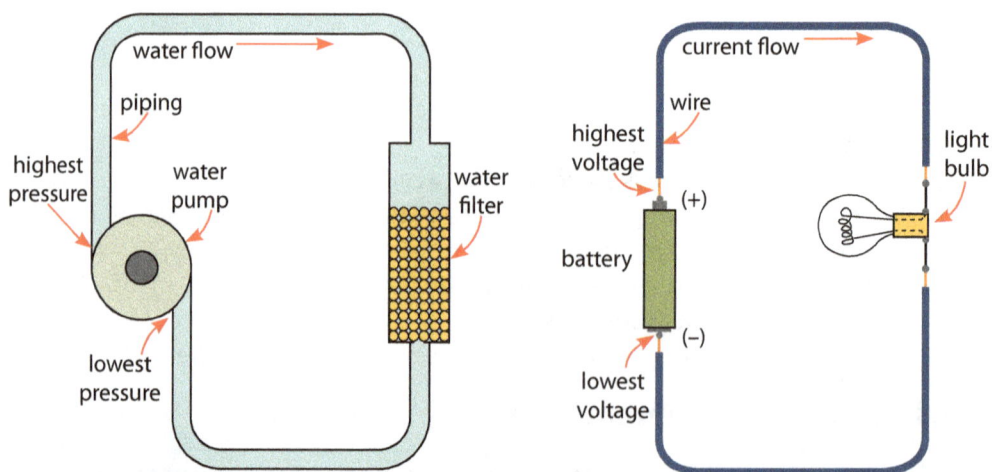

Figure 13.10. The water circuit as an analogy for the electrical circuit.

white hot—giving off light. But we don't really need to worry about those details at present.) The filament is a thin wire compared to the larger wires in the rest of the circuit, and the battery must force the flow of electrons through this wire. Just as placing the water filter in the water piping makes it more difficult for the pump to force the water through, placing a light bulb in the electrical circuit makes it more difficult for the battery to push the electric current through. For this reason, in the world of electrical circuits, any device in the circuit that makes it harder for the current to flow is called a *resistor*.

Now let's discuss what it is making the water and the electric current flow in these two circuits. Looking again at Figure 13.10, you see that the highest and lowest pressures in the water circuit are at the connections to the pump. Think back for a moment to the discussion in Chapter 9 on pressure. Recall from that discussion that for a fluid to flow, there must be a difference in pressure, and that the fluid flows from high pressure to low pressure. So in the water circuit, the water is pushed by the pressure of the pump and it flows around the circuit toward the low pressure end of the pump.

Just as the pressure difference forces water to flow, in the electrical circuit it is a *voltage* difference that forces the charge to flow. As you see in Figure 13.10, the highest voltage is at the end of the battery labeled (+), and the lowest voltage is at the end of the battery labeled (−). We refer to these places as the positive (+) and negative (−) ends of the battery.

Figure 13.11 illustrates in more detail the ways pressure and voltage cause flow. In the upper part of the figure, the high pressure pushes on a section of water, forcing it to flow toward the lower pressure region. The lower part of the figure shows the analogy for the flow of electric charge. The red arrows indicate the lines of the electric field produced by the rows of positive and negative charge. As you know from Chapter 5, the field lines point from positive to negative, and they indicate the direction of the force that will be present on a positive charge placed in the field.

Figure 13.11. Just as a difference in pressure forces water to flow, a difference in voltage causes charge to flow.

Without getting into the details of defining exactly what voltage is, we can say that the voltage difference is proportional to the strength of the electric field; the stronger the electric field is, the greater the voltage difference is, and the more charge flows.

A pressure difference makes fluid flow, and a voltage difference makes charge flow.

I hope you noticed something strange about Figure 13.11. Is there something bothering you about that diagram?

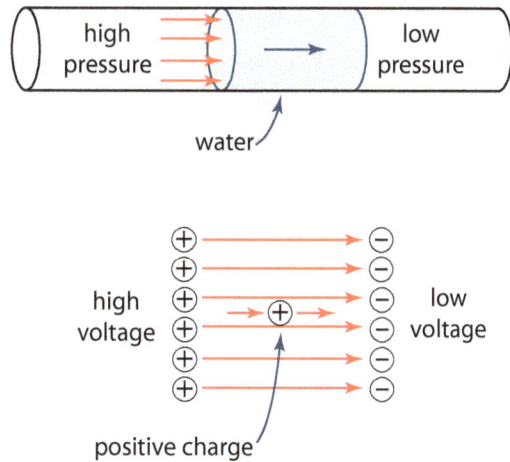

The diagram shows a *positive* charge being pushed to the right by the voltage difference. But I have been saying that in electric circuits we have negative charges—electrons—flowing in the wires. If you noticed this problem, good. You noticed the big difference between the water circuit and the electric circuit.

IF positive charge were flowing in our electric circuits, then the water circuit and the electric circuit would be basically just alike. But in fact, in the electric circuit we don't have positive charges flowing, we have negative charges flowing, and the negatives charges go in the opposite direction! But Figure 13.10 shows the electric current flowing in the direction positive charges would flow. Is this just a trick?

Yes, actually, it is. Here's the deal. Back in the early days of electrical theory people noticed that mathematically, negative charges flowing toward the positive terminal of a battery is exactly the same as positive charges flowing toward the negative terminal of the battery. And we know that physics is all about modeling nature with mathematics. But if we do calculations with negative charges, we are going to have a bunch of negative signs in our equations.

Electrons flowing toward positive is mathematically identical to positive charges flowing toward negative.

So the Powers That Be decided that when we talk about electric circuits and do calculations with them, we will just pretend that positive charge is flowing. Then all the negative signs go away from the equations and everything is simpler. Mathematically, the results are exactly the same. But we all have to remember that really, physically, what's happening in the wires is that negative charges are flowing the other way. Figure 13.12 illustrates this point.

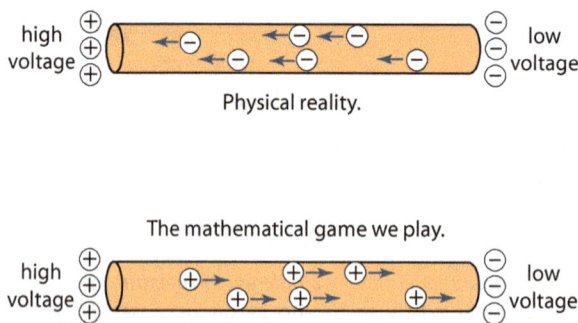

Physical reality.

The mathematical game we play.

Figure 13.12. In electric circuits, electrons flow in the wire toward the positive terminal of the battery. In our math, we pretend positive charges flow the other direction.

To conclude this section, let's consider the units for electric current and how they relate to the actual charges that make up the current. Recall from Chapter 8 that the SI System unit for current is the ampere, or amp for short. In fact, the ampere is one of the seven SI System base units. Now, compared to the charge on an individual electron, an amp of current represents a *lot* of flowing charges. Just as huge numbers of water molecules flowing together form a current of water, huge numbers of electrons flowing together form an electrical current.

Electric charge is measured in units called coulombs (C). One ampere of current flowing in a wire is equal to one coulomb of charge per second passing through the wire. To give you an idea of the numbers of electrons we are talking about in an electrical current, if you have a wire with a current of one ampere (1A) flowing in it, 6,250,000,000,000,000,000 electrons flow past a given point in that wire each second! Once again, we see that the numbers of particles involved at the atomic level with quite ordinary things is almost scary.

Learning Check 13.3

1. What is electric current?
2. Pretend you are going to explain electric current to a friend who knows nothing about it. Write a couple of paragraphs in your own words, using the analogy between flowing water and flowing electric current to explain electricity to your friend.
3. Explain the difference between the physical reality of current in a wire and the way we handle current mathematically.

13.4   DC Electric Circuits and Ohm's Law

Our next adventure into the mathematics of physics is to learn how to compute voltages and currents in simple DC electric circuits. In general, there are two general types of electrical systems: *DC* systems and *AC* systems. The power distribution system that brings electrical power to the buildings we live and work in is an AC system. "AC" stands for *alternating current*. In an AC system, the current does not flow steadily in one direction. Instead, it oscillates back and forth at a frequency of 60 Hz. This type of system has many advantages for power distribution, but is more difficult to analyze mathematically.

All the electric circuits in our music equipment, mobile devices, computers, and televisions are DC systems. "DC" stands for *direct current*. In a direct current system, the current flows steadily in one direction, making the mathematics very simple. Portable DC electrical systems such as mobile phones and flashlights are powered by batteries. Stationary electronic systems like computers and televisions have circuitry inside that converts the AC power coming to the building into DC, which is used by the electronics. To keep things simple, we are going to look at only the simple case of a single battery connected to a single device, such as a light bulb or "resistor."

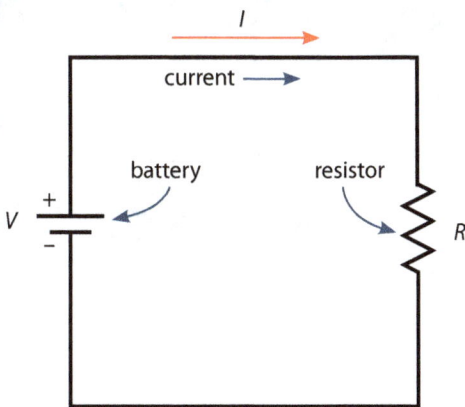

Figure 13.13 shows a *schematic diagram*, the standard way of representing electric circuits. This diagram represents the exact same circuit as the one shown in Figure 13.10. A schematic diagram shows only how the electrical components are connected together; the diagram bears no relation at all to what the circuit actu-

Figure 13.13. A schematic diagram for a simple DC circuit with a battery and one resistor.

Schematic diagrams are the standard way of representing electric circuits.

ally looks like. The wires could be one micrometer long in a computer chip, or 12 inches long in a flashlight; they are shown the same way in either case.

In an electric circuit, the battery supplies the voltage, V, that makes the current, I, flow. (Yes, the symbol for current is I, not C. C is used for something else that we are not going to get into.) You see in the diagram that the current flows in the direction of imaginary positive charge, as we discussed in the previous section. The symbol for a battery has two short parallel lines, one longer than the other. By convention, the longer line always represents the positive terminal of the battery.

Also shown in the figure is the symbol for a *resistor*. In the electric circuit of a flashlight, the resistor is actually a light bulb. The light bulb resists the flow of electric current. The variable we use to specify how much the light bulb resists the current is called the *resistance*, R. Most other DC circuits are a lot more complex than a flashlight. A great deal of electronic circuitry is involved to provide all the functionality of device like a computer or mobile phone. In complex circuits like this, special devices called "resistors" are used to control the flow of current in various parts of the circuit. Figure 13.14 illustrates what these devices look like when they are installed on *printed circuit boards* in electronic components. Throughout the 1960s, 70s, and 80s, resistors looked like those in the upper image. In the 1990s, newer manufacturing technologies led to miniaturized devices that can be installed and connected more easily in automated assembly lines. Examples of these are shown in the lower photo.

Whether the resistance in a circuit is a light bulb or a resistor, the math is the same. The mathematical relationship between the voltage, current, and resistance in a circuit is called *Ohm's law*, named after the man who discovered it, German scientist Georg Ohm (Figure 13.15). Ohm was working as a high school teacher (hooray!), conducting research into the nature of electricity using equipment he made himself. In 1827, at the age of 36, Ohm published the results of his work and Ohm's law was brought into the world. I suppose the three

older style electronic resistors

newer style electronic resistors

Figure 13.14. Older and newer styles of electronic resistors, such as those found in electronics components.

most well-known equations in all of physics are Newton's second law of motion ($a = F/m$ or $F = ma$), Einstein's law of mass-energy equivalence ($E = mc^2$), and Ohm's law. Here is Ohm's law:

$$V = I \cdot R$$

Ohm's law

This equation states that the voltage difference, V, provided by the battery is equal to the product of the current, I, and the resistance, R. As we have seen, the unit for current is the *ampere* (A), one of the SI base units. The unit for voltage is the *volt* (V), and the unit for resistance is the *ohm* (Ω), obviously named after the discoverer of Ohm's

Figure 13.15. German scientist Georg Simon Ohm.

law. The symbol for the ohm, Ω, is the upper case Greek letter omega, the last letter in the Greek alphabet. To help students compare the names and units for the variables in electricity, I always like to put them all together in a table, and I have done so in Table 13.1.

Variable	Symbol	Unit	Symbol
voltage	V	volt	V
current	I	ampere	A
resistance	R	ohm	Ω

Table 13.1. Electrical variables, units, and their symbols.

As we have done with other equations, let's use just a bit of algebra to put Ohm's law into an alternate form. When we divide both sides of the equation by the resistance, R, we have

$$V = I \cdot R$$

$$\frac{V}{R} = \frac{I \cdot R}{R}$$

$$\frac{V}{R} = I$$

which gives us

$$I = \frac{V}{R}$$

Ohm's law, form for computing current

With this form of the equation, we can calculate the current flowing in a circuit, given the voltage and the resistance. We now look at a few example calculations. As you have seen before, we must use conversion factors to get everything into standard units before we calculate the result. For DC circuit problems, always begin by converting the units of measure into volts, amperes, and ohms.

EXAMPLE 13.1

An electric lantern uses a 6-volt battery to illuminate a light bulb. The resistance of the bulb is 5.5 Ω. Determine the current flowing through the light bulb.

Begin by listing the given information and the unknown.

$V = 6 \text{ V}$

$R = 5.5 \ \Omega$

$I = ?$

The units are already in the form we need them to be in, so we can proceed to write down the equation and compute the result.

$$I = \frac{V}{R} = \frac{6 \text{ V}}{5.5 \ \Omega} = 1.09 \text{ A}$$

Checking over the work, the denominator of the fraction is a bit less than the numerator, so I expect the result to be a bit more than 1, and it is.

In the next example, we get more into the unit conversions. It is common in today's world of microelectronics for currents to be in the range of microamps. This is the result of resistances that are in the kilohm (kΩ) or megohm (MΩ) range. (Notice that when the terms kilohm or megohm are written out, the *o* or *a* at the end of the kilo– or mega– prefix is dropped.) Always convert everything into volts, amps, and ohms before inserting values into an equation.

EXAMPLE 13.2

The current flowing in a certain microcircuit is 3.58 µA. If the resistance in this circuit is 245 kΩ, determine the voltage of the power supply running the circuit.

Begin by writing down the given information.

$I = 3.58 \ \mu\text{A}$

$R = 245 \text{ k}\Omega$

$V = ?$

This time we must perform two unit conversions, so we proceed to these next.

$$I = 3.58 \ \cancel{\mu A} \cdot \frac{1 \, A}{1,000,000 \ \cancel{\mu A}} = 0.00000358 \text{ A}$$

$$R = 245 \ \cancel{k\Omega} \cdot \frac{1,000 \ \Omega}{1 \ \cancel{k\Omega}} = 245,000 \ \Omega$$

$$V = ?$$

Now we can proceed, using the Ohm's law equation in the form that allows us to solve for the voltage. Write down the equation first, then insert the values with units. Finally, compute the result and write it down with the correct units of measure.

$$V = I \cdot R = 0.00000358 \text{ A} \cdot 245,000 \ \Omega = 0.877 \text{ V}$$

As a final check, our answer is in volts as required. The multiplication involves one very large value and one very small value, so we expect the result to come out neither large nor small.

For one final example, we will look at a problem that requires us to state our result in units other than the standard volts, amps, and ohms. When given a DC circuit problem like this, always first compute the result using the standard V, A, Ω units. Then at the end, convert your result to the units the problem requires. This is a different procedure than we have followed so far, where we performed the unit conversions so the units in the given information match what we are required to show in our result. But in the previous cases, we could see the length, mass, and time units and could verify that everything cancelled correctly. If you multiply or divide non-standard electrical units, there is a good chance you won't know what units to place in the result. So for these, use the standard units to perform the computation, then convert to nonstandard units at the end if you need to.

EXAMPLE 13.3

The power supply in a computer operates at 5.5 V. If this power supply is connected to a resistance of 87 kΩ, calculate what the resulting current is. State your result in microamps.

Begin by writing down the given information.

$$V = 5.5 \text{ V}$$

$$R = 87 \ k\Omega$$

$$I = ?$$

Next, convert the resistance to ohms.

$V = 5.5$ V

$R = 87 \text{ k}\Omega \cdot \dfrac{1{,}000 \ \Omega}{1 \text{ k}\Omega} = 87{,}000 \ \Omega$

$I = ?$

Now write the appropriate equation, insert the values, and compute the result.

$I = \dfrac{V}{R} = \dfrac{5.5 \text{ V}}{87{,}000 \ \Omega} = 0.00006322$ A

The final step is to convert this value into microamps, as required by the problem.

$I = \dfrac{V}{R} = \dfrac{5.5 \text{ V}}{87{,}000 \ \Omega} = 0.00006322 \text{ A} \cdot \dfrac{1{,}000{,}000 \ \mu\text{A}}{1 \text{ A}} = 63.22 \ \mu\text{A}$

Double checking all the math, and the unit conversion factor I used, everything looks good.

Learning Check 13.4

1. What is the difference between DC and AC electricity?
2. What is a schematic diagram?
3. Use the circuit diagram in Figure 13.13 as a reference for the following questions.

   a. The current flowing in a resistor is 5.22 mA. If the resistor value is 970 Ω, determine the voltage of the battery.

   b. A 12-volt battery is keeping the overhead light on in a car, even though the car's engine is off. If the resistance of the overhead light bulb is 125 ohms, determine the current flowing in the circuit.

   c. A 1.5-volt battery is used to power the motor in a small clock. If the resistance of the motor is 11.6 kΩ, determine the current flowing in the circuit. State your answer in milliamps.

   d. A current of 25.5 μA flows through a resistance of 1.1 MΩ. Determine the voltage driving this circuit.

e.   A 5.2-volt power supply in a computer is running a circuit with a resistance of 2.96 MΩ. Determine the current flowing in the circuit, and state your answer in microamps.

Answers to #3:	b.	0.096 A	d.	28.05 V
a. 5.06 V	c.	0.129 mA	e.	1.76 μA

13.5   Series and Parallel Circuits

Most of the electrical and electronic gadgets in the world have more than just one resistor. Many electronic devices have thousands of resistors. So to conclude this chapter, we take a brief look at the two main ways resistors can be connected together to form a *multi-resistor network.*

One way to connect resistors together is to connect them in *series.* Series-connected resistor networks are shown in Figure 13.16. For all three diagrams in this figure, the dots at the ends of the wire indicate where we would connect a battery to the network.

On the left in this figure is the single resistor we have already seen. In the center is a resistor network with two resistors connected in series. The key thing to notice is that the resistors are located one after the other in a single loop of wire. Also, with more than one resistor in a network, we can no longer just label it R. We need a way to distinguish the resistors from one another. This is done with subscripts, so in the two-resistor series network the resistors are labeled R_1 and R_2. On the right of the figure is a network containing three resistors connected in series. As before, notice that all three resistors are in a single loop of wire, one after the other. Subscripts are used to distinguish the resistors as R_1, R_2, and R_3.

In a series network, all resistors are in the same loop of wire, one after the other.

The other way to connect multiple resistors together is to connect them in *parallel.* Figure 13.17 illustrates resistors connected this way. On the left for comparison is the single resistor. In the center is a network with two resistors connected in parallel, and on the right is a network with three resistors in parallel. The key thing to notice about the parallel networks is that each resistor is located in its own

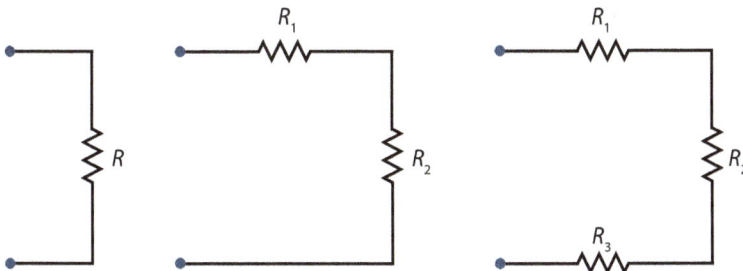

Figure 13.16. A single resistor (left), compared to two resistors in series (center), and three resistors in series (right).

285

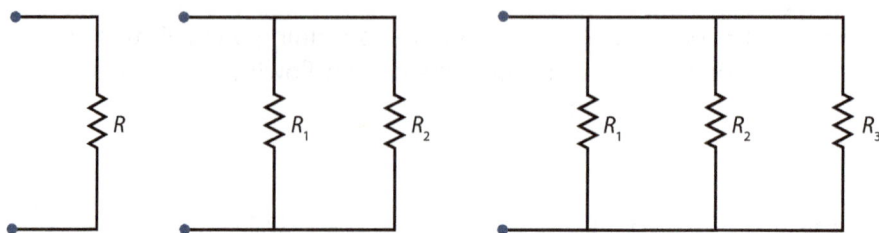

Figure 13.17. A single resistor (left), compared to two resistors in parallel (center), and three resistors in parallel (right).

branch or loop of the network. All the branches are connected together at one end to a wire that would go to the positive terminal of a battery. All the other ends are connected together to a wire that would go to the other terminal of the battery. As with the series networks, when we have more than one resistor, we use subscripts to distinguish the resistors from one another.

In a parallel network, each resistor is in its own branch and the branches are connected together at the ends.

Figure 13.18 illustrates what happens when a battery is connected to two-resistor and three-resistor series networks. As you see, in each of these networks there is only one loop for the current to flow in, so the same current, labeled I, is flowing in every part of the network.

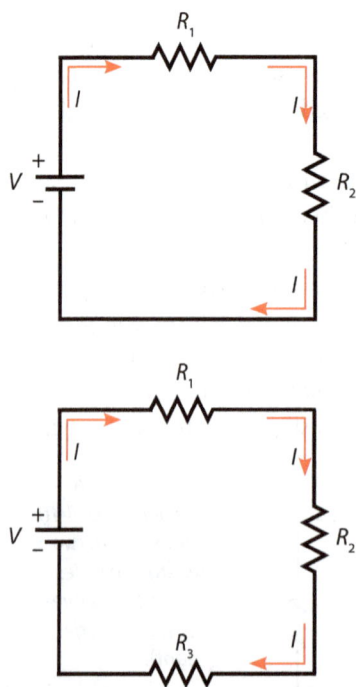

Always remember that current flowing in a wire in an electric circuit is like water flowing in a pipe in a water circuit. If water is flowing in a pipe with two or three water filters, one after the other, the same water flow rate is present in every part of the system. How could it be otherwise? Where else can the water go? If a certain amount of water is coming out of the pump, this same amount of water must be flowing through each water filter, and the same amount of water must be entering the pump at the low-pressure point.

The same thing happens here. Whatever current is flowing out of the battery, that same amount of current must be flowing through each of the resistors, and that same amount of current must also be entering the battery at the low-voltage point (the negative end of the battery).

We have a different situation entirely with the two-resistor and three-resistor parallel circuits shown in Figure 13.19. In these two diagrams, I labeled the current coming out of the battery as I_1. Follow I_1 as it goes around the loop. After leaving the battery, the first thing that happens to this cur-

Figure 13.18. Series circuits with two and three resistors.

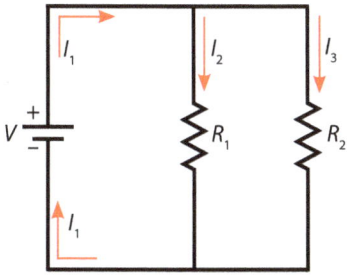

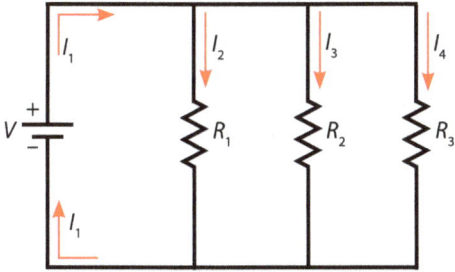

Figure 13.19. Parallel circuits with two and three resistors.

Figure 13.20. Branching streams as an analogy for branching currents in a parallel circuit.

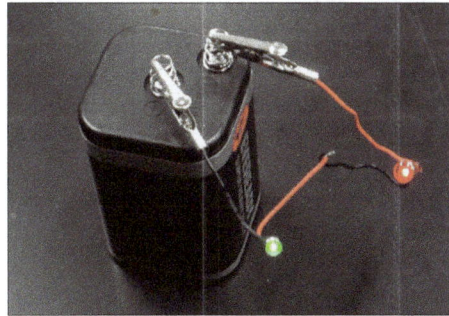

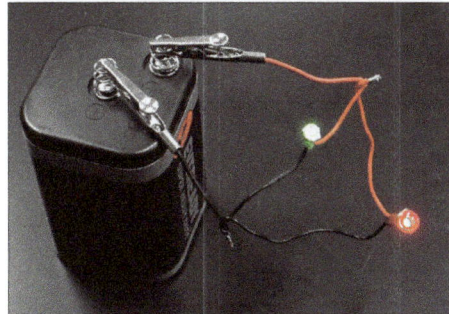

rent is that it comes to a junction or "fork in the road"—a place where two wires are connected together.

Again, think about what water in a pipe does in this situation. Clearly, some of the water goes one way in one branch and the rest goes the other way in the other branch. In the two-resistor circuit, I labeled these two currents I_2 and I_3. In the three-resistor circuit there are two junctions. Part of the current, I_2, takes the first branch that flows through resistor R_1. Whatever current is left branches again at the second junction, part of the current, I_3, flowing through R_2, and the remainder, I_4, flowing through resistor R_3.

As another analogy, Figure 13.20 illustrates parallel circuits with branching currents in stream of water. Here I have labeled the water currents flowing as W_1, W_2, and so on. As you see, after the currents go in their individual branches and come back together again, they all add back together to give us the same water flow we had to begin with,

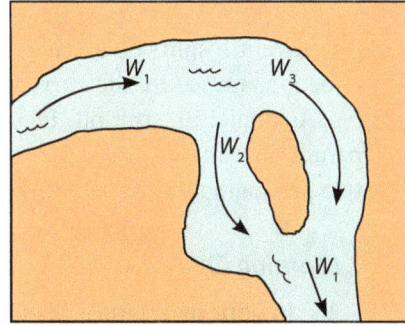

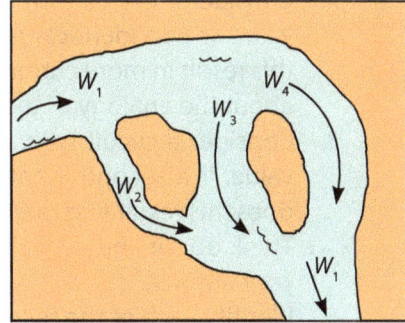

Figure 13.21. Actual circuits: A two-LED series circuit (top), and a two-LED parallel circuit (bottom).

W_1. The same thing happens with electric current in a parallel network. The original current I_1 splits up at each junction, but all the currents combine back together to be I_1 again before the currents get back to the battery.

As a final illustration, Figure 13.21 shows actual examples of both series and parallel circuits of red and green LED lights connected to a battery. Compare these two photographs with the schematic diagrams shown Figures 13.18 and 13.19.

Learning Check 13.5

1. In your own words, describe series and parallel resistor networks.
2. Consider a circuit with a certain battery voltage and a certain resistor value. If two identical resistors are added in series with the first, does this result in more current flowing out of the battery or less? (Hint: Think about the analogy of the water circuit.) Explain your answer.
3. Consider a circuit with a certain battery voltage and a certain resistor value. If a second identical resistor is added in parallel with the first, does this result in more current flowing out of the battery or less? (Hint: Think about the analogies with streams and pipes of water.) Explain your answer.
4. Identify each of the following resistor networks as a series network or a parallel network.

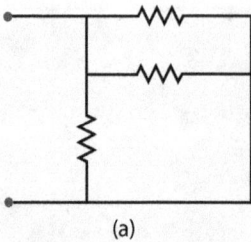

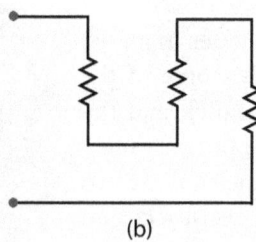

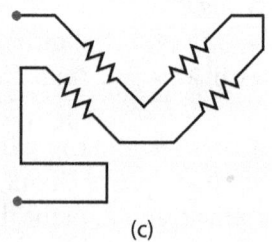

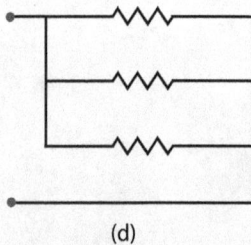

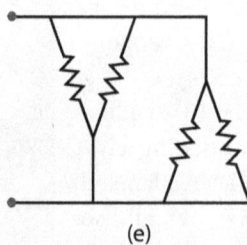

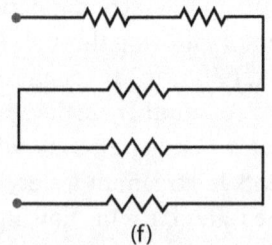

(a)   (b)   (c)

(d)   (e)   (f)

Scientists, Experiments, and Technology

What do you get when electricity meets chemistry? *Electrochemistry*, of course. Without electrochemistry, none of us would enjoy the mobile digital environment we now have, because electrochemistry is the science behind modern day "batteries."

I put the term battery in quotes, because the little batteries we use in flashlights and cameras are really not batteries; they are *dry cells*. In battery history, the dry cell, which contains no liquid, was preceded by the *wet cell*. If you chain several cells together, then you have a *battery*, a group of cells working together. Car batteries are made of six wet cells connected together. Each cell produces 2.2 V, and when six of them are placed in series, the combined voltage is 13.2 volts (even though they are called 12-volt batteries).

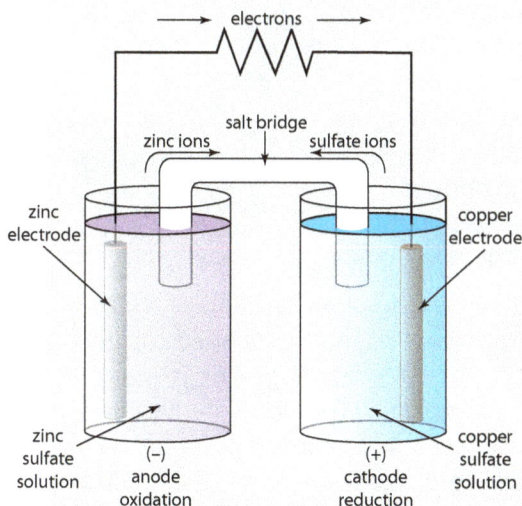

Italians Luigi Galvani and Alessandro Volta were behind the first device that could produce current like a battery, the *Voltaic pile* of 1800. But the science of electrochemistry took a big step forward with the invention of the *Daniell cell* by English chemist John Daniell in 1836. The Daniell cell, shown to the left, is a "wet cell"; it uses solutions to run redox reactions, and these reactions produce the voltage and current the cell provides.

The Daniell cell is made of two containers, each containing a metal electrode and a metal salt solution. The *cathode* is the electrode where the metal is being reduced (gaining electrons)—the copper electrode in the diagram. In the copper container, copper ions (Cu^{2+}) from a copper sulfate solution are reduced by attaching to the copper cathode. The electron supply for this reduction reaction is coming from the anode container, where the zinc *anode* is being oxidized (losing electrons). As the atoms in the zinc anode are oxidized, they form ions (Zn^{2+}) and go into solution. So as the cell operates, supplying electric current to whatever is connected in the external electron pathway, copper ions reduced to neutral copper atoms are plating onto the copper cathode, and zinc atoms oxidized to zinc ions are coming off the zinc anode into the solution. If this runs long enough, the zinc anode eventually dissolves away to nothing.   (Continued next page.)

This wet cell is rechargeable. If you take out the resistor and connect in a DC power supply or battery with its positive terminal connected to the copper cathode, the electrons are forced to flow in the opposite direction. This puts oxidized Cu^{2+} ions back into the copper sulfate solution and puts reduced zinc metal back onto the zinc electrode.

But electrochemistry can do more than run batteries. The same process that reduces copper ions and plates them onto the copper cathode can just as easily plate one metal onto another metal. This is called *electroplating*. A copper electroplating process using the same copper sulfate solution as before is shown to the right. Here there is only one solution. The cathode is made of the metal to be plated (Me in the figure). Copper ions from the copper sulfate solution are reduced at the cathode to become copper metal, plated onto the cathode material. The anode is copper, and copper atoms oxidizing at the anode go into the solution, replenishing the supply of Cu^{2+} ions.

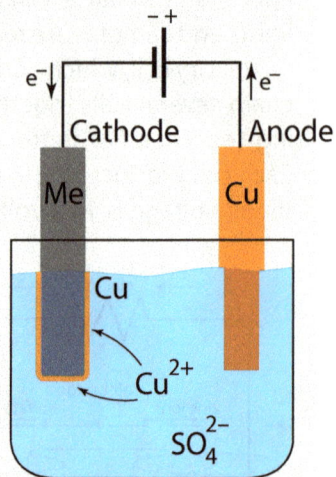

This is the exact plating process used for making pennies. Pennies are zinc disks covered with copper plating. (Prior to 1982 they were solid copper.) There are also many other applications for electroplating. Zinc is electroplated onto steel in a process called *galvanizing*. (The name derives from Luigi Galvani's name.) Galvanized steel doesn't rust, so metal parts can have the strength of steel while being protected from rusting by the electroplated zinc coating. Chromium is also electroplated onto steel. This is called *chrome plating*. Chrome plating not only protects the steel from rust, but is also pretty and shiny. Just the thing for nice bumpers and trim parts on cars and trucks. (Although, not many cars have chrome-plated bumpers any more; they are mostly plastic now.) Gold and brass are also often plated onto other metals.

In the late 1930s, Richard Feynman, a future winner of the Nobel Prize in Physics, developed processes for plating metal onto plastic. These processes are now used for everything. Just looking around on my desk, I see my calculator and a camera lens both have metal plated plastic parts on them, as do the loudspeakers for my stereo. And it's all done with redox chemistry combined with DC electricity.

Chapter 13 Exercises

Answer each of the questions below as completely as you can. Write your responses in complete sentences.

1. Explain the difference between conduction and induction as ways static electricity forms.
2. Explain what happens to the electrons when a negatively charged object is grounded.
3. Does lightning travel from a cloud to ground or from the ground to the cloud? Explain your answer. (Be careful—this is a tricky question.)
4. Describe the similarities between pressure in a water system and voltage in an electrical circuit.
5. What makes series resistor networks different from parallel networks?
6. A chemist intends to run a current through a beaker of water. The resistance between two electrodes (wires) inserted into the water is 2.25 kΩ. What voltage is required to produce a current of 35 mA?
7. A toy car uses four batteries (dry cells, actually) in series to get an operating voltage of 6 volts. If the car's motor has a resistance of 6.55 kΩ, what current flows? State your answer in milliamps.
8. What voltage is required to deliver a current of 82.9 μA to a resistance of 160 kΩ?
9. In an aluminum smelting facility, a voltage of 450 V is applied to a resistance of 4.09 mΩ. Determine the current that flows.
10. Determine the voltage needed to run a current of 105.5 mA through a resistance of 75 μΩ. State your answer in microvolts.
11. An ordinary lightning strike can deliver a current of 35 kA. If the resistance of the ionic channel is 128 Ω, determine the voltage between the ground and the cloud. State your answer in megavolts.
12. A voltage of 45.4 mV is applied to a resistance of 27 kΩ. Determine the resulting current and state your answer in microamps.
13. An inverter is a device that converts DC electricity from a battery into AC. A battery system is designed to operate at a voltage of 175 V, delivering current into an inverter with a resistance of 155 mΩ. Determine the current that flows into the inverter. State your answer in kiloamps.

Answers	8.	13.26 V	11.	4.48 MV
6. 78.75 V	9.	110,024 A	12.	1.68 μA
7. 0.916 mA	10.	7.91 μV	13.	1.129 kA

Experimental Investigation 11: Series and Parallel Circuits

Overview

- *The goal of this experiment is to observe the behavior of resistor networks connected in series and in parallel.*
- Measure the resistance of a small lamp. Then connect two lamps in series and measure the combined resistance. Repeat for three lamps in series.
- Connect two lamps in parallel and measure the combined resistance. Repeat for three lamps in parallel.
- Connect three lamps in series, and then connect the lamp network to a battery pack. Measure the voltage difference across each of the three lamps individually. Remove one lamp from its base and observe the effect on the rest of the lamps in the network. Replace the lamp in its base and make the same test for the other two lamps in the network.
- Connect three lamps in parallel, and then connect the lamp network to a battery pack. Measure the voltage difference across each of the three lamps individually. Remove one lamp from its base and observe the effect on the rest of the lamps in the network. Replace the lamp in its base and make the same test for the other two lamps in the network.

Basic Materials List

- small incandescent lamps with bases (3)
- alligator clip connecting leads (6)
- digital multimeter
- 6-volt lantern batteries (2)

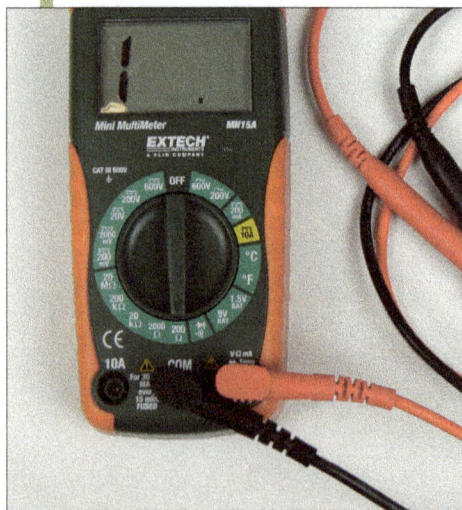

In this experiment, we use small incandescent lamps (common in flashlights before LEDs took over) as resistors. We connect the bulbs first in series and then in parallel to see how these combinations affect the total resistance of the network. Then we try out these two kinds of networks by connecting them to a power supply (the battery pack).

For making measurements with electric circuits, we use a tool called a digital multimeter (DMM). A DMM measures voltages, currents, and resistances. The way the test leads connect to the DMM depends on model of DMM. For this experiment, you use a DMM to measure resistance and voltage, so insert the black test lead into COMMON terminal, and insert the red lead into the VOLT/OHM terminal.

We begin our measurements by measuring the resistance of lamps connected in different ways. First, measure the resistance of a single lamp. To do this, insert the lamp in a base. Set your DMM to measure resistance, using a range selection of 100 Ω to 200 Ω (depending on your DMM), as shown in the first photo. (When the DMM reads "1 ." as in the photo, this means the resistance is too high to measure in the current range setting. You need to set the range to a higher setting. In the photo, the DMM is on but not connected to anything.) Touch the test leads to the terminals on the lamp base, read the resistance, and record the result in your lab journal. (It doesn't matter which test lead is connected to which terminal.) Insert your other two lamps into bases, and perform the same test. If the three resistances are different, compute the average (mean) resistance for the three.

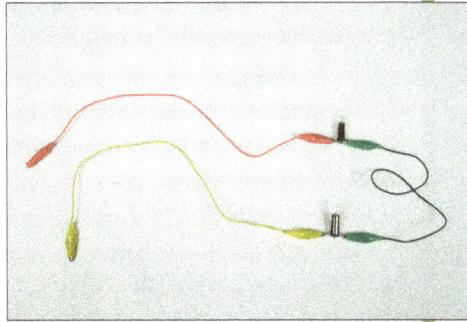

Using wires with alligator clips on each end, connect two lamps together in series, as illustrated in the first photo on this page. Measure the resistance of this series network by touching the DMM test leads to the ends of the network. Then add a third lamp to the chain and measure the resistance of a series network with three lamps. The second photo shows the DMM connected to a series network for the resistance measurement.

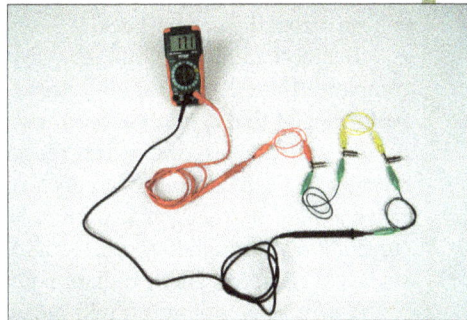

Now connect two lamps in parallel as shown in the third photo. Measure the total resistance of this combination. Then connect a third resistor in parallel with the first two and measure the resistance of this three-lamp parallel network.

Next, we measure the voltages in circuits formed by connecting the lamps—in series and in parallel—to a battery pack. Our battery pack consists of two, 6-volt lantern batteries connected in series to give us a 12-volt battery, as shown in the fourth photo.

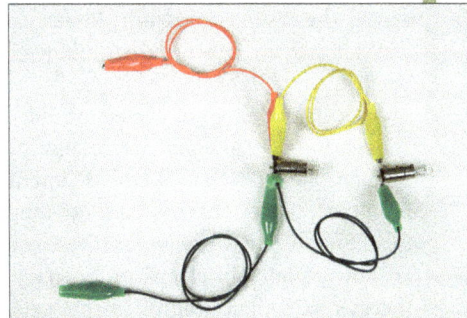

Switch your DMM over to the setting for measuring DC voltages in the range of 10 to 20 V (depending on your DMM).

For each circuit arrangement, record the following measurements:

- the voltage across the battery pack
- the voltage across each of the lamps

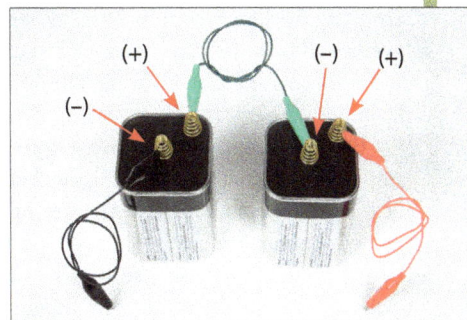

There are five separate circuit arrangements to test: three for series connections and two additional ones for parallel connections. (Note that when a single lamp is in the circuit, the series and parallel connections are the same, since there is only one way to connect a single lamp to the battery pack.) Measure and record all the voltages for these circuit arrangements:

- one lamp connected to the battery pack
- two lamps connected in series to the battery pack
- three lamps connected in series to the battery pack
- two lamps connected in parallel to the battery pack
- three lamps connected in parallel to the battery pack

The first two photos below show the setup connected together for the three-lamp series circuit and the three-lamp parallel circuit. The third photo illustrates a measurement of the voltage across the battery pack.

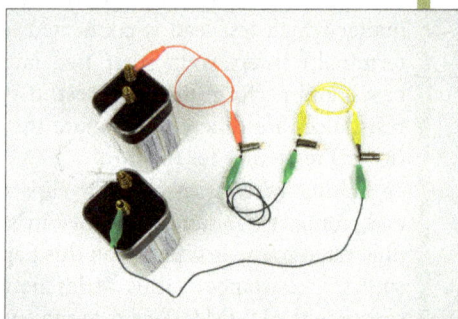

The measurements with the DMM are now complete. There is one more series of tests to perform, and that is to observe how series and parallel circuits respond when the circuit is interrupted. A light bulb contains a tiny wire, called a *filament*, that gets extremely hot when current flows through it. As you recall from our studies of electromagnetic radiation in Chapter 3, all warm objects emit infrared radiation, but a really hot object emits light at wavelengths across the visible spectrum. When all visible wavelengths are produced simultaneously, the result is white light.

As light bulbs are cycled on and off, the stresses on the filament eventually cause the filament to fail—that is, the filament breaks into pieces and current can no longer flow through it. When this happens, we say the lamp is "burned out." In electrical language, we say there is an "open circuit." In this experiment, we can simulate a lamp burning out by simply removing the lamp from its base. When a lamp is removed from its base, an open circuit is created and current can no longer flow through the lamp filament.

For the last set of tests, we begin with the lamps connected in parallel, since that is how they are connected for the last set of voltage tests. The lamps are probably hot by now from the previous testing, so use a tissue or paper towel to prevent burning your fingers. While the three lamps are connected to the battery and illuminated, remove one of the lamps from its base and observe what the other lamps do. Reinsert this lamp into its base and perform the same test on each of the other two lamps in turn. Record all your observations in your lab journal.

Finally, connect the lamp network back into a series circuit with three lamps (the first photo on the previous page). With the lamps all illuminated, remove each lamp one by one from its base, observe what the other lamps do, and replace the lamp in its base. Record all your observations in your lab journal.

Analysis

In your report, you need to work out a way to display all your resistance and voltage measurements in a table. It may help to make separate tables for the series connections and the parallel connections. In your report, address the following items.

1. Study your resistance data for a single lamp and for two or three lamps connected in series. From your data, determine the general mathematical rule for the combined resistance of several resistors connected in series. Use your rule to predict the combined resistance for five resistors connected in series, assuming that each of them has a resistance of 1,500 Ω. (Note: The rule you determine works even if the resistor values are not the same.)
2. Study your resistance data for a single lamp and for two or three lamps connected in parallel. Determine the general rule for the combined resistance of several identical resistors connected in parallel. Use your rule to predict the combined resistance of five resistors connected in parallel, assuming that each of them has a resistance of 1,500 Ω. (Note: The rule you determine for parallel resistors may only be valid for the case when the resistor values are identical. There is a more general rule that you may learn in a later course. Or you can look it up if you are curious.)
3. Study your voltage data for a single lamp connected to the battery pack and for two or three lamps in series connected to the battery pack. Determine the general rule for what the voltage across a single resistor is when several identical resistors are connected in series to a battery. Use your rule to predict the voltage across a single resistor if four 250-Ω resistors are connected in series to a 12-V battery.
4. Study your voltage data for a single lamp connected to the battery pack and for two or three lamps in parallel connected to the battery pack. Determine the general rule for what the voltage across a single resistor is when several identical resistors are connected in parallel to a battery. Use your rule to predict the voltage across a single resistor if four 250-Ω resistors are connected in parallel to a 12-V battery.
5. Write a description of how a series circuit of lamps behaves if one of the lamps burns out. Explain why the circuit behaves this way.
6. Write a description of how a parallel circuit of lamps behaves if one of the lamps burns out. Explain why the circuit behaves this way.
7. As you know, Christmas lights for trees or houses have 50 or 100 little lamps wired together in a single string of lights. Back in the early 1970s, the light bulbs for Christmas lights were not as reliable as they are now, and people always kept a few extra lamps on hand to replace those that burned out. Back then, Christmas lights were wired as series circuits. Nowadays, they are always wired as parallel circuits. Based on your experimental results, you should be able to explain why parallel wiring replaced series wiring for Christmas tree lights.

Chapter 14
Magnetism and Electromagnetism

Across the center of this image is a glass plate, and under the glass plate is a magnet. On top of the plate is a blob of fluid with magnetic properties, known as a ferrofluid. *In the presence of a magnetic field, the ferromagnetic particles in a ferrofluid align themselves with the magnetic field lines, and the surface of the fluid forms points oriented with the magnetic field lines. If the magnet is removed, the fluid relaxes to a small blob of liquid with a consistency similar to that of liquid soap.*

Magnetic fields have been put to many different uses in technology. We look at some of them in this chapter.

OBJECTIVES

After studying this chapter and completing the exercises, you should be able to do each of the following tasks, using supporting terms and principles as necessary.

1. Explain how magnetism arises in ferromagnetic materials.
2. Explain how information can be stored in the domains of magnetic media.
3. State Ampère's law and describe the circumstances behind its discovery.
4. State Faraday's law of magnetic induction.
5. Describe how Ampère's law is used to make DC motors turn.
6. Choose an example of how Ampère's law is used in another contemporary technology and explain how Ampère's law makes it work.
7. Choose an example of how Faraday's law is used in a contemporary technology and explain how Faraday's law makes it work.

VOCABULARY TERMS

You should be able to define or describe each of these terms in a complete sentence or paragraph.

1. Ampère's law
2. coil
3. crystallite
4. diamagnetic
5. domain
6. Faraday's law of magnetic induction
7. ferromagnetic
8. grain
9. lodestone
10. magnetic media
11. magnetite
12. paramagnetic

14.1 Magnetism and Its Cause

Magnetism has been known since ancient times. The ancients' name for naturally occurring magnetic stones was *lodestone*. Nowadays, this substance is more commonly known as the mineral *magnetite*. A chunk of this material is shown in Figure 14.1. By the 12th century, it was known that a piece of lodestone suspended from a string aligns itself to point

Figure 14.1. The mineral magnetite, also known as lodestone, is a naturally occurring magnet.

north, and thus was born the first magnetic compass and the use of the compass for navigation.

The basic form of magnetism we are familiar with in magnets and ferrous materials is called *ferromagnetism*. Materials such as magnetite and iron that retain permanent magnetism are ferromagnets. Many materials respond very weakly to magnetic fields without being able to retain any magnetism themselves. Such materials are called *paramagnetic* if they are attracted by a magnetic field and *diamagnetic* if they are repelled by a magnetic field. Liquid oxygen, pictured in Table 6.1, is an example of a strongly paramagnetic material.

Magnetism originates from the motion of electrons in atoms. You may recall from Chapter 5 that electric current creates a magnetic field, an effect described by Ampère's law. We look at Ampère's law in more detail in the next section. The point here is that *any* moving charge is an electric current, as we saw in the previous chapter. Since electrons in atoms are whirling around in the atoms all the time, magnetism emerges right down at the atomic level.

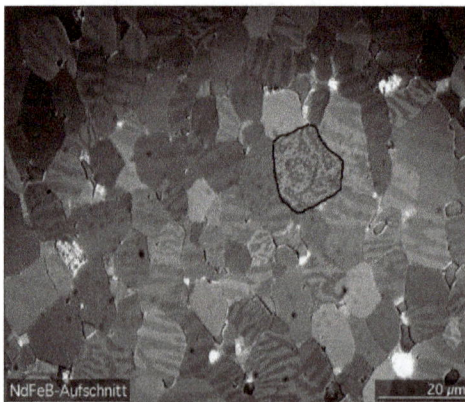

Figure 14.2. Metal crystal grains, with domains visible as light and dark stripes. The image shows a region of metal 0.1 mm wide.

As you know, metals are crystals. But in fact, a metal is not structured as just one big continuous crystal lattice. When the structure of these crystals is examined at the microscopic level, we see that the crystal is actually divided up into small regions called *crystallites* or *grains*. The grains are regions within the crystal lattice that are misaligned with respect to one another. The photograph in Figure 14.2 shows the grains in a metal—they look like countries on a map.

In zinc, the grains are visible to the naked eye. You may recall from the previous chapter that zinc is the metal

The crystal structure in metals is divided up into grains—chunks of crystal lattice misaligned with respect to each other.

used in galvanizing, and if you look at a polished galvanized surface, you can easily see the domains in the crystal structure.

Figure 14.3. Grains in a ferromagnetic material, with alternating domains of aligned magnetism indicated by the arrows.

Within the grains of a ferromagnetic metal, there is an even finer structure. In a ferromagnetic material, the magnetic effects of atoms line themselves up with each other, and the atoms align in tiny regions that all have their magnetic north and south poles pointing in the same direction. These interesting regions of aligned atoms are called

298

domains. In the photograph of Figure 14.2, the domains are visible as light and dark stripes within the grains.

In the sketch of Figure 14.3, the domains are represented by alternating colored arrows indicating the direction of the magnetism for each domain. This sketch represents the ferromagnetic material when it is *not* magnetized, since within each grain are several domains oriented in opposite directions that cancel each other out.

Regions of atoms with their magnetic fields aligned in the same direction are called domains.

Figure 14.4 shows what happens to the domains in the material in the presence of an external magnetic field. As you see, the domains that lie within the magnetic field actually align themselves with the field. In this figure, the field lines from the magnet are shown in red with the arrows pointing from north to south. Once an external magnetic field appears, the process of domain realignment takes a few seconds in a large sample of material. Then the material has been magnetized.

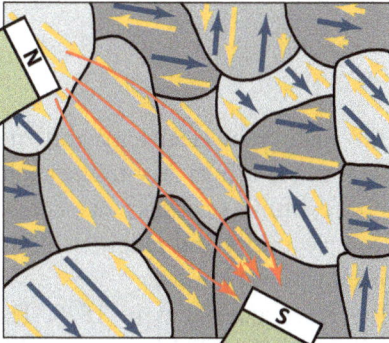

Figure 14.4. Alignment of magnetic domains in the presence of an external magnetic field.

The magnetization and alignment of domains in a ferromagnetic material is how information is stored on recording tape, floppy disks, and hard drives. The tape or disk is made of a plastic base material covered with a polished, ferromagnetic oxide coating. Figure 14.5 shows a side view of a section of tape used in Compact Cassettes, popular from the 1970s until Compact Disks (CDs) arrived in the 1990s.

The recording head produces a magnetic field that orients the domains in patterns that represent the data or signal being recorded. Magnetization of individual domains on a contemporary magnetic hard drive occurs at breathtaking speeds. On the external backup drive for my computer, a 300-megabyte (300 MB) backup takes less than one minute. A "byte" is a string of data requir-

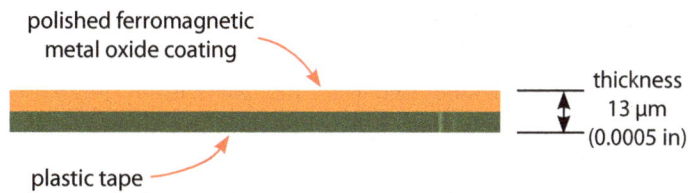

polished ferromagnetic metal oxide coating

thickness
13 μm
(0.0005 in)

plastic tape

Figure 14.5. Recording media, such as tapes and disks, have an oxide coating on a plastic base material. Dimensions shown are for the tape used in Compact Cassettes, a recording medium that nearly vanished until a nostalgia craze in the 2000s stimulated a comeback.

ing eight little domains to record it. This means that backing up 300 MB of data requires 8 × 300 × 1,000,000 = 2,400,000,000 magnetizations of domains on the disk. Doing this in 60 seconds means that each domain magnetization occurs in 0.000000025 seconds (0.025 microseconds)!

Learning Check 14.1

1. What is a lodestone?
2. What is the source of magnetism down at the atomic level?
3. Explain what grains and domains are.
4. How does a magnetic recording medium such as a hard drive store information?

14.2 Ampère's Law

Figure 14.6. Danish physicist and chemist Hans Oersted.

The history of electricity and magnetism goes back—as many things do—to the ancient Greeks. By the 19th century, scientists suspected a connection between electricity and magnetism, but the connection itself was not known.

The story goes that in 1820, Danish scientist Hans Oersted (Figure 14.6) was walking one morning to the University of Copenhagen, where he was a professor. He had already been thinking about electricity and magnetism for several years, but that morning it just dawned on him that electric current must produce a magnetic field. So he hurried into the room where his students were waiting for the morning class. In front of them, he connected a voltage source (a Voltaic pile, in fact) to a circuit of wire with a switch. He place a magnetic compass near the wire. When he switched on the current, the compass needle deflected, demonstrating that a magnetic field was being produced by the current in the wire.

Hans Oersted first demonstrated that a current-carrying wire produced a magnetic field.

Apparently Oersted's students did not realize what they had just witnessed, and were not particularly impressed. But Oersted himself knew how important this discovery was, and he immediately published it. The news hit the scientific community like a ton of bricks. The connection between electricity and magnetism had been found!

Just a few months later, French scientist and mathematician André-Marie Ampère (Figure 14.7) heard about Oersted's discovery. He immediately went to work experimenting with wires and Voltaic piles of his own. He found that two parallel current-carrying wires would repel or attract one another like magnets, depending on whether their currents were flowing in the same direction or in opposite directions. After six years of mathematical analysis, Ampère published what is now known as *Ampère's law*. A formal mathematical statement of Ampère's law requires calculus. But in layman's language, Ampère's law may be stated as follows:

> ### Ampère's Law
>
> A current-carrying wire produces a magnetic field around it, and the strength of the magnetic field is directly proportional to the current in the wire.

Figure 14.7. French scientist and mathematician André-Marie Ampère.

We looked at Ampère's law briefly back in Chapter 5. Figure 5.11 shows the shape of the magnetic field around a wire. Figure 5.12 shows what happens when a current-carrying wire is formed into a coil—you get an electromagnet.

The reason Ampère's law is named after André-Marie Ampère instead of Hans Oersted is that the law is not simply a statement that current-carrying wires produce magnetic fields. Ampère's law is a precise mathematical statement that correctly predicts the strength and direction of the magnetic field. Ampère's law includes the prediction that when a wire is wound into a coil, the strength of the magnetic field produced is proportional to the number of wire loops—turns—in the coil.

It is hard to overstate the importance of the electromagnetic *coil*. The simple coil of wire that can switch on and off a magnetic field by switching on and off an electric current has affected our lives more than anyone can describe. I'll just pick a few random examples to illustrate. Just for fun, rather than taking examples from industry, we'll stick to common devices you can find in your own home.

Remember the loudspeakers I mentioned in Section 12.2 when we were discussing longitudinal waves? Figure 14.8 is a diagram showing the location of the *voice coil* inside the speaker. The electric signal from a stereo is connected to the voice coil. The coil is mounted on the back of the speaker cone, suspended in a hole inside a permanent magnet. The electrical current passing through the coil is a wave that matches the sound wave to be produced by the speaker. As the current passes through the coil, the coil's own magnetic field—produced according to Ampère's law—pushes against the magnetic field of the permanent magnet, forcing the speaker cone to move back and forth. As the speaker moves back and forth, it produces a sound wave in the air.

Figure 14.9 is a picture of some slot cars on a slot car track. When I was 13, I enjoyed racing slot cars. These little cars have DC motors in them that operate according to Ampère's law. In the cut-away view of Figure 14.10, you can see two coils of wire (shown in green) wrapped on a steel core mounted on a rotating axle. Surrounding the rotating parts are two curved permanent magnets (red and blue). The magnetic field produced by the coils pushes against the magnetic field of the magnets, turning the motor shaft.

Without a way to change the current in the coils, the forces

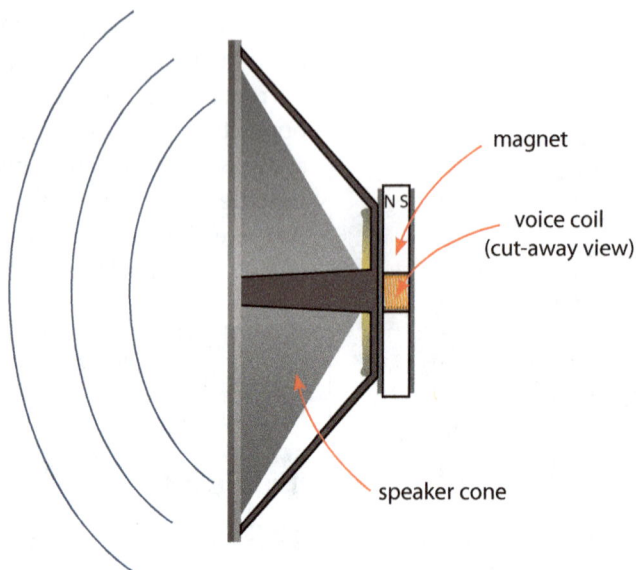

magnet

voice coil
(cut-away view)

speaker cone

Figure 14.8. Diagram of a loudspeaker, showing the location of the voice coil inside the magnet.

Figure 14.9. These toy slot cars run on DC motors, operating according to Ampère's law.

turning the motor would only turn the motor half way around. When the south pole of the coils got around to the blue magnet it would stay there because of the magnetic attraction between the south pole of the coils and the north pole of the magnets. The two copper wires marked (+) and (−) from the battery connect to the rotating coils at the copper ring on the axle. These contact points are called *brushes*, and they press against the rotating copper ring supplying current to the coils as they rotate. The copper ring is split into two pieces, and as the coils turn, the brushes switch the flow of current through the coils back and forth so that the magnetic forces always push the coils in the same direction, keeping the motor going.

A more recent technology is the *brushless* DC (BLDC) motor. If you are

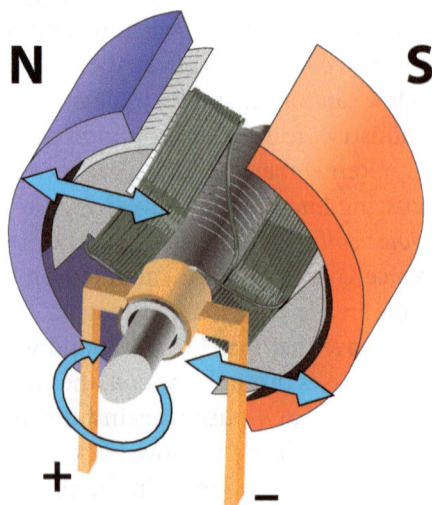

Figure 14.10. DC motor.

a radio-controlled car hobbyist or a computer buff, you may already know a bit about BLDC motors. But regardless of whether you know about them, they are common these days. They power the cooling fans in computers, they power electric RC (radio-controlled) cars, which I'm sure you have seen, and a they are found in scores of other applications.

Loudspeakers, motors, and read/ write heads in hard drives all operate according to the principle of Ampère's Law.

Figure 14.11 shows a computer cooling fan taken apart to show the BLDC motor inside. In the brushless design, the coils are stationary and the permanent magnets do the rotating. The fan in the figure has ring of small permanent magnets mounted around the inside. The computer rapidly switches the current to the four coils on and off so that the magnetic fields they produce are always pushing against the magnetic fields of the permanent magnets to keep the fan spinning.

Figure 14.11. The brushless DC motor inside a computer cooling fan.

As a final example, let's go back again to the computer hard drive. On page 94 is a picture of the read/write head in a hard drive. Figure 14.12 shows a close-up of the tip of the arm, where the head (coil) is. In this close-up, it is easier to see where the actual arm is; the lower portion of what appears to be part of the arm is actually just a reflection off the highly polished surface of the magnetic disk.

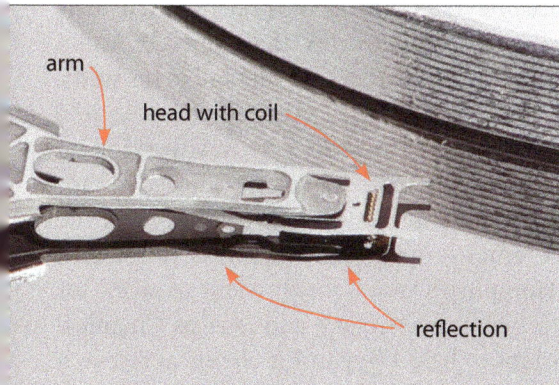

Figure 14.12. The read/write head in a computer hard drive.

At the tip of the arm is the head, which contains a tiny coil of wire. To write data onto the disk, the current to the coil is switched back and forth, making the end of the coil either a north pole or a south pole. The coil is suspended only a few nanometers above the magnetic surface of the disk, as shown in Figure 14.13. The coil is wrapped around a ferromagnetic core that is

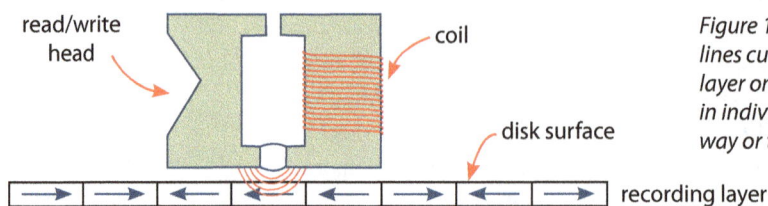

read/write
head

coil

disk surface

Figure 14.13. The magnetic field lines curve down into the recording layer on the disk, aligning the fields in individual domains to be one way or the other.

recording layer

designed to force the magnetic field lines from the coil down into the magnetic recording layer on the disk. This magnetizes the individual domains on the disk so they point one way or the other, each one representing one bit of data. As the disk spins, the coil is switched on and off so fast that it can record 40,000,000 bits each second. Oh, one more thing: the arm holding the head is itself positioned back and forth by a magnetic coil. Ampère's law strikes again.

Learning Check 14.2

1. Write a sentence stating Ampère's law.
2. Write a paragraph describing how the discovery of Ampère's law came about.
3. Explain how the magnetic principle in Ampère's law is used to make DC motors turn.
4. Choose another technology based on Ampère's law and write a brief description of how it works.

14.3   Faraday's Law of Magnetic Induction

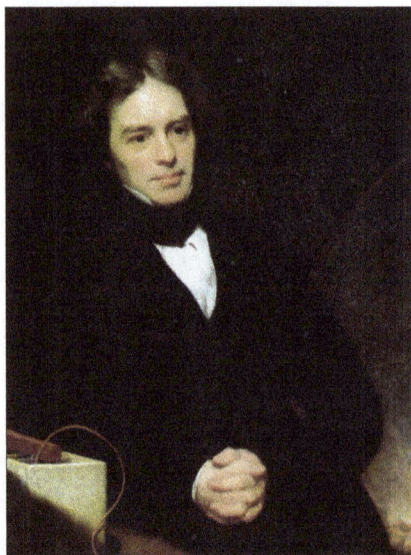

Figure 14.14. English scientist Michael Faraday.

I love the rags-to-riches story of English scientist Michael Faraday (Figure 14.14). He was from the lower class in English society and received little formal education. He was apprenticed in a bookbinder's shop at the age of 14. But he became one of the most influential scientists in history, so noteworthy that Albert Einstein kept a picture of Faraday on the wall in his study.

Faraday became interested in scientific study by reading the books he was working on in the shop. When he was 20, he finished his apprenticeship and made contact with one of the most famous, upper-class scientists of the day, Sir Humphry Davy. Faraday went to work for Davy, first as a secretary, and later as Chemical Assistant to help Davy in his chemical research

Figure 14.15. Images of Michael Faraday's lab in London.

in the areas of elements and gases. Some images from Faraday's lab, now the Faraday Museum in London, are shown in Figure 14.15.

Once Faraday had the opportunity to begin working as a scientific researcher, his scientific genius began to take off. As Davy's Chemical Assistant, he was a major contributor to the early study of electrochemistry (see pages 293–294). But Faraday had a deep interest in electricity and magnetism, and Davy couldn't keep Faraday from fooling around with electromagnets. Faraday coined the term *diamagnetism* that we encountered earlier. His work led him to the discovery in 1831 of the law that now bears his name: Faraday's law of magnetic induction. The law may be stated this way:

> ## Faraday's Law of Magnetic Induction
>
> **A changing magnetic field near a coil of wire induces a current in the coil.**

This law is illustrated in Figure 14.16. A coil of wire is connected to a resistor and is passed by the moving permanent magnet. As the magnet passes the coil, the magnetic field of the magnet penetrates into the coil, and when it does so, a blip

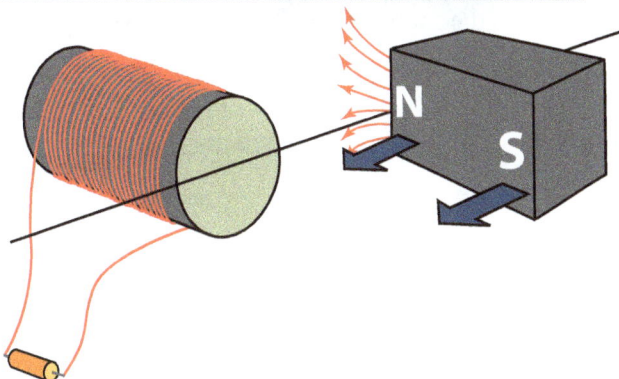

Figure 14.16. As the magnet passes the coil, a brief pulse of current is induced in the coil.

of current is induced in the coil. Other electronic circuits connected to the resistor can sense this current.

It is important to note that Faraday's law stipulates that the external magnetic field passing through the coil must be *changing* in order for the current to be induced. If the magnet in the figure stops in front of the coil, the induced current ceases.

Ampère's law was a big deal because it led to the electric motor. Faraday's law was an even bigger deal because it led to the electric generator.

If Ampère's law was a big deal because it led to the invention of the electric motor, Faraday's law of magnetic induction was an even bigger deal because it led to the invention of the electric generator. With the invention of the generator, an engine could be used to generate electricity from scratch. A generator works just like a motor running backwards. Look again at the sketch of a motor in Figure 14.10. In that case, electricity enters the coils, producing a magnetic field in the coils. The new magnetic field pushes against the surrounding magnetic field and turns the motor.

In a generator, things work the other way around. An engine, powered by a fuel of some kind, is used to turn the coils. Now, the magnetic fields of the surrounding magnets will sometimes be passing through the coils and sometimes not, depending on where the coils are in their rotation. This continuously changing magnetic field passing through the coils induces an electric current in the coils, as long as the engine keeps turning the coils.

As we did with Ampère's law, here we look at some other (and more fun) examples of how Faraday's law has been put to use in common devices we have around us. Figure 14.17 illustrates the inner parts of a basic microphone. Attached to the front where the sound waves arrive is a membrane or diaphragm that stretches back and forth in response to the sound waves. (Your ear drum does the same thing.) Attached to the back of the diaphragm is a small, light coil of wire. Inside the coil, but not touching it, is a permanent magnet. The magnet is mounted inside the microphone housing and does not move. As the coil vibrates back and forth from the sound waves, its position relative to the magnet changes in a pattern that matches the sound waves. These changes of the magnetic field passing through the

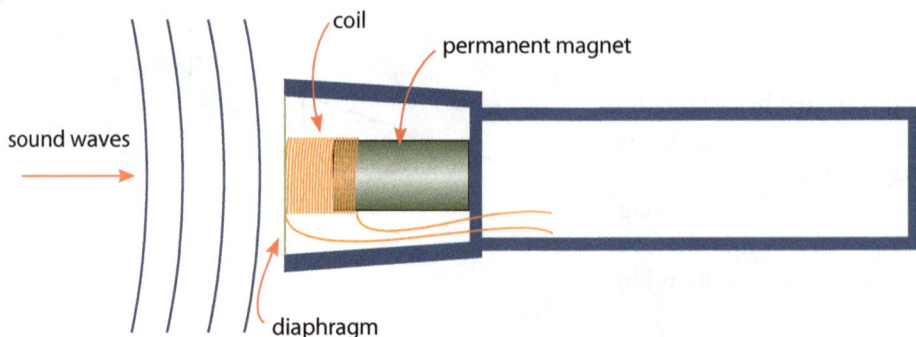

Figure 14.17. A basic microphone, operating on the principle of Faraday's law of magnetic induction.

coil induce a current in the coil that is exactly analogous to the sound waves that cause it.

If it occurred to you that a microphone works a lot like a loudspeaker, very good! A microphone is essentially a loudspeaker working in reverse. Recall that motors and generators are basically the same, and that a generator is just like a motor working in reverse. Motors use electricity and Ampère's law to turn the motor shaft. Generators use a turning shaft and Faraday's law to produce electricity.

The same thing holds for loudspeakers and microphones. Loudspeakers use an electrical signal and a coil inside a magnet to vibrate the speaker cone. They operate according to Ampère's law because the electrical signal in the coil produces a field and this field pushes against the permanent magnet to make the speaker cone vibrate.

Microphones are like loudspeakers working in reverse, just as generators are like motors working in reverse.

Microphones use Faraday's law of magnetic induction. The coil vibrates from the sound waves, and this motion in the field of the permanent magnet induces an electric current in the coil.

Our next example is again about music. The magnetic pickups on electric guitars work according to the same principle—Faraday's law of magnetic induction. If you play the electric guitar, you have Michael Faraday to thank for the basic idea behind the technology. And maybe you don't play an electric guitar; you just listen to electric guitars in music (and who doesn't?). Either way, it's Faraday we have to thank.

Figure 14.18 shows a front view of the pickups on an electric guitar. One pickup is one of the oblong, white objects with the six metal circles inside. The guitar shown actually has four separate pickups. This allows the musician to get different sounds out of the guitar. The metal dots under each of the strings are permanent magnets, each one mounted inside of—you guessed it!—a coil of wire. Figure 14.19 is a diagram of the arrangement of the magnet, the coil, and the gui-

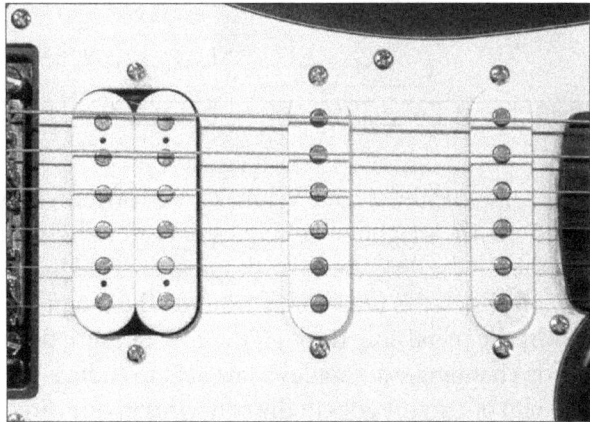

Figure 14.18. Magnetic pickups on an electric guitar.

tar string. As you see, the field lines from the magnet extend up into the area where the string is. The strings are made of steel—a ferromagnetic material. As a string vibrates, the field surrounding it is disturbed. The pickup coil is very sensitive to such changes because it is made of several thousand turns of very thin wire. The changing magnetic field near this coil induces a current in the coil that is exactly analogous to the vibration of the string.

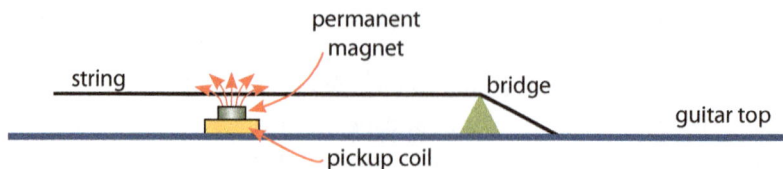

Figure 14.19. Magnet, coil, and guitar string on a guitar pickup.

As a final example, let's return one more time to the read/write head on the computer hard drive. You can see what's coming, right? The process of writing data onto the hard drive made use of Ampère's law. So what's going to happen when we need to read the data? The process of writing in reverse! Ampère's law and Faraday's law are mirrors of each other. To write data, the drive uses Ampère's law. A current in the coil of the read/write head produces a magnetic field that magnetizes the domains in the ferromagnetic recording layer on the top of the disk. Figure 14.20 shows the diagram again, but this time illustrating how data is read from the disk.

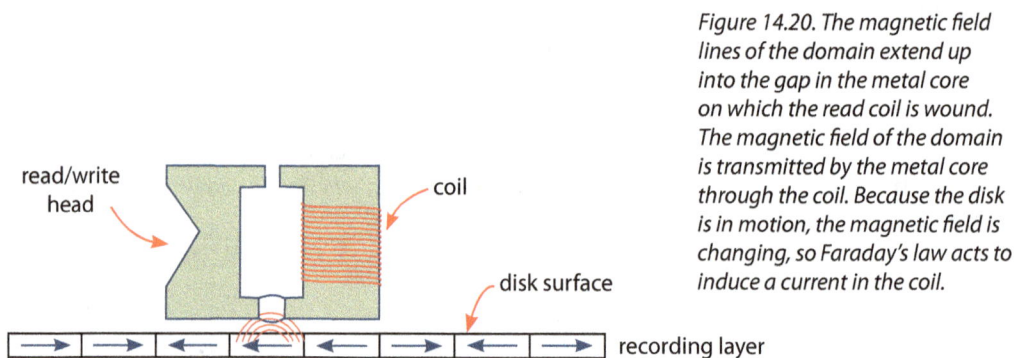

Figure 14.20. The magnetic field lines of the domain extend up into the gap in the metal core on which the read coil is wound. The magnetic field of the domain is transmitted by the metal core through the coil. Because the disk is in motion, the magnetic field is changing, so Faraday's law acts to induce a current in the coil.

To read the data, we make use (of course) of Faraday's law of magnetic induction. The magnetic field lines of each individual domain extend up above the surface of the disk. As the disk spins, the field lines come into the gap in the metal core on which the read coil is wound. The magnetic field of the domain is transmitted by the metal core through the coil. Because the disk is in motion, the magnetic field is changing, so Faraday's law acts to induce a tiny blip of current in the coil. This blip of current goes in the coil wires in one direction or the other, depending on the orientation of the magnetization of the domain, allowing each domain to be read as a "1" or a "0," the two digits used to record all digital information.

It is hard to imagine that this arrangement of coils and domains can be used to read and write huge amounts of data at high speed, but obviously every computer hard drive is proof that it can. Like every other technology, this one has improved dramatically over the past few years. The home computer I had in 1995 had a hard drive in it that could store 40 megabytes (MB) of data. The hard drive in the computer I use now can store 2 terabytes (TB) of data—50,000 times as much!

Learning Check 14.3

1. Write a sentence stating Faraday's law of magnetic induction.
2. Explain how the magnetic principle in Faraday's law is used to read data from a computer hard disk.
3. Choose another technology based on Faraday's law and write a brief description of how it works.

Chapter 14 Exercises

Answer each of the questions below as completely as you can. Write your responses in complete sentences.

1. Pretend your grandmother is watching a science program that talks about "ferromagnetic materials" and "grains and domains." She isn't sure what grains and domains are, but knows that since you are studying this in school you can probably explain it to her. Give it a shot. Explain these terms in a couple of paragraphs that even your grandmother can understand.
2. We have seen repeatedly that Ampère's law and Faraday's law are mirrors of each other. Examples of technologies illustrating this include, a) motors and generators, b) loudspeakers and microphones, and c) writing data and reading data on a hard disk. Choose two of these pairs and write explanations that illustrate how Ampère's law and Faraday's law are put to use. (Write these so your grandmother can understand them, too. Writing these explanations is challenging, and even a bit tricky, so take your time to think this through.)
3. Explain specifically how the grains and domains in a ferromagnetic material are involved in recording information on a recording tape or hard disk.

Experimental Investigation 12: Magnetic Field Strength

Overview

- Set up a simple DC circuit with a 6-V battery connected to a 10-foot loop of wire. With the power to the wire off, place a magnetic compass under the wire so the wire is in line with the compass needle.
- Connect the circuit, and measure the compass needle deflection. Raise the wire a few millimeters, and repeat. Do this at several different heights from 10 mm up to 250 mm.
- *The goal of this experiment is to determine how the strength of the magnetic field around a current-carrying wire varies with the distance from the wire.*

Basic Materials List

- magnetic map compass
- insulated wire with alligator clips (10 ft)
- 6-volt lantern battery
- masking tape
- wooden stands (from the Kinetic Energy experiment)
- plastic work clamps (2)
- plastic clips (2)
- table (non-metallic), carpenter's level
- 1 × 4 pine, 24 to 36 inches in length
- rule, metric, 18 inch

As we have seen in Chapters 5 and 14, Ampère's law states that a current-carrying wire produces a magnetic field circling around it. The strength of this magnetic field is proportional to the current flowing in the wire. In this experiment, we use the deflection of a magnetic compass to determine how the strength of the magnetic field around a wire varies with respect to the distance from the wire.

The experimental setup is shown in the photo below. After you make sure that your table is level, tape a magnetic compass to a 1 × 4, and clamp the 1 × 4 to the work table with plastic work clamps. Adjust the scale on the ring around the compass so the compass needle is in line with the 0° mark. Make a large loop of insulated copper wire, about 10 feet long, with alligator clips on each end. Clip one end of the wire to the battery. The other end will be clipped on and off the battery as needed. Using the two adjustable supports from the Kinetic Energy experiment, arrange the loop of wire so you can adjust it to different heights above the compass and so it passes directly over the compass needle. Run the rest of the wire in a big loop, as far away from the compass as possible.

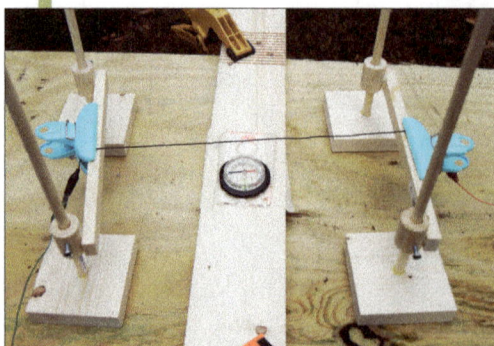

Here is a very important detail you must attend to: any ferrous metal, such as steel, that is close to the compass will affect the position of the compass needle. This goes for the steel studs often used in building construction, metal furniture, and so on. For this reason, it is best to perform this experiment outdoors, far from any buildings, cars, or metal structures. The larger any nearby metal structure is, the farther you need to be from it. If your school meets in a metal building, you should setup your table for this experiment at least 100 ft from the building. You must also take care that your experimental setup doesn't make use of any steel clamps, tools, or tables.

To begin, adjust the wire supports down as low as they will go. Clip the wire to the support arms with plastic clips, such as are used to close bags of potato chips. Adjust the support stands so the wire is tight, horizontal, and perfectly aligned with the compass needle, as shown in the photo to the right. Use a rule marked in millimeters to measure the height of the wire above the 1 × 4 supporting the compass. By placing the compass on a 1 × 4 instead of directly on the table, the first height measurement may be only 10 mm or so. Record the height measurement in a table in your lab journal. Now clip the other end of the wire to the battery, and read the deflection in degrees on the compass. As soon as you have this reading, disconnect the wire from the battery, and record the compass deflection for that height in your lab journal.

Raise the supports 10 millimeters or so, and adjust them so the wire is horizontal, tight, right over the compass needle, and perfectly aligned with it. Measure and record the wire height. Then briefly connect the power and read the needle deflection. Repeat this procedure until your support stands are as high as they can go. Make sure you have measurements for at least 15 different heights. When you reach the point where your deflection measurements are only changing by a degree or so, then double the distance you raise the wire for the next measurement (e.g., raise it 20 mm instead of 10 mm, etc.).

Analysis

We are using the deflection of the compass needle as an indicator of the magnetic field strength around a current-carrying wire. Prepare a graph of compass deflection in degrees (vertical axis) versus wire height in millimeters (horizontal axis). Your instructor will help you format the axes and scales properly. In your report, address the following questions.

1. Describe the shape of your graph. As the height increases, does the strength of the magnetic field from the wire decrease in a linear fashion, or in some other way?
2. Explain how the results of this experiment would change if we used two 6-V batteries in series to get a 12-V battery, as we did in Experimental Investigation 11.
3. In your own words, explain why turning on the current causes a compass needle to deflect.
4. Imagine that we used 20 feet of wire instead of 10 and made the wire into a double-wrapped loop so that we had two wires passing over the compass instead of one. How would you expect this to change the results of this experiment?

Glossary

absolute zero	The lowest temperature theoretically possible; the temperature at which all molecular motion ceases.
acceleration	Changing velocity; the rate at which an object's velocity is changing.
acid	A compound in aqueous solution that neutralizes a base to form a salt and water.
acid rain	Rain water that has a pH below 7; caused by emissions of sulfur dioxide, nitrogen oxide, and other pollutants.
alkali	A base that contains the hydroxide ion.
alternating current	Oscillating electric current. In the U.S., the frequency of alternating current (AC) is 60 Hz.
ampere	The SI System unit for electric current, representing a flow of charge of one coulomb per second; one of the seven SI System base units.
Ampère's law	The principle developed by André Ampère (1826) that an electric current generates a magnetic field that is proportional to the strength of the current. If the current is flowing in a coil of wire, the strength of the magnetic field is also proportional to the number of turns of wire in the coil.
amplitude	The height of a wave, measured from the centerline of the wave to the top of a peak.
aqueous solution	A solution in water.
arc	The passage of electric current through the air by means of ionization of the air. In the process, the ionized particles in the air gain energy and emit an intense violet light. The violent process of ionization produces shock waves in the air that we hear as a snapping, crackling, or zapping sound.
atmosphere	The layer of gas surrounding the earth, composed of 78% nitrogen, 21% oxygen and small amounts of other gases.
atom	The smallest part of an element, composed of a nucleus of protons and neutrons surrounded by a cloud of electrons located in orbitals according to their energies.
atomic mass	The average number of nucleons (protons and neutrons) in the nuclei of the atoms of a particular element.

atomic number	The number of protons in the nuclei of the atoms of a particular element; the elements are listed in the Periodic Table of the Elements according to atomic number.
barometer	A device for measuring atmospheric pressure.
barometric pressure	The local value of atmospheric pressure.
base	A compound in aqueous solution that neutralizes an acid to form a salt and water.
base unit	One of the seven fundamental units of measure in the SI system of units. The base units are meter (length), kilogram (mass), second (time), ampere (electric current), candela (luminous intensity), kelvin (temperature), and mole (amount of substance).
beta decay	A nuclear reaction that occurs in the nucleus of an atom when a neutron spontaneously converts into a proton, emitting an electron and a neutrino in the process (see nuclear decay).
Big Bang	The theoretical origin of the universe as an immense explosion of energy from a single point; occurred 13.8 billion years ago according to widely accepted theory.
boiling point	The temperature at which a substance enters a phase transition from liquid to gas (or vapor), or vice versa.
brittleness	A physical property that describes a substance's tendency to shatter.
caloric theory	The now discarded theory that heat is a weightless gas inside hot objects that flows to cooler objects when released, as in a fire.
caustic	Synonym for alkaline, which is a descriptor for a base that contains the hydroxide ion (an alkali).
charge	An electrical property of all protons and electrons. Protons have positive charge, and electrons have an identical amount of negative charge; accompanied by an electric field.
chemical equation	An equation written with the chemical formulas of the compounds involved in a chemical reaction that shows the reactants and the products, their ratios, and (possibly) the forms they are in.
chemical formula	An expression of the elements in a compound and their ratios in which the elements are denoted by their chemical symbols.
chemical potential energy	The energy stored in the chemical bonds of a compound; can be released by processes such as combustion or digestion.
chemical property	A property that describes chemical reactions a substance will or will not engage in.
chemical symbol	The symbol for one of the 118 elements listed in the Periodic Table of the Elements.

circular wave	A wave on water formed by the wind; referred to as circular because a floating object moves in a circular pathway as water waves pass beneath it.
coefficient	A numeral written in front of a chemical formula in a chemical equation; indicates the proportions of the reactants and products in a chemical reaction.
coil	Turns or loops of electrically conductive wire, usually wrapped around a ferromagnetic core.
combustion	A chemical reaction in which a substance combines with oxygen to produce heat and flame.
compound	Two or more different elements chemically bonded together.
compressive strength	A physical property that describes the ability of a substance to withstand crushing forces.
conduction	In electricity: When electric charge moves in an electrically conductive substance, such as a metal. In heat transfer: The process of heat transfer by the propagation of vibrations of atoms in a solid.
conduction electrons	Electrons in a metal that are free to move around in the metallic crystal lattice.
conductor	A substance that conducts electricity (electrical conductor) or heat (thermal conductor), such as a metal.
conservation of energy	The principle that energy cannot be created or destroyed, only changed from one form to another.
convection	The process of heat transfer by molecular motion in fluids. As fast molecules mingle with slower molecules, they collide and energy is transferred from the faster molecules to the slower molecules.
corrosion	The destructive oxidation of a metal.
covalent bond	The chemical bond that forms between nonmetal elements by means of sharing electrons. Hydrogen also forms covalent bonds.
crystal	The atomic structure of a substance held together by ionic bonds, characterized by the orderly arrangement of atoms in a rigid lattice structure.
crystal lattice	The structural framework in which the atoms are arranged in a crystal.
crystallite	Synonym for *grain*.
crystallography	The study of crystal structures by means of X-Ray diffraction.
Cycle of Scientific Enterprise	The process of forming a theory to account for a body of facts, then formulating testable hypotheses from the theory, then testing the hypotheses experimentally and analyzing the

	results. If a hypothesis is confirmed, the theory is strengthened, otherwise the theory is weakened.
density	The ratio of mass to volume for a substance; a physical property that describes the amount of mass in a given volume.
derived unit	A unit of measure in the SI system based on a combination of various of the seven base units.
diamagnetism	The magnetic property of a substance that is repelled by magnetic fields but which cannot itself be magnetized.
diffraction	When waves encounter corners or obstructions and bend and spread out as a result.
direct current	Electric current that flows steadily in one direction.
dissociation	When the ions in a crystal lattice come apart and go into solution during the process of dissolving.
dissolving	When a solute mixes with a solvent so that the particles of solute and solvent are mixed together uniformly all the way down to the molecular level.
domain	A microscopic region within one of the grains of a ferromagnetic material, where the magnetism of all the atoms is aligned in the same direction.
Doppler effect	The change of pitch perceived when the pitch is produced by a moving object and the object passes by the observer. The pitch is higher than normal as the moving object approaches because the motion compresses the wavelength of the sound waves, and lower than normal after the object passes because the motion stretches out the wavelength of the sound waves.
ductility	A physical property, typical of many metals, that indicates that a substance can be drawn into a wire by pulling the substance through a die (a metal block with a hole in it).
echolocation	The process used by bats, dolphins, and some other animals to navigate and hunt by means of emitting sounds and interpreting the echoes that come back.
elasticity	A physical property that describes a substance's ability to stretch without breaking or becoming permanently deformed.
electric current	Moving electric charge.
electric field	A field around electric charges; produces a force on other charges present in the field.
electrical conductivity	A physical property that describes the ability of a substance to conduct electricity. Metals typically have high electrical conductivity.
electricity	When the electrical effects of charged particles are made evident through electric current or static electricity.

electromagnet — A magnet formed by winding a wire into a coil and passing a current through it.

electromagnetic force — A force caused by an electric or magnetic field.

electromagnetic radiation — Waves (or photons) of pure energy at any wavelength in the electromagnetic spectrum.

electromagnetic spectrum — The entire range of wavelengths of electromagnetic radiation from long wavelength radio waves to short wavelength X-Rays and gamma rays; includes microwaves, infrared radiation, visible light, and ultraviolet radiation.

electron — One of the three fundamental particles inside atoms; located in orbitals by energy around the nucleus; possesses negative charge, and has a mass of 1/1,836 the mass of a proton.

electrostatic force — A force between objects that are charged with static electricity.

element — A substance characterized by being composed of atoms all possessing the same number of protons in the nucleus.

energy — Holds everything together and enables every process to happen. One of the basic constituents of the universe, the others being matter and order.

evaporation — When a substance undergoes the phase transition from liquid to vapor without first being heated to the boiling point.

experiment — A scientific test of a hypothesis.

Faraday's law of magnetic induction — The principle discovered by Michael Faraday (1831) that a changing magnetic field passing through a coil of wire induces a current in the coil.

ferromagnetism — The magnetic property of a material that can retain magnetism and is affected by magnetism.

ferrous metal — A metal containing iron.

field — A region in space around an object in which other objects experience a force if they are susceptible to the particular type of field involved.

field lines — Arrows in a graphical representation of a field that indicate the direction of the force that is present on an object placed in the field, assuming that the object is the type to be affected by the type of field in question.

fine-tuning — The fact that throughout nature many physical constants are set at just the right values to enable complex life to exist on earth; if any of these values were different by even a small amount, life could not exist.

fission — A nuclear chain reaction in which a neutron shatters the nucleus of a heavy element, producing nuclear fragments and additional free neutrons that shatter other nuclei.

flammable	The chemical property that describes a substance's ability to burn. Combustible.
fluid	A substance that flows; a liquid or a gas.
force	A push or pull.
frequency	In an oscillating system, the number of cycles completed per second.
friction	A force that converts the kinetic energy of motion into heat.
fusion	A nuclear reaction in which two hydrogen nuclei (protons) fuse together to form a helium nucleus, releasing energy in the process.
galaxy	A large spiral structure in outer space consisting of some 100 billion stars. There are about 100 billion galaxies in the known universe.
general theory of relativity	Albert Einstein's theory (1915) of gravity; an explanation of gravity in terms of acceleration, and of the curvature of the geometry of spacetime around a massive object.
grain	In metals, small regions within the crystal lattice that are not aligned with adjacent regions.
gravitational field	A field around an object possessing mass; produces a force on other massive objects; also affects electromagnetic radiation.
gravitational force	The weakest of the four fundamental forces; the force of attraction between two objects due to their gravitational fields. The force is so weak that it is typically only noticeable in the case of very large masses.
gravitational potential energy	Stored energy due to an object's location in a gravitational field.
ground	An electrical connection to the earth so that the earth may be used as a zero-voltage reference for electrical systems.
halide	A compound formed with a halogen.
halogen	One of the elements in Group 17 of the Periodic Table of the Elements.
hardness	A physical property that describes a material's ability to resist scratching.
heat	Energy in transit from a warm body to a cooler one, by means of infrared radiation, thermal conduction, or convection.
heat of fusion	The energy required to melt a solid substance while keeping its temperature the same.
heat of vaporization	The energy required to vaporize (boil) a liquid substance while keeping its temperature the same.

heat transfer	When energy is transferred from a warm object to a cooler one by means of infrared radiation, thermal conduction, or convection.
heterogeneous mixture	A mixture in which the presence of at least two different substances is visible to the eye, possibly with the aid of a microscope.
homogeneous mixture	A mixture with a composition that is uniform all the way down to the molecular level. Also called a solution.
hydrocarbon	Any molecule consisting of only hydrogen and carbon atoms, typically fossil fuels and other compounds derived from them.
hydroelectric power	The generation of electricity by damming water and allowing the water behind the dam to fall through a turbine-generator system.
hydrogen ion	A proton; the nucleus of a hydrogen atom; possesses a charge of +1.
hydroxide	A polyatomic ion consisting of one atom of oxygen and one atom of hydrogen, and possessing a charge of –1.
hypothesis	A testable, informed prediction, based on a theory, of what will happen in certain circumstances.
incident	A term that describes a ray or wave as it approaches the boundary of a new medium, where it reflects and/or refracts.
incompressibility	A property of solids and liquids, since they typically do not change volume under pressure (at least not much).
induction	In electricity: When the presence of nearby charges causes movable charges in a metal to accumulate. In magnetism: See Faraday's law of magnetic induction.
inertia	A property of all matter that causes matter to prefer its present state of motion; quantified by the variable *mass*.
inflammable	Flammable.
insulator	In electricity: A substance that does not conduct electricity. In heat: A substance that resists the flow of heat.
interaction	An alternative way of understanding forces in terms of subatomic particle exchanges.
internal energy	The sum of all the kinetic energies possessed by all the particles in a substance.
International System of Units	The system of units of measure administered in France and accepted by most of the countries in the world; the metric system.
inverse-square law	An inverse proportion in which one quantity decreases as a second quantity increases; the first quantity decreases in proportion to the square of the second quantity; generically represented as $y = k/x^2$.

ion	An atom that has gained or lost one or more electrons, and has thus acquired a net electric charge.
ionic bond	A chemical bond between metal and nonmetal atoms in which one or more electrons from each metal atom are transferred to the nonmetal atoms; this process results in ions that are electrically attracted to one another and so form a crystal lattice.
isotope	Any variety of the atoms of an element, differing in the number of neutrons the atoms have in their nuclei.
kilogram	The SI System base unit for mass; one of the seven base units in the SI System.
kinetic energy	Energy an object possesses by virtue of its motion.
law of universal gravitation	Isaac Newton's theory (1687) that all objects with mass attract all others, with a force that is inversely proportional to the square of the distance between them.
laws of motion	Isaac Newton's three laws (1687) summarizing the behavior of objects with mass. When no net force is present on such an object, the object retains its present state of motion (first law); when a net force is present, the object accelerates in proportion to the force, and in inverse proportion to its mass (second law); and any force always produces an equal and opposite reaction force (third law).
laws of nature	The orderly and very mathematical system of rules that objects in nature obey; a manifestation of the order found throughout nature.
liter	A unit of volume equal to 1,000 cubic centimeters; not an official SI System unit, but commonly used in scientific study.
lodestone	Magnetite.
longitudinal wave	A wave with the characteristic that the oscillation producing the wave is moving parallel to the direction in which the wave propagates.
luster	Shininess; a physical property.
magnetic field	A field around a magnet or electromagnet; causes a force on other magnets, electromagnets, current-carrying wires, and magnetic materials.
magnetic media	Disks or tapes covered with a polished, ferromagnetic oxide coating, and used for storing information in the patterns of alignment in the magnetic domains of the ferromagnetic material.
magnetite	A naturally occurring mineral possessing permanent magnetism.

magnetosphere	The magnetic field surrounding the earth, and protecting the earth from the high-energy protons and electrons in the solar wind.
malleability	A physical property that indicates a substance's ability to be hammered into thin sheets; typical of many metals.
mass	The variable used to quantify the amount of inertia in an object.
mass-energy equivalence	Einstein's theory (1905) that mass and energy are different forms of the same thing, that amounts of mass and energy may each be expressed in terms of the other; the theory can be used to calculate the energy released when a quantity of matter is converted into energy in a nuclear reaction.
matter	Anything that has mass and volume; one of the three basic things the universe is composed of, the other two being energy and order.
mechanical equivalent of heat	The theory that heat and kinetic energy are different forms of the same thing (energy) and may be converted from one form to the other.
mechanical wave	Any wave requiring a medium in which to propagate; all waves other than electromagnetic waves.
medium	Matter of some kind through which the energy of mechanical waves can propagate.
melting point	The temperature at which a substance undergoes the phase transition from solid to liquid or vice versa.
meniscus	The curved shape that forms on the surface of most liquids when they are place in a glass tube such as a graduated cylinder.
mental model	A theory.
metalloid	A group of seven elements located in between the metals and the nonmetals in the Periodic Table of the Elements, possessing some of the properties of both.
meter	The SI system base unit for length; one of the seven base units in the SI System.
metric prefix	Prefixes added to units of measure in the SI System to indicate multiples or fractional multiples. For example, the prefix kilo– indicates a multiple of 1,000, so a kilometer is 1,000 meters. The prefix micro– indicates a fractional multiple of 1/1,000,000, so a micrometer is 1/1,000,000 of a meter.
metric system	The common term in the U.S. for the International System of Units.
Milky Way	The name for our galaxy.

mixture	A combination of substances that occurs without any chemical reaction. Substances in the mixture retain their own properties and may be separated from one another by physical means such as filtration and boiling.
molecule	A chemically bonded cluster of atoms; the smallest particle in compounds formed by covalent bonding between nonmetals (including hydrogen).
monopole	A single pole of a magnet (north or south), which cannot exist by itself.
multi-resistor network	A network of resistors in an electric circuit consisting of more than one resistor.
neutralization	When an acid (pH < 7) and a base (pH > 7) react to produce a salt in water, with a pH of 7 (neutral).
neutron	One of the three basic particles inside atoms; located in the nucleus; possesses no electric charge, and has a mass very close to the mass of a proton (but slightly greater).
normal line	A line perpendicular to the boundary between two media, used as a reference for measuring angles involved in wave reflection and refraction.
nuclear decay	A naturally occurring nuclear reaction in which an atom of a radioactive element spontaneously changes the composition of its nucleus by the emission of particles and/or electromagnetic radiation.
nuclear energy	The use of controlled nuclear fission to produce steam which can then power a turbine and generate electricity.
nucleus	The center of an atom where the protons and neutrons are.
ohm	The SI System unit for resistance.
Ohm's law	The principle discovered by Georg Ohm (1827) that in an electric circuit, the voltage drop across a resistor is equal to the current flowing through the resistor multiplied by its resistance.
orbital	An energy region in an atom where up to two electrons may reside.
order	The fine mathematical structure observed everywhere in nature; one of the three basic things the universe is made of, the other two being energy and matter.
oscillation	Continuous, periodic movement back and forth or up and down.
oxidation	When an atom loses electrons to become a positively charged ion.
parallax error	A measurement error caused by not having your line of sight perpendicular to the scale you are measuring with.

parallel resistance	Two or more resistors connected together at both ends.
paramagnetism	The property of a substance that is attracted by magnetic fields, but which cannot itself be magnetized.
period	The length of time required for an oscillating system to complete one full cycle of its oscillation.
pH	A scale measuring how acidic or basic a substance is. The scale runs from 0 to 14, 7 being neutral. The farther below 7 the pH is, the more acidic the substance. The farther above 7 the pH is, the more basic the substance.
pH indicator	A substance or set of chemically colored papers that can be used to indicate the pH of other substances.
phase diagram	A graph of temperature vs. energy for a particular substance, showing at least one phase transition.
phase transition	The process of a substance changing phase from solid to liquid, liquid to gas, gas to liquid, liquid to solid, solid to gas, etc.
photon	A single quantum of electromagnetic energy.
photovoltaic cell	A semiconductor device that converts light into electrical signals.
physical property	A non-chemical characteristic of a substance.
plasma	An ionized gas, recognized as one of the four phases of matter, the other three being solid, liquid, and gas (or vapor).
polar	When the electrons in a molecule are not balanced, so one end or region of the molecule is more positively charged and another region is more negatively charged.
pole	One end of a magnet, identified as either north or south.
polyatomic ion	A group of nonmetal atoms (which may include hydrogen) covalently bonded together and possessing a net electrical charge.
potential energy	Energy that is stored and can be released by conversion into a different form of energy.
precipitate	As a verb: To form a solid and come out of solution. As a noun: A solid that has formed and has come out of solution during a chemical reaction.
pressure	The amount of force present per unit area on a surface.
product	The compounds formed by a chemical reaction.
proton	One of the three basic particles inside atoms; located in the nucleus; possesses positive charge and has a mass very close to the mass of a neutron (slightly less).
pure substance	An element or a compound.
quantized	Divided into discrete lumps or packets; non-continuous.

quantum	Albert Einstein theorized in 1905 that all energy is divided into discrete packets called *quanta*. A single packet of energy is called a *quantum* of energy.
radiation	Electromagnetic waves. In heat transfer, refers specifically to electromagnetic waves in the infrared region of the electromagnetic spectrum. Not to be confused with nuclear radiation, which can involve massive particles such as protons, electrons and neutrons.
radioactive	A term for substances that spontaneously undergo nuclear decay, producing nuclear radiation in the process (see radiation).
reactant	One of the compounds going into a chemical reaction.
redox reaction	A chemical reaction in which one substance is reduced and another substance is oxidized. Since oxidation and reduction always occur together, the combined reactions are called redox reactions.
reduction	When a positive ion gains electrons to become less positively charged, or when a neutral atom gains electrons to become a negatively charged ion.
reflection	When a wave bounces off the boundary between two media, obeying the law of reflection in the process.
refraction	When a wave passes through the boundary separating two media, changing velocity and direction in the process.
resistance	Electrical resistance to the flow of electric current; a property of resistors.
resistor	A device in electric circuits used to regulate the flow of electric current.
right circular cylinder	An object with a circular base and a curving side that is perpendicular with the base.
right rectangular solid	An object with a rectangular base and rectangular sides, in which all angles are right angles.
salt	A compound formed by the neutralization reaction between an acid and a base. May also be described as a metal bonded to a halogen, or a metal bonded to certain polyatomic ions such as sulfate, nitrate, or phosphate.
schematic diagram	An abstract representation of the devices in an electric circuit and how they are connected.
science	The process of using experiment, observation, and reasoning to develop mental models (theories) of the natural world.
scientific fact	A scientific statement supported by a great deal of evidence, which is correct so far as we know.
second	The base unit of time in the SI System; one of the seven base units.

series resistance	Two or more resistors joined together one after the other in a single electrical pathway.
shear strength	A physical property that describes a substance's ability to withstand shearing forces (forces that tend to cut an object into two pieces).
shell	A layer in an atom containing one or more orbitals where the atom's electrons reside.
SI System	The International System of Units administered in France and accepted by most of the countries in the world; commonly known in the U.S. as the metric system.
solar cell	A semiconductor device that converts the energy in sunlight into electricity.
solar system	The sun and the planets, along with the planets' moons and other smaller bodies.
solar wind	A stream of high-energy charged particles (protons and electrons) continuously emitted by the sun.
solute	A solid, liquid, or gas that is dissolved into a fluid.
solution	A homogeneous mixture formed when one substance (a solute) dissolves into another substance (the solvent).
solvent	A fluid into which a solute is dissolved to form a solution.
sonar	A technology in which echoes of sound waves are used to locate objects under water.
speed	How fast an object is moving.
state of motion	The state of an object characterized by its possessing a particular velocity; that is, possessing a particular speed in a particular direction; includes the state of being at rest (a speed of 0 m/s).
static discharge	The release of a static accumulation of charge produced when an electrostatic voltage is high enough to cause the surrounding air to ionize, forming a conductive channel in which charge flows; produces an arc (see arc).
static electricity	An accumulation of charge.
steady-state universe	The discarded theory that the universe is neither expanding nor contracting, and had no beginning, but has always been here.
strong nuclear force	One of the four fundamental forces; the strongest of all forces; binds the protons together in the nucleus of atoms.
subatomic particle	Mainly protons, electrons, and neutrons, although there are approximately 60 less commonly known particles.
sublimation	A phase transition in which a substance transitions from a solid to a gas without going through the liquid phase.
telos	A Greek word meaning purpose, goal, or end.

tensile strength	A physical property that indicates a substance's ability to withstand stretching forces.
theory	A model, representation, or explanation that seeks to account for the related facts and provide means for producing new hypotheses.
thermal conductivity	A physical property that indicates the ability of a substance to conduct heat.
thermal energy	Energy an object possesses after being heated.
thermal equilibrium	A state when two objects (or an object and its environment) are at the same temperature and heat flow between them ceases.
thermal properties	Properties that describe how a substance acts as it is heated.
time interval	A specific period of time.
transverse wave	A wave with the characteristic that the oscillation producing the wave is perpendicular to the direction in which the wave is propagating.
truth	The way things really are; known to us by the direct and immediate testimony of our senses, by reasoning with valid logic based on true premises; or, according to many faith traditions, revealed to us by God; not a part of the language of scientific claims.
turbine	A system of blades mounted on a rotating shaft such that when steam, water, or air flows past the blades the shaft turns.
ultrasonic	Any sound wave with a frequency higher than about 20 kHz, the upper limit of human hearing.
uniform acceleration	A situation caused by the presence of a constant net force, in which the velocity of an object increases or decreases by the same amount each second.
unit conversion factor	An expression with a value of one, written by placing equivalent quantities expressed in different units of measure in the numerator and denominator of a fraction.
vapor	The gaseous phase of matter; a term applied to describe the gaseous state of substances that are solids or liquids at room temperature
vaporization	The process of transitioning to the gas or vapor phase; boiling.
velocity	A quantity that describes how fast an object is moving and in which direction.
visible spectrum	The portion of the electromagnetic spectrum that we can sense with our eyes through the colors red, orange, yellow, green, blue, and violet.
volt	The SI System unit of measure for voltage.

voltage	The electrical "pressure" that forces charge to flow in an electrical circuit.
volume	The variable used to quantify the amount of space an object takes up.
wave	A disturbance in space and time that carries energy from one place to another.
wave packet	A synonym for photon.
wavelength	The spatial separation between the peaks of a wave.
weak nuclear force	One of the four fundamental forces; the agent involved in beta decay.
work	A mechanical process whereby energy is transferred from one object to another by a force pushing through a certain distance.

Appendix A
Making Accurate Measurements

Making accurate measurements in experiments requires care. It also requires learning some practices that help you minimize error. In this appendix, we review some of these practices.

A.1 Parallax Error and Liquid Meniscus

Two common measurement issues involving special technical terms are avoiding *parallax error* and working correctly with the *meniscus* on liquids. These terms both have to do with using analog instruments with measurement scales that must be correctly aligned for an accurate measurement.

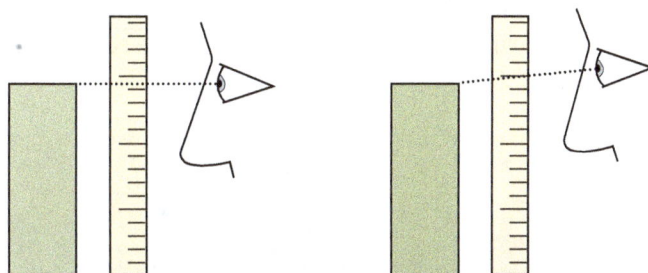

Figure A.1. In the sketch on the left, the measured object, measurement instrument, and viewer line of sight are correctly aligned. In the sketch on the right, the misalignment of the viewer is causing parallax error in the measurement.

Parallax error occurs when the line of sight of a person taking a measurement is at an incorrect angle relative to the instrument scale and the object being measured. As shown in Figure A.1, the viewer's line of sight must be parallel to the lines on the scale and perpendicular to the scale itself. Misalignment of the viewer's line of sight results in an inaccurate measurement due to parallax error.

When a liquid is placed in a container, the surface tension of the liquid causes the liquid to curve up or down at the walls of the container. In most liquids, the surface of the liquid curves up at the container wall, but a well-known example of downward curvature occurs when liquid mercury is placed in a glass container. We concentrate here on the common upward curvature exhibited by water.

If the container is tall and narrow, as with a graduated cylinder, then the curving liquid at the edges gives the liquid surface an overall bowl shape. This bowl-shaped surface is called

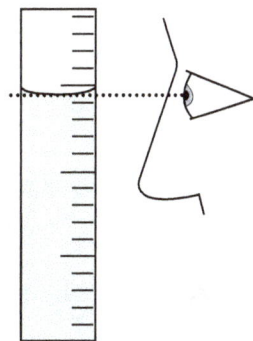

Figure A.2. Correctly reading a liquid volume in a graduated cylinder. The viewer's line of sight is at the bottom of the meniscus, and is perpendicular to the scale to avoid parallax error.

a *meniscus*. The correct way to read a volume of liquid is to read the liquid at the bottom of the meniscus. Figure A.2 illustrates the correct way to read a volume of liquid in a graduated cylinder by reading the liquid level at the bottom of the meniscus, avoiding parallax error.

A.2 Measurements with a Meter Stick or Rule

1. For maximum accuracy, avoid using the end of a wooden rule. The end is usually subject to a lot of pounding and abrasion, which wears off and compresses the wood on the end.

2. As indicated in Figure A.3, arrange the rule against the object to be measured so the marks on the scale come in contact with the object being measured. This helps minimize parallax error.

3. As indicated in Figure A.4, use a straight-edge to assure the end of a metal rule is accurately aligned with the edge of an object being measured.

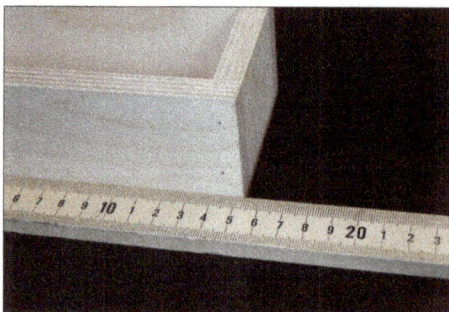

incorrect   *correct*

Figure A.3. Proper placement of a rule or meter stick.

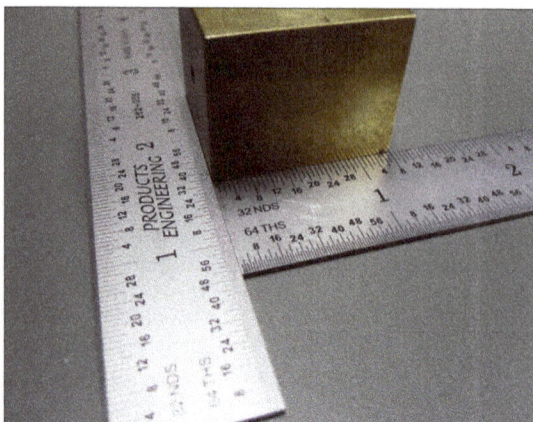

Figure A.4. Use of a straightedge for proper alignment of the end of the rule with the object being measured.

A.3   Measurements with a Triple-Beam Balance

1.   Calibrate the balance before making measurements. This is accomplished by turning the calibration weight under the pan until the scale's alignment marks are perfectly aligned.

2.   Make sure the 10-g and 100-g weights are locked into a notch on the beam. Otherwise, the measurement will not be correct.

3.   As shown in Figure A.5, when adjusting the position of the gram weight, it is good practice to slide this weight with the tip of a pencil held below the beam instead of with your finger. If done carefully, this technique allows the gram weight to be manipulated into position without disturbing the balance of the beam as the balance point is approached.

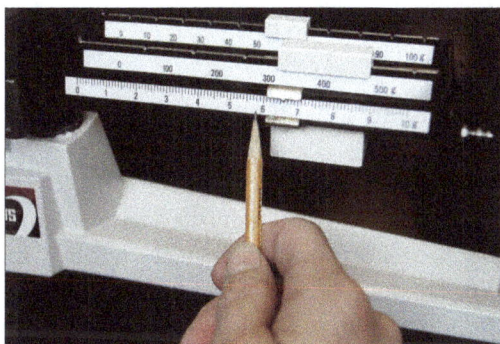

Figure A.5. Move the gram slider with a pencil to prevent disturbing the beam as the balance point is approached.

A.4   Measurements with an Analog Thermometer

1.   Mercury thermometers are more accurate than spirit thermometers. However, if a mercury thermometer breaks, you have a real problem cleaning up and disposing of the spilled mercury. Thus, for student use, spirit thermometers have replaced mercury thermometers almost completely.

2.   When measuring temperatures, be sure to notice that the thermometers have a mark indicating the proper degree of immersion for the most accurate reading.

3.   Thermometer accuracy can be severely compromised if gaps get into the liquid. Always store spirit thermometers vertically in an appropriate rack to help prevent gaps from getting into the liquid.

One of the conventional calculations in secondary school science labs is the so-called "experimental error." This experimental error is typically defined as the difference between the predicted value and the experimental value, expressed as a percentage of the predicted value, or

$$\text{experimental error} = \frac{|\text{predicted value} - \text{experimental value}|}{\text{predicted value}} \times 100\%$$

(The absolute-value sign in this equation guarantees that the experimental error is always a positive value.)

From the perspective of the average student, this use of "experimental error" makes perfect sense. After all, students are studying well-established theories and the goal of the experiment is to learn about the theory, not to validate or refute it. In the world of science, however, experiments are the gold standard by which theories are judged. When there is a mismatch between theory and experiment, it is often the theory that is found wanting. That is how science advances.

Over the years, research and discussions with practicing scientists have led me to the conclusion that the phrase "experimental error"—despite its common use in secondary science classes—is misguided. Used in this way, the term *error* implies that the theory is *correct* and that the error in the experiment may be summarized by this difference equation. However, the difference between prediction and experimental result may not be caused by deficiencies in the experiment. In more general scientific practice, the *theory* may not be correct. Referring to the overall difference between prediction and experimental result as "experimental error" is thus a bad habit to get into.

Consider this case: an experimental measurement of velocity produces a value that is consistently less than the predicted value. Most likely this is because the predictions did not take air resistance into account. Is this an experimental error? It is more correct to say that the theory is inaccurate because we made the unrealistic assumption (and used it to form our theoretical prediction) that there would be no significant air resistance. Such causes of differences between predictions and measurements are quite common, and it is well if our future scientists (maybe you) can understand that this is not an error in the experiment or the measurements.

As a result of these considerations, I have adopted different terminology. I now use the phrase "percent difference" to describe the value computed by the above equation. When quantitative results can be compared to quantitative predictions, students should compute the percent difference as

$$\text{percent difference} = \frac{|\text{predicted value} - \text{experimental value}|}{\text{predicted value}} \times 100\%$$

Scientists do use the phrase "experimental error" and they work hard to estimate the experimental error associated with every data point. But it is important that this error estimate is *independent* of the theory being tested. The experimental error must be strictly related to the uncertainty in the measurements, and not to any weakness in the theory. Errors in experiments creep in through measurement instruments, variations in environmental conditions, variations in the samples under test, and so on. Together, these errors are quantified (expressed mathematically) with an *uncertainty* value. (Often, the uncertainty is expressed with a calculation from statistics called the *standard deviation*).

After scientists have estimated the error in the data and know the uncertainty in the experimental result, they compare the experimental error to the percent difference to see if the theory matches the experimental result.

CHAPTER 4

I first discovered many of the ideas in Section 4.3 from viewing *The Privileged Planet*, a produced by Illustra Media (illustramedia.com).

CHAPTER 6

Atomic mass values for the Periodic Table of the Elements were taken from *A Guide to the Elements*, Albert Stwertka, Oxford, 2012. Arrangement of elements 57, 89, 71 and 103 is based on *The Periodic Table, A Very Short Introduction*, Eric R. Scerri, Oxford, 2011.

The basic information for the Growing Crystals experimental came from waynesthisandthat.com.

CHAPTER 9

Data values in Figure 9.5 are from coppercananda.ca.

Reference density values for Experimental Investigation 6 are from engineeringtoolbox.com.

CHAPTER 11

Reference pH values for Table 11.2 are from engineeringtoolbox.com.

The images attributed to the Creative Commons were obtained from commons.wikimedia.org. The science pages at en.wikipedia.org are well maintained and cite their own scholarly sources. These were used as sources for some details and for fact checking.

Image Credits

ii–5, John D. Mays. 6, 1.2: John D. Mays. 6, 1.3: Giovanni Piranesi, public domain. 7, gold: Erwinrossen; public domain. 7, nanotubes: Mstroeck, licensed under CC-BY-SA-3.0. 7, nanotube: STM Taner Yildirim (The National Institute of Standards and Technology—NIST); public domain. 8, Kurzon; public domain. 9, 1.5: Hendrick Bloemaert, public domain. 9, 1.6: Charles Turner; public domain. 10, 1.7: Unknown; Source: firstworldwar.com; public domain. 10, 1.8: John D. Mays. 11, 1.9: Unknown; public domain. 11, 1.10: AB Agrelius and Westphal, public domain. 12–13, John D. Mays. 14, Leaflet, public domain. 16, The History of the Universe.jpg via https://commons.wikimedia.org/wiki/File:The_History_of_the_Universe.jpg, public domain. 17, Eagle: NASA/ESA Hubble, public domain. 17, Cat's Eye: NASA/Hubble ST, public domain. 17, Andromeda: NASA/ESA/Herschel/PACS/SPIRE/J, public domain. 18, NASA, public domain. 19, John D. Mays. 20, bodoklecksel, licensed under CC-BY-CA-3.0. 21, 2.5: Sky Wing Sky, licensed under CC-BY-SA-3.0. 21, 2.6: F. Schmutzer, 1921, public domain. 23, Author: MB, licensed under CC-BY-SA-2.5 Generic. 24, Gringer, public domain. 25, John D. Mays. 26, Martin Pettitt from Bury St. Edmunds, UK, licensed under CC-BY-SA-2.0. 27, 2.12: Adam Kliczek / Wikipedia, licensed under CC-BY-SA-3.0. 27, 2.13: Bureau of Reclamation photographer, public domain. 28, ubirajararodrigues51, licensed under CC-BY-SA-2.0. 29, top: Decumanus, licensed under CC-BY-SA-3.0. 29, center: Trevor MacInnis, licensed under CC-BY-SA-2.5. 29, bottom: Fletcher6, licensed under CC-BY-SA-3.0. 30, top: Summi, licensed under CC-BY-SA-3.0. 30, center: David Pfister-Senz, licensed under CC-BY-SA-3.0. 30, bottom: Peter Van den Bossche from Mechelen, Belgium, licensed under CC-BY-SA-2.0. 31, NASA, public domain. 32, Rjglewis, licensed under CC-BY-SA-3.0. 33, John D. Mays. 34, Tennessee Valley Authority; public domain. 36–43, John D. Mays. 44, Statue: Caspar von Zumbusch; Photo: Daderot; public domain. 46–47, John D. Mays. 50, top: John D. Mays. 50, center: NASA/IPAC; public domain. 50, bottom, NASA/IPAC; public domain. 51–52, John D. Mays. 53, Bow hunter: Unknown; public domain. 53, Kids/car: Roger Price, from Hong Kong, Hong Kong, licensed under CC-BY-SA-2.0. 53, White car/red car: Bidgee, licensed under CC-BY-SA-3.0 Australia. 53, Crane: John D. Mays. 54, John D. Mays. 55, Left: Daderot; public domain. 55, Right: Laurence "Green-Reaper" Perry, licensed under CC-BY-SA-3.0. 56, John D. Mays. 58, Top: John D. Mays. 58, Center: FutureNJGov, licensed under CC-BY-SA-3.0. 58, bottom: NASA/Apollo 8 crewmember Bill Anders; public domain. 59–61, John D. Mays. 62, Dr. Richard Feldmann (Photographer), public domain. 64, U.S. Navy photo by Photographer's Mate 2nd Class Aaron Peterson, public domain. 66, Vassil, public domain. 67, 4.7: Lsmpascal, licensed under CC-BY-SA-3.0. 65, 4.8: Dartmouth College, public domain. 67, 4.9: Uwe Kils, aka Kils at en.wikipedia, licensed under CC-BY-SA-3.0. 68, 4.10: Photo by NEON ja, colored by Richard Bartz, licensed under CC-BY-SA-2.5. 68, 4.11: Nick Hobgood, licensed under CC-BY-SA-3.0. 69, 4.12: NASA, public domain. 70, 4.13: NASA, public domain. 70, 4.14: NASA, public domain. 72, Christian Müller; Stephan Kemperdick; Maryan Ainsworth; et al, Hans Holbein the Younger: The Basel Years, 1515–1532, Munich: Prestel, 2006, public domain. 73, 4.16: NASA (Apollo 12 crew), public domain. 73, 4.17: NASA, public domain. 74, NASA/Crew of Expedition 22, public domain. 76, Zureks, public domain. 78, 5.1: Godfrey Kneller, 1689, public domain. 78, 5.2: John D. Mays, based on Johnstone, licensed under CC-BY-SA-3.0. 79–81, John D. Mays. 82, top: Unknown, public domain. 82, bottom: André Castaigne, 1903, public domain. 84, John D. Mays. 85, NASA, public domain. 86, John D. Mays. 87, Top, John D. Mays. 87, Bottom: Initi, public domain. 88–90, John D. Mays. 91, Top, John D. Mays. 91, Bottom: Copyright 1970, CERN. 92, Top: Jan Ainali, licensed under CC-BY-SA-3.0. 92, Center: Yosemite, licensed under CC-BY-SA-3.0. 92, Bottom: Christian Jansky, licensed under CC-BY-SA-3.0. 94–95, John D. Mays. 96, left: Ben Mills, public domain. 96, right: Unknown, public domain. 98, 6.1a left: Sakurambo, public domain; right: Ben Mills, public domain. 98, 6.1b left: Ingvald Straume, licensed under CC-BY-SA-3.0; right: Benjah-bmm27, public domain. 98, 6.1c left: Jynto, public domain; right: Jynto, public domain. 98, 6.1d left: Ben Mills, public domain; right: Ben Mills, public domain. 98, 6.1e left: Ben Mills, public domain; right: Ben Mills, public domain. 99, 6.1f left: Ben Mills, public domain; right: Ben Mills, public domain. 99, 6.1g: Adam Redzikowski, public domain. 99, 6.1h: Yinch, licensed under CC-BY-SA-3.0. 99, 6.1i left: Ben Mills, public domain; right: Ben Mills, public domain. 99, 6.1j: Unknown, public domain. 100, 6.2a: Ben Mills, public domain. 100, 6.2b: John D. Mays, generated using CrystalMaker®, CrystalMaker Software Ltd, Oxford, England (www.crystalmaker. com). 100, 6.2c: Calcite-unit-cell-3D-vdW.png from http://commons.wikimedia.org/wiki/ File:Calcite-unit-cell-3D-vdW.png. Author Benjah-bmm27, public domain. 100, 6.2d: Rob Lavinsky, iRocks.com, licensed under CC-BY-SA-3.0. 101, John D. Mays. 102, 6.2e: Rob Lavinsky, iRocks.com, licensed under CC-BY-SA-3.0. 102, 6.2f: Periodictableru, licensed under CC-BY-SA-3.0. 103, Nikolai Yaroshenko, public domain. 104–106, John D. Mays. 107, calcium: Calcium unter Argon Schutzgasatmosphäre.jpg via https://commons. wikimedia.org/wiki/File:Calcium_unter_Argon_Schutzgasatmosph%C3%A4re.jpg. Author: Matthias Zepper, public domain. 107, oxygen: Dr. David Hillier, licensed under the GNU General Public License. 107, nitrogen, Jurii, licensed under CC-BY-SA-3.0. 107, sulfur: Ben Mills, public domain. 107, sodium: John D. Mays. 107, chlorine: W. Oelen, licensed under CC-BY-SA-3.0. 108, 6.9 rear, middle: John D. Mays; front: Ben Mills, public domain. 111–113, John D. Mays. 114, top: John D. Mays; bottom: RudolfSimon, licensed under CC-BY-SA-3.0. 116–117, John D. Mays. 118, Communication Specialist 1st Class Chad J. McNeeley, U.S. Navy, public domain. 120, Divinemerce, licensed under CC-BY-SA-3.0. 123, John D. Mays. 126–129, John D. Mays. 133, Feynman: The Nobel Foundation, 1965, public domain; Curie: Unknown, public domain; Oersted: Unknown, public domain; Collins: Bill Branson, NIH, public domain; Planck: AB Lagrelius & Westphal, public domain; Roentgen: Unknown, public domain; Crick: Marc Lieberman, licensed under CC-BY-SA-2.5; Watson: public domain; Einstein: F. Schmutzer, 1921, public domain. 138, Greg L, licensed under CC-BY-SA-3.0. 140, 8.1: Unknown, licensed under CC-BY-SA-3.0. 140, 8.2: Ana al'ain, licensed under CC-BY-SA-3.0. 141, Unknown, public domain. 143, 8.4/8.5/8.6: John D. Mays. 143, 8.7: Unknown, public domain. 146–147: John D. Mays. 157–163, John D. Mays. 164, Magnus Hagdorn, licensed under CC-BY-SA-2.0. 166, 9.1: David Shankbone, licensed under CC-BY-SA-3.0. 166, 9.2: Photograph by Jastrow, public domain. 166, 9.3: John D. Mays. 167, 9.4: Seyhan668, licensed under CC-BY-SA-3.0. 167, 9.5: John D. Mays. 170, NASA/ESA, public domain. 171, John D. Mays. 173, top: Sze Ning, licensed under CC-BY-SA-2.0; bottom: John D. Mays. 175–177, John D. Mays. 178, Justinc, licensed under CC-BY-SA-2.0. 179, Sarathly,

Index

Periodic Table of the Elements

Legend: ■ liquid at room temperature ■ radioactive

1	2	3	4	5	6	7	8	9	10	11	12	13	14	15	16	17	18
1 **H** Hydrogen 1.0079																	2 **He** Helium 4.0026
3 **Li** Lithium 6.941	4 **Be** Beryllium 9.0122											5 **B** Boron 10.811	6 **C** Carbon 12.011	7 **N** Nitrogen 14.0067	8 **O** Oxygen 15.9994	9 **F** Fluorine 18.9984	10 **Ne** Neon 20.1797
11 **Na** Sodium 22.9898	12 **Mg** Magnesium 24.3050											13 **Al** Aluminum 26.9815	14 **Si** Silicon 28.0855	15 **P** Phosphorus 30.9738	16 **S** Sulfur 32.066	17 **Cl** Chlorine 35.4527	18 **Ar** Argon 39.948
19 **K** Potassium 39.098	20 **Ca** Calcium 40.078	21 **Sc** Scandium 44.9559	22 **Ti** Titanium 47.88	23 **V** Vanadium 50.9415	24 **Cr** Chromium 51.9961	25 **Mn** Manganese 54.9380	26 **Fe** Iron 55.847	27 **Co** Cobalt 58.9332	28 **Ni** Nickel 58.6934	29 **Cu** Copper 63.546	30 **Zn** Zinc 65.39	31 **Ga** Gallium 69.723	32 **Ge** Germanium 72.61	33 **As** Arsenic 74.9216	34 **Se** Selenium 78.96	35 **Br** Bromine 79.904	36 **Kr** Krypton 83.80
37 **Rb** Rubidium 85.468	38 **Sr** Strontium 87.62	39 **Y** Yttrium 88.9059	40 **Zr** Zirconium 91.224	41 **Nb** Niobium 92.9064	42 **Mo** Molybdenum 95.94	43 **Tc** Technetium 98.9072	44 **Ru** Ruthenium 101.07	45 **Rh** Rhodium 102.9055	46 **Pd** Palladium 106.42	47 **Ag** Silver 107.8682	48 **Cd** Cadmium 112.411	49 **In** Indium 114.82	50 **Sn** Tin 118.710	51 **Sb** Antimony 121.76	52 **Te** Tellurium 127.60	53 **I** Iodine 126.9045	54 **Xe** Xenon 131.29
55 **Cs** Cesium 132.905	56 **Ba** Barium 137.327	71 **Lu** Lutetium 174.967	72 **Hf** Hafnium 178.49	73 **Ta** Tantalum 180.9479	74 **W** Tungsten 183.85	75 **Re** Rhenium 186.207	76 **Os** Osmium 190.2	77 **Ir** Iridium 192.22	78 **Pt** Platinum 195.08	79 **Au** Gold 196.9665	80 **Hg** Mercury 200.59	81 **Tl** Thallium 204.3833	82 **Pb** Lead 207.2	83 **Bi** Bismuth 208.9804	84 **Po** Polonium 208.9824	85 **At** Astatine 209.9871	86 **Rn** Radon 222.0176
87 **Fr** Francium 223.0197	88 **Ra** Radium 226.0254	103 **Lr** Lawrencium 262.11	104 **Rf** Rutherfordium 261.11	105 **Db** Dubnium 262.114	106 **Sg** Seaborgium 263.118	107 **Bh** Bohrium 262.12	108 **Hs** Hassium (265)	109 **Mt** Meitnerium (266)	110 **Ds** Darmstadtium (281)	111 **Rg** Roentgenium (281)	112 **Cn** Copernicium (285)	113 **Nh** Nihonium (284)	114 **Fl** Flerovium (289)	115 **Mc** Moscovium (288)	116 **Lv** Livermorium (293)	117 **Ts** Tennessine (294)	118 **Og** Oganesson (294)

Lanthanides:

57 **La** Lanthanum 138.9055	58 **Ce** Cerium 140.115	59 **Pr** Praseodymium 140.9077	60 **Nd** Neodymium 144.24	61 **Pm** Promethium 144.9127	62 **Sm** Samarium 150.36	63 **Eu** Europium 151.965	64 **Gd** Gadolinium 157.25	65 **Tb** Terbium 158.9253	66 **Dy** Dysprosium 162.50	67 **Ho** Holmium 164.9303	68 **Er** Erbium 167.26	69 **Tm** Thulium 168.9342	70 **Yb** Ytterbium 173.04
89 **Ac** Actinium 227.0278	90 **Th** Thorium 232.0381	91 **Pa** Protactinium 231.0359	92 **U** Uranium 238.0289	93 **Np** Neptunium 237.0482	94 **Pu** Plutonium 244.0642	95 **Am** Americium 243.0614	96 **Cm** Curium 247.0703	97 **Bk** Berkelium 247.0703	98 **Cf** Californium 251.0796	99 **Es** Einsteinium 252.083	100 **Fm** Fermium 257.0951	101 **Md** Mendelevium 258.10	102 **No** Nobelium 259.1009

www.ingramcontent.com/pod-product-compliance
Lightning Source LLC
Chambersburg PA
CBHW080903220326
41598CB00034B/5458